Thermodynamics of Seawater

Marine Science

Series Editor

Donald W. Hood
Institute of Marine Science
University of Alaska
Fairbanks, Alaska

Volume 1 **Bioactive Compounds from the Sea**
Edited by Harold J. Humm and Charles E. Lane

Volume 2 **Internal Gravity Waves in the Ocean**
Jo Roberts

Volume 3 **Thermodynamics of Seawater**
as a multicomponent electrolyte solution
(in two parts)

Part 1 **Entropy, Volume, Expansibility, Compressibility**
J. V. Leyendekkers

Additional Volumes in Preparation

Thermodynamics of Seawater

as a multicomponent electrolyte solution

(in two parts)

PART 1

Entropy, Volume, Expansibility, Compressibility

J. V. LEYENDEKKERS

MARCEL DEKKER INC. New York and Basel

MARCEL DEKKER, INC.

270 Madison Avenue, New York, New York 10016

LIBRARY OF CONGRESS CATALOG CARD NUMBER: 76-18422

ISBN: 0-8247-6486-2

Current printing (last digit):
10 9 8 7 6 5 4 3 2 1

PRINTED IN THE UNITED STATES OF AMERICA

FOREWORD

Seawater is a solution, which means that it is a very complex system requiring for its adequate study many branches of mathematical physics; statistical mechanics, electrostatics, hydrodynamics, and thermodynamics. This monograph (Parts I and II) deals with the latter branch, and makes a valuable contribution to the characterisation of seawater as a solution. The authors have not simply made an abstract of the literature, but have developed new theory and ideas for inclusion together with a critical selection of already published work. The authors have been able to do this from the standpoint of those who have already made important contributions to the subject.

Early marine scientists regarded the ocean as simply an inorganic solution. This idea was quickly dispelled as analytical chemists wrestled with the difficulties of performing seemingly simple quantitative determinations. After nearly a century an immense volume of information has accumulated, yet little attention has been given to the detailed study of seawater as a unique solution. Research and understanding still suffer from the lack of authoritative and stimulating texts. Understanding even of the basic concept of salinity, fundamentally important for comparing waters from different parts of the globe, is so deficient that governments have found it necessary to sponsor special working-groups of marine chemists, so that a theoretical, thermodynamic basis for world tracking of water masses could be obtained. The amount of research potential is enormous. Water comprises nearly 97% of the mass of a sample of seawater and the study of this liquid still continues with great enthusiasm. Even now, no unequivocally accepted model of liquid water has yet been developed. Moreover, the small percent of solute converts the liquid to a multicomponent electrolyte solution. In regard to an exact theory, this type of solution has long remained intractable. In recent years much progress has been made, however, and it is now quite feasible to predict the chemical potentials of seawater components. This is an important achievement since all measurements of systems in equilibrium reduce to the determination of this single thermodynamic function. The chemical potentials of seawater components and the variations with temperature and pressure form the basis of this work.

It does not seem necessary to point out in detail the value of understanding the thermodynamic properties of seawater as a solution, any more than it would seem necessary to point out the value of knowing the ordinary physico-chemical properties. It is, in fact, the interrelationships between the master variables, temperature, pressure, volume and composition, that are the important determinants of all types of oceanographic processes. Thermodynamics defines these relationships very precisely, in the form of fairly straightforward mathematical equations. The agreement or otherwise of calculated and observed thermodynamic properties provides the best check on the reliability of both experimental data and theoretical assumptions.

The effect of the bulk solute in seawater, in regard to its modification of the properties of water, is well brought out. As well as the physico-chemical changes, such modification of the properties of water are important from the biological aspect. Many interesting correlations between the seawater and pure water properties emerge. There is much to stimulate further research, not only in the specific field treated but in other more or less unrelated fields sharing only the common ground of oceanographic relevance.

G. F. Humphrey

Chairman
Intergovernmental
Oceanographic Commission
UNESCO, Paris

ACKNOWLEDGEMENTS

I thank Dr. D. R. Kester and the publishers for initiating this work. I am greatly indebted to the inspiration and help of Dr. G. F. Humphrey over a period of many years and for the guidance and encouragement given so generously by Professor R. J. Hunter. I should like, also, to thank The Trustees, Science and Industry Endowment Fund, CSIRO, and the Australian Research Grants Commission for financial help.

J. V. Leyendekkers

CONTENTS

Thermodynamics of Seawater

CHAPTER 1

FUNDAMENTALS

Some Thermodynamic Concepts and Equations--the Partial Molal Quantities

1.1 INTRODUCTION.

The purpose of this monograph is to give a systematic discussion of the thermodynamics of seawater (treated as a multicomponent electrolyte solution), and as such, all the partial molal quantities for seawater components (water plus the major electrolytes) will be covered in some detail in Part I (entropy, volume, expansivity, compressibility) and Part II (free energy, enthalpy, specific heat). Information on many other electrolytes is included for completeness. This allows useful comparative trends to be illustrated, and provides data for "salt pollution" studies.

In general, seawater is considered as a unique solution, the components of which are in fixed relative compositions. This "artificial seawater" does, in fact, closely approach the real solution in the open ocean. But, even if it did not, it is obvious that a systematic description of the thermodynamics of seawater only becomes really feasible on the basis of such an "ideal" (if you like) solution. The problem of variations in composition has been anticipated and can be overcome by simple extension of the equations and principles outlined. This is because the individual components of seawater have been characterised in the multicomponent electrolyte solution in a general way, and referred as well to their own binary solutions, at the same ionic strength as the solution.

Thermodynamics is basically an experimental science and this is the main reason why it is so useful. It is not at all necessary to understand the apparently "abstract" concepts of thermodynamics in order to be able to use the equations. These equations are well established and relate the familiar quantities of pressure, temperature, volume, energy and composition. Thermodynamics provides useful relationships between these properties and their derivatives. Since some quantities are easier to measure than others, thermodynamics enables the experimenter to choose a simpler and more accurate experimental approach. For example, even data of the highest accuracy is severely strained by differentiation and it is often adviseable to use the experimental data for the derived quantity itself. A good example of this is given by the communication of Larson (1970) concerning the ionization of water as a function of temperature versus the calorimetrically measured heat of ionization (see Chapter 3, Section 3.75). In the following chapters, especially Chapter 6, the flexibility and usefulness of thermodynamics in this respect will no doubt become apparent to the reader, if it is not already so.

In this first chapter some of the basics of thermodynamics are outlined very briefly. This is to illustrate where the various quantities characterising seawater fit in the general scheme. A more detailed or less selective (emphasis on solutions) treatment is considered outside the scope of this text. A few general references are given at the end of this chapter for further reading. Chapter 2 gives some mathematical guidelines (for those who need them) and includes the relevant units to be used.

1.2 THERMODYNAMIC LAWS

The following outline of the laws is along the lines given by McGlashan (1966). A few explanatory notes are given on the useful quantities, enthalpy, work-content and the Gibbs free energy. Reliance is placed on the summarising of the thermodynamic concepts by mathematical equations. This seems the most practical approach.

The distinguishing basic quantity (cf. mass, length, time, electric current) of thermodynamics is temperature. This quantity indicates the direction of transfer of heat. Such a transfer of heat from one system to another changes the energy of each system. The energy of a system will also change if work $\underline{w}$ is done on it (from the "probably" more familiar law of conservation of energy of electromechanics). The first law of thermodynamics is just the energy conservation law expanded to include the effect of temperature. Thus, changes in the thermodynamic energy, U, are given by

$$\Delta U = w + q \qquad (1.2.1)$$

where w is the work done on the system and q is the heat flow into the system. The additional quantity q is experimentally familiar in terms of thermometry and calorimetry. In calorimetry, changes of temperature and amounts of work are measured and the first law (eq 1.2.1) is used to calculate the energy change. It is not feasible to assign absolute values to energies - only changes in energy are measured. The quantities w and q are not properties of a system but are defined only for a given path of change. That is, their values depend on the way in which the process is carried out. However, their sum (ΔU) is uniquely determined by the initial and final states of the system.

When the work, w, is the work of expansion against an external pressure and the expansion occurs in such a way that no heat enters or leaves the system, the process is ADIABATIC, i.e.

$$\Delta U = w = -\int_{V_A}^{V_B} PdV$$

where V_A and V_B are the initial and final volumes and P the external pressure.

The other limiting case for eq 1.2.1 is when w = 0, i.e.

$$\Delta U = q$$

Here no work is done by the surroundings (e.g. a chemical reaction occurring in a constant volume calorimeter).

The <u>second law</u>. A system with uniform intensive properties (e.g. density, specific volume, temperature) is called a phase. A body of seawater, for example, with the same temperature, density, dielectric constant and so on, is a phase. A column of seawater for which one or more of the intensive properties will vary continuously from the bottom to the surface may be regarded as a vertical pile of separate phases.

If it is assumed that any work done is pressure-volume (PV) work, then for any phase the second law may be expressed

$$dU = -PdV + Td\cancel{S} + \Sigma_i \mu_i dn_i \qquad (1.2.2)$$

(if other work is done, the appropriate electromechanical term xdz is simply added to the R.H.S. of eq 1.2.2, e.g. for reversible electrical work the term $(\partial U/\partial e)de$ is added, e representing the charge, or, for increase in interfacial area, A_s, the additional term (σdA_s) is added, where σ is the surface tension).

In eq 1.2.2, T is the thermodynamic temperature, and $\cancel{S}$ is the entropy (the S is crossed to distinguish entropy from salinity which will be symbolised by the familiar S). The chemical potential, and number of moles of the substance i, are represented by μ_i and n_i, respectively. The sum covers all the substances present. The familiar quantities pressure, P, and volume, V, are measured by ordinary mechanical methods, and n_i by the usual chemical methods. The quantities $\cancel{S}$ and μ will be discussed in following sections. Together with T, they are completely and independently defined by eq 1.2.2, since a way of measuring each of them is derivable from this equation via suitable experimental techniques. Differences of entropy or of chemical potential are derived from calorimetric, thermometric, or electromechanical measurements. Ratios of T can be measured via the gas thermometer (this uses the ideal thermometric substance, a gas at very low pressure).

The fundamental equation 1.2.2 may be transformed to more convenient relationships by defining the quantities

H, the enthalpy

A, the work-content function or Helmholtz function

and G, the free energy, (Gibbs free energy), viz

$$H = U + PV \qquad (1.2.3)$$

$$A = U - T\underline{S} \qquad (1.2.4)$$

$$G = U - T\underline{S} + PV = H - T\underline{S} \qquad (1.2.5)$$

Note that G bears the same relation to A that H bears to U.

<u>H, the Enthalpy</u>. In most calorimetric measurements the pressure, rather than the volume is kept constant. The only work done is that due to the volume change ΔV, thus

$$\Delta U = q - p\Delta V \quad \text{(from eq 1.2.1)}$$

or

$$q = (U_C + PV_C) - (U_B + PV_B) \qquad (1.2.3\underline{a})$$

where B and C represent the initial and final states. Thus

$$q = H_C - H_B = \Delta H \quad \text{(from eq 1.2.3)}$$

No attempt is made to give an absolute value to the enthalpy, H, since changes in this quantity are readily measured. Whilst, in this case (the constant-pressure calorimeter) ΔH and q are equal, this is naturally not the case when the process involves pressure changes.

Thus, for the process

$$\overset{B}{H_2O \text{ (liq, 1 atm)}} \rightarrow \overset{C}{H_2O \text{(gas, 0.01atm)}}$$

ΔH is given by

$$\Delta H = (U_C + P_CV_C) - (U_B + P_BV_B) \qquad (1.2.3\underline{b})$$

The value of ΔH is determined only by the initial and final states and not by the mechanisms of the process (i.e. the special way in which the process is carried out), as can be seen from eq (1.2.3b).

The enthalpy is important in dealing with flow processes. The Joule-Thomson coefficient (Chapter 6) derives from experiments on a flow process. The characteristic of this special case is that $\Delta H = 0$.

<u>A, the work-content function</u>. The symbol A comes from the German word Arbeit (work). Occasionally it is of interest to consider a system under constant volume and temperature conditions. Then the work-content function is used. Although, since most experiments are carried out at

constant T or P, the quantities H and G are more generally useful than the corresponding quantities U, and A. For a reversible, isothermal (constant temperature) process with no entropy production, the amount of work that must be done on the system is equal to the increase in A, the work content. If a closed system is undergoing a change in state at constant temperature, then the change in A, ΔA, is a measure of the maximum work that can be performed. In an actual isothermal process the work performed by the system is less, owing to friction and other degradation processes, so that ΔA is a measure of maximum efficiency. A is a property of the system, determined by its <u>state</u> and independent of its history.

<u>G, The free energy</u>. The work done against the constant pressure of the atmosphere is equal to $p\Delta V$, (ΔV being the volume change) and this work must always be done when a process occurs at constant pressure whether this process is reversible or irreversible. Thus, in general, the maximum work that could be done by a system undergoing a change in state is less than the total maximum work $-\Delta A$, by the expansion work. That is, the net work performed by a system, $-w'$, is the total work done by a system less the expansion work. w' can be related to the free energy

$$-w' = -\Delta G = G_B - G_C \qquad (1.2.5a)$$

where B is the initial state and C the final state. Thus, in any process occurring at constant temperature and pressure, $G_B - G_C$ represents the maximum of work, in addition to expansion work, which can be obtained from a given process. It is for this reason that G is known as the free energy. The quantities G and A, or, more adequately, the molal or partial molal values of these quantities, are sometimes called thermodynamic potentials on account of certain rough analogies to mechanics. As noted above, it can be seen that G bears the same relation to A that H bears to U.

These new definitions can be used to transform eq 1.2.2 to any of the following, more convenient equations

$$dH = VdP + Td\underline{S} + \Sigma_i \mu_i dn_i \qquad (1.2.3\underline{c})$$

$$dA = -PdV - \underline{S}dT + \Sigma_i \mu_i dn_i \qquad (1.2.4\underline{a})$$

$$dG = VdP - \underline{S}dT + \Sigma_i \mu_i dn_i \qquad (1.2.5\underline{b})$$

Any system may be regarded as being made up of one or more phases. Such a system may or may not be in internal equilibrium (a one-phase system can be in an internal non-equilibrium state only because a chemical reaction is taking place spontaneously in the phase). When the thermodynamic energy U and the volume V are constant the entropy $\underline{S}$ of the system either increases (non-equilibrium) or remains constant (equilibrium). In mathematical terms

$$d\underline{S} \ (= \Sigma d\underline{S}) \geq 0 \quad (U,V \text{ constant}) \qquad (1.2.2\underline{a})$$

This condition completes the statement of the second law. Obviously, the idea of entropy is necessitated by the existence of irreversible processes.

Equation 1.2.2 describes the condition for any particular phase of the system while (1.2.2a) defines the state of the complete system, under the specified conditions. Together with eq 1.2.1 these eqs summarise thermodynamics as a whole. The application from then on reduces to the derivation of useful relations using rather uncomplicated mathematics, (see Chapter 2).

From eq 1.2.2a it follows:

$$d\underline{S} \geq 0 \quad (H \text{ and } P \text{ constant})$$

$$dA \leq 0 \quad (T \text{ and } V \text{ constant})$$

$$dG \leq 0 \quad (T \text{ and } P \text{ constant})$$

(See eqs 1.2.3c, 1.2.4a, 1.2.5b)

Third law. This law (perhaps more correctly considered as a statistical mechanical result rather than a thermodynamic law) has provided the basis for a large fraction of the numerical data on chemical reactions. A satisfactory statement of the law is generally taken as:

Every substance has a finite positive entropy, provided the entropy of each element in some crystalline state be taken as zero at the absolute zero of temperature. At this temperature the entropy of a perfect crystalline substance becomes zero, and may become zero for other substances.

This principle is of great value in the prediction of chemical equilibrium constants (represented by $\underline{K}$) and consequently of the direction of chemical reactions.

From pure thermodynamics

$$-RT\ln\underline{K}(T,P^*) = \Delta H(T,P^*) - T\Delta\underline{S}(T,P^*)$$

where ΔH is the enthalpy change and $\Delta\underline{S}$ is the entropy change which occur when the equilibrium reaction takes place, at the standard pressure P^* and the temperature T. The ΔH can be measured either directly or indirectly from calorimetric measurements. The difference in entropy

$$\underline{S}(T,P^*) - \underline{S}(T \to 0)$$

can be measured for any substance by making calorimetric measurements of heat capacity and enthalpy of transition from a very low temperature, usually about 10°K, on the solid, up through any change to liquid or gas, to the temperature T and pressure P^*. Then by simple arithmetic,

$$\Delta\underline{S}(T,P^*) \quad -\Delta\underline{S}(T \to 0) \quad \text{is obtained.}$$

The third law provides the value of $\Delta\underline{S}(T \to 0)$, and thus $\Delta\underline{S}(T,P^*)$, so that K can be calculated.

1.3 REVERSIBLE PROCESSES AND EQUILIBRIUM.

Most of the equations given throughout this text will apply to systems in equilibrium. As will be seen, this enables a great deal of information to be catalogued concerning the properties of seawater. The concept of equilibrium is therefore a very important one and, in fact, underlies most of the measurements done in the laboratory, or even in the field under controlled conditions. Properties such as vapor pressure lowering, freezing point lowering, boiling point rise, solubility, specific heat, enthalpy, surface tension and the potentials of galvanic cells (to name a few) are all determined by measurements on systems in equilibrium. In contrast, properties such as viscosity, diffusion and electrical conductance are derived from measurements on systems not in equilibrium.

It should be noted that CLOSED systems are those in which matter cannot transfer across the boundaries, but energy can be exchanged with the surroundings. On the other hand, both energy and matter can be exchanged with the surroundings for OPEN systems.

The second part of the second law, eq 1.2.2a, gives the criterion for reversibility or equilibrium. When the total entropy change is zero, then the process is reversible. Otherwise the entropy is greater than zero. The entropy can change in a reversible process, but there is only a transfer of entropy and no entropy production. In other words, the entropy change in any system is equal and opposite to the entropy change in the other systems involved in the reversible process. For any *reversible* process the increase in entropy of a system or part of a system is equal to the heat which it absorbs, divided by the absolute temperature, thus,

$$d\mathcal{S} = \delta q_{rev}/T \qquad (1.3.1)$$

(entropy has the same dimensions as heat capacity and may be expressed in calories per degree).

For an irreversible process the entropy produced as a result of irreversible processes within the system must be included, so that $d\mathcal{S}$, the increase in entropy of the system must include $\delta q_{rev}/T$ (the entropy transferred from the surroundings) plus $d\$_{irr}$, (the irreversible part), to give

$$d\$ = \delta q_{rev}/T + d\$_{irr} \tag{1.3.2}$$

Thus, the entropy increase of a system undergoing an irreversible process is the sum of entropy transferred from other systems and entropy produced by the irreversible processes within the system. For equilibrium, the last term of eq 1.3.2 is reduced to zero, and eq 1.3.1 is obtained.

As an example of an irreversible process we can consider a flow of heat δq from a body _1_ to body _2_ (at lower temperature), the net entropy increase is

$$d\$_{irr} = \delta q(1/T_2 - 1/T_1) \tag{1.3.3}$$

$d\$_{irr}$ approaches zero as T_1 and T_2 approach each other, i.e. the process approaches equilibrium. Or, when heat flows between two systems at different temperature, the process is more nearly reversible the _smaller the temperature difference_. Although the net entropy increase may not be zero, it can become negligible in comparison with $\delta q_{rev}/T$ and the process is then considered reversible. For any infinitesimal rise of temperature in a system, with heat capacity C, (irrespective of how it is produced)

$$\delta q_{rev} = CdT$$

so that by eq 1.3.1 $\quad d\$ = C(dT/T)$

If the heating is done at constant volume then

$$d\$ = C_v(dT/T) = C_v d\ln T \tag{1.3.4}$$

or $\quad (\partial\$/\partial T)_V = C_V/T \qquad (1.3.4a)$

and at constant pressure $\quad d\$ = C_p(dT/T) = C_p d\ln T \qquad (1.3.5)$

or $\quad (\partial\$/\partial T)_p = C_p/T \qquad (1.3.5a)$

where C_V and C_P represent the heat capacity at constant volume and constant pressure, respectively.

Thus, we can derive

$$\partial^2 S/\partial T \partial V = (1/T)(\partial C_v/\partial V)_T = (1/T)(\partial^2 U/\partial T \partial V) \quad (1.3.6)$$

$$\partial^2 S/\partial T \partial P = (1/T)(\partial C_p/\partial p)_T = (1/T)(\partial^2 H/\partial T \partial P) \quad (1.3.7)$$

These equations perhaps give more substance to the concept of entropy, since they relate it to such familiar things as heat and temperature. See also eqs 1.6.1, 1.8.2 and 1.8.6.

It is not considered very useful to expound detailed explanations of the "meaning" of such thermodynamic functions as entropy (or even enthalpy or the free energy). This is left to the statistical-mechanics experts. They might say that the entropy increase in an "isolated body" that moves spontaneously towards equilibrium is simply related to the increase in the number of independent eigenstates accessible to such a body. However, this number is simply related to the purely geometrical or spatial "disorder" only for three special cases, viz. mixtures of isotopes, perfect gases, or of crystals at very low termperatures (near 0°K). (McGlashan 1966).

The idea of entropy representing the state of disorder is not generally very accurate for most "practical" systems. Entropy is a criterion for irreversibility, in fact as already noted the idea of entropy is needed because irreversible processes exist.

Usually, in the present text, the free energy and its derivatives will be of primary interest. As noted above, the minimum amount of work, other than expansion work, required to bring about a given isothermal change of state is given by the increase in free energy. However, in any actual process more work would be required. Similarly, the maximum amount of work, other than expansion work, that can be done by the system is given by the decrease in free energy. An actual process yields less work. For isothermal processes where no work other than expansion work is done

$$0 = dG + Td\$_{irr} \qquad (1.3.8)$$

such a process can take place spontaneously, with resultant entropy production, only if the free energy decreases. If, for the change in state, there is no change in free energy, no change can take place spontaneously and the two states will be in thermodynamic equilibrium.

The condition that entropy is not produced, only transferred, in a reversible process, has its counterpart for free energy. In any reversible process at constant T and P, free energy can be transferred from the system to the surroundings but is not lost. Thus, for a reversible process with work being done on adjacent systems (the surroundings), the decrease in free energy of the system is just equal to the increase of the free energy of the surroundings (the work "stored" in the surroundings). When an irreversible process takes place (at constant temperature and pressure) resulting in entropy production, there will be a corresponding free-energy destruction. But if the system concerned can perform zero net work - in other words, if the system is subject to no external forces except a constant pressure - then the net work is zero and no free energy can be transferred. Thus with respect to any process occurring at constant temperature and pressure, the state of equilibrium is defined by the equation

$$dG = 0 \qquad (1.3.9)$$

It is therefore a general criterion of equilibrium, with respect to every possible change, that the free energy remain unchanged in any infinitesimal process occurring at constant temperature and pressure.

1.4 SOLUTIONS

The above paragraphs constitute a general outline of the structure of thermodynamics. However, the main concern here is with seawater, which will be treated as an aqueous mixture of electrolytes. Thus, from here on the emphasis will be on solutions, on the basis of reversible thermodynamics (i.e. covering systems in equilibrium).

It is assumed, unless otherwise stated that the extensive properties (volume, mass, entropy, free energy, enthalpy and so on) of a solution are determined by pressure, temperature, and the total amount of each component. Thus, the volume, mass and so on, of 1 kg of seawater of salinity 35‰, say, are different to the corresponding quantities for 2 kg of seawater, of the same salinity, at the same temperature and pressure.

A component is defined as any substance in the solution that can be independently varied, e.g. water or electrolytes. The ions are not components, the term constituents will be reserved for such dependent entities. An ion such as Cl-, for example, can only be added (without an enormous increase in energy of the solution) by adding a salt of this anion.

The intensive properties (density, dielectric constant and the partial molal quantities - see sections 5, 6) are determined by pressure, temperature and the relative amounts of the several components (i.e. composition). Although we cannot vary the concentrations of individual ions independently, it is possible to combine the ions in different ways to obtain different "recipes" for seawater (see Tables 4.16 and 4.17). The composition is fixed, as long as the moles of the salts for a given recipe are constant, relative to the moles of water (usually 55.51 moles of water (1 kg) are taken as the relative value, the moles of salts are then equivalent to their molality or moles per kg water). The moles of an individual salt could, however, be changed by choosing a different recipe without changing the composition, provided the recipe chosen was equivalent to that of seawater (i.e. had the same moles of each ion as before). The salinity, on the other hand, could remain constant and yet the composition could change. The salinity could be defined in terms of a set recipe, that is, it could be transformed into a suitable quantity to represent composition. It has been found convenient in this text to choose a recipe based on the mixture of salts given by Kester *et al* (1967). This is discussed in Section 1.10.

Generally, solutions are studied at constant temperature and pressure. The key quantity in this case is the partial molal free energy from which the other partial molal quantities (the partial molal differentials of volume, enthalpy and entropy) are derived by differentiating with respect to pressure and temperature. Further derivations then give other partial molal quantities such as heat capacity, expansibility and compressibility (see Figure 1.1).

In the following chapters these quantities will be treated in detail. However, a brief outline of the character of the partial molal free energy and the general derivations of the other partial molal quantities will be given here. It should be noted that the bar notation and the term *partial molal*, such as the partial molal free energy, $\bar{G}_i$, always refer to partial derivatives at constant pressure and temperature.

1.5 THE CHEMICAL POTENTIAL

The concept of the distribution of thermal energy is easily grasped. Thermal equilibrium exists in a system when the temperature is the same in all parts of that system. This is a useful local criterion in distinction to the entropy criterion of the second law (eq 1.2.2a), which focuses on changes in the whole of a complicated system. In regard to thermal equilibrium, this predicts that any infinitesimal flow of heat within the system is associated with zero total entropy change.

Since temperature indicates the direction of transfer of heat it might be considered as a measure of the escaping tendency of heat. That is, if two parts of a system are in thermal contact and at different temperatures, heat will escape (or flow) to the cooler part.

The same concepts as given for heat apply to the distribution of a material substance through any system. If the escaping tendency of each substance is constant throughout the system then a condition of equilibrium exists. If the escaping tendency of a particular substance is greater in one part of a system than another, then it will move towards the part where the escaping tendency is less. For example, if a solution of KCl is supersaturated the escaping tendency of potassium chloride in solution is greater than that of solid potassium chloride. On the other hand, the escaping tendency is less if the solution is undersaturated. When the solution is saturated, the escaping tendencies in the two phases (solution and solid) are equal and there will be a state of equilibrium.

The property which gives a quantitative measure of this escaping tendency of a material substance (from one phase or part of a system to another) is the chemical potential. If a chemical substance is free to move from one place to another, then it will move spontaneously to the state of lower chemical potential.

When there is only a single component in a phase, e.g. water (in the liquid phase) or ice (in the solid phase) or water vapor (in the gas phase), the molal free energy, G, represents the chemical potential. However, the corresponding quantity for a <u>solution</u> is the <u>partial molal free energy</u>. This is the partial derivative of the free energy with respect to the number of moles of the component, at constant temperature, pressure and moles of other components. For example, in an aqueous solution of sodium chloride (a binary, or two-component solution - salt plus water) the chemical potentials, μ, (or partial molal free energies,

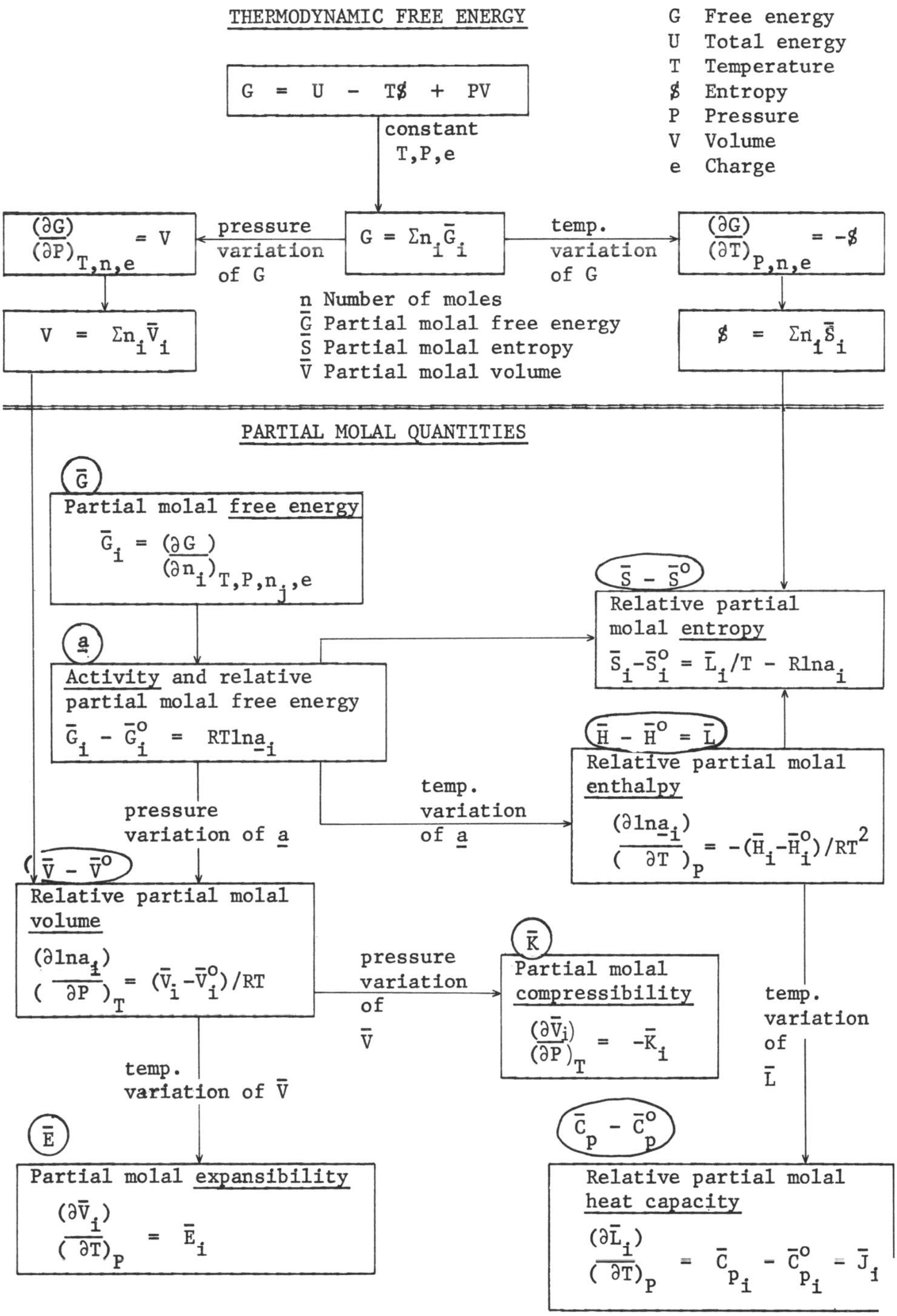
THERMODYNAMIC FREE ENERGY
G Free energy
U Total energy
T Temperature
$ Entropy
P Pressure
V Volume
e Charge
G = U - T$ + PV
constant T,P,e
(∂G)/(∂P)T,n,e = V
pressure variation of G
G = Σn_i Ḡ_i
temp. variation of G
(∂G)/(∂T)P,n,e = -$
V = Σn_i V̄_i
n Number of moles
Ḡ Partial molal free energy
S̄ Partial molal entropy
V̄ Partial molal volume
$ = Σn_i $̄_i
PARTIAL MOLAL QUANTITIES
Ḡ
Partial molal free energy
Ḡ_i = (∂G)/(∂n_i)T,P,n_j,e
S̄ - S̄°
Relative partial molal entropy
S̄_i - S̄°_i = L̄_i/T - Rlna_i
a
Activity and relative partial molal free energy
Ḡ_i - Ḡ°_i = RTlna_i
H̄ - H̄° = L̄
Relative partial molal enthalpy
(∂lna_i)/(∂T)P = -(H̄_i - H̄°_i)/RT²
pressure variation of a
temp. variation of a
V̄ - V̄°
Relative partial molal volume
(∂lna_i)/(∂P)T = (V̄_i - V̄°_i)/RT
pressure variation of V̄
K̄
Partial molal compressibility
(∂V̄_i)/(∂P)T = -K̄_i
temp. variation of L̄
temp. variation of V̄
Ē
Partial molal expansibility
(∂V̄_i)/(∂T)P = Ē_i
C̄p - C̄°p
Relative partial molal heat capacity
(∂L̄_i)/(∂T)P = C̄p_i - C̄°p_i = J̄_i

FIGURE 1.1

pmfe's) of the two components are given by

$$\mu_{water} = \bar{G}_{water} = (\partial G/\partial n_{water})_{P,T,n_{NaCl}} \quad (1.5.1)$$

$$\mu_{NaCl} = \bar{G}_{NaCl} = (\partial G/\partial n_{NaCl})_{P,T,n_{water}} \quad (1.5.2)$$

where n represents the number of moles of the component. G is the total free energy of the solution, so that the total differential is, see Figure 1.1,

$$dG = (\partial G/\partial P)dP + (\partial G/\partial T)\,dT + \Sigma_i(\partial G/\partial n_i)dn_i$$

$$= VdP - \underline{S}dT + \Sigma_i \mu_i dn_i$$

which is eq 1.2.5b.

At constant temperature and pressure dT and dP are zero and the partial molal free energies are the chemical potentials, governing the direction of transfer of the chemical substances.

It should be noted that, since

$$dA = -PdV - \underline{S}dT + \Sigma_i \mu_i dn_i \quad (1.2.4a)^*$$

$$\mu_i = (\partial A/\partial n_i)_{T,V} \text{ as well as } (\partial G/\partial n_i)_{T,P} \quad (1.5.3)$$

At equilibrium, with T, V constant, dA = 0.

Thus the same chemical potential, μ_i, controls the direction of transfer of chemical substances for constant volume and for constant pressure conditions.

Even in the more complex problems involving solutions in centrifugal or gravitational fields the partial molal free energy remains a measure of the escaping tendency of any substance, *i*, that is present. In other words, the constancy of $\bar{G}_i$ or the activity, a_i, throughout the system, remains the criterion of equilibrium of distribution of *i*. For example, consider a solution in a centrifugal field. If the component, *i*, is a distance r from the axis of rotation, in a radial column of solution that is in equilibrium.

$$d\bar{G}_i = 0$$

so that

$$(\partial\bar{G}_i/\partial r)dr + (\partial\bar{G}_i/\partial p)dp + (\partial\bar{G}_i/\partial n_i)dn_i = 0$$

and since

$$(\partial\bar{G}_i/\partial r)_{P,T,n} = -M_i\omega^2 r$$

where M_i is the molecular weight, and ω the angular velocity, and

$$(dp) = \rho\omega^2 r(dr)$$

where ρ is the density of the solution, the useful eq

$$(\rho\bar{v}_i - M_i)\omega^2 r(dr) + (\partial\bar{G}_i/\partial n_i)dn_i = 0$$

is obtained, where $\bar{v}_i$ is the partial molal volume of _i_.

The _pmfe_ is obviously the master function in studies of chemical transfer in solutions at constant temperature and pressure. It may be of interest, before going on to other partial molal quantities, to illustrate how the _pmfe_ is used. That is, to consider a simple process during which one component is transferred to or from a solution, i.e. from one phase to another. The general equation for such a process is developed and the result for equilibrium illustrated.

Consider a solution of two non-volatile salts, say sodium chloride and magnesium sulfate in water, at 25°C and 1 bar pressure. The salts have practically zero vapor pressure so that the gas in equilibrium with the solution is pure water vapor. This would not be the case for the hydrogen halides, for example. With the system at equilibrium, dG = 0 for an infinitesimal transfer of water from the gas to the solution phase. The total free energy of the system, gas and solution, is just the sum of the contributions of each phase, i.e.

$$G = G(\text{gas}) + G(\text{solution}) \qquad (1.5.4)$$

The free energy of the n_g moles of water vapor (gas being denoted by g) is $n_g G_w(g)$, where $G_w(g)$ is the molal free energy (G per mole), i.e. the chemical potential, since there is only one component in the gas phase.

The free energy of the solution depends on the relative amounts of the salts and water, as well as the total amount. This means that G(solution) is a function of n_w and n_{NaCl} and n_{MgSO_4}, i.e. the number of moles of the components (salts and water), as well as the temperature and pressure. The total differential is, with j, k representing NaCl and $MgSO_4$, respectively,

$$dG(\text{solution}) = (\partial G/\partial T)_{P,n_w,n_j,n_k} dT + (\partial G/\partial P)_{T,n_w,n_j,n_k} dP$$
$$+ (\partial G/\partial n_w)_{P,T,n_j,n_k} dn_w + (\partial G/\partial n_j)_{P,T,n_w,n_k} dn_j$$
$$+ (\partial G/\partial n_k)_{P,T,n_w,n_j} dn_k \qquad (1.5.5)$$

If all the variables, except n_w are held constant, i.e. dT, dP, dn_j, dn_k are zero

$$dG(\text{solution}) = (\partial G/\partial n_w)_{P,T,n_j,n_k} dn_w \qquad (1.5.6)$$

Then, when dn_w moles of water transfer from the gas to solution

$$dG = dG(\text{gas}) + dG(\text{solution})$$
$$= -G_w(\text{gas})dn_w + (\partial G/\partial n_w)_{P,T,n_j,n_k} dn_w \qquad (1.5.7)$$

If the system of gas and solution are in equilibrium

$$dG = 0 \text{ and}$$
$$G_w(\text{gas}) = (\partial G(\text{solution})/\partial n_w)_{P,T,n_j,n_k}$$
$$= \bar{G}_w \qquad (1.5.8)$$

The same principles apply if a solute is taken as the component being transferred (from the solid phase into solution in this case), e.g. solid calcium carbonate in equilibrium with a solution saturated with calcium carbonate.

1.6 DERIVATIVES OF THE CHEMICAL POTENTIAL - OTHER PARTIAL MOLAL QUANTITIES.

The derivative of the pmfe with respect to temperature at constant pressure and composition (moles of components) is given by

$$(\partial \bar{G}_i/\partial T)_P = \partial((\partial G/\partial n_i)_{P,T})/\partial T$$
$$= \partial((\partial G/\partial T)_P)/\partial n_i$$
$$= -(\partial \$/\partial n_i)_{P,T} = -\bar{s}_i \quad (1.6.1)$$

(for details on the mathematics see Chapter 2)

This is the partial molal entropy (the symbol \$ is obviously not necessary in this case) which is related very simply to the partial molal enthalpy, $\bar{H}_i$, and the pmfe, viz.

$$\bar{H}_i = \bar{G}_i + T\bar{s}_i$$
$$= \mu_i + T\bar{s}_i \quad (1.6.2)$$

Similarly, the derivative of the pmfe with respect to pressure at constant temperature and constant moles of the components is given by

$$(\partial \bar{G}_i/\partial P)_T = \partial((\partial G/\partial n_i)_{P,T})/\partial P$$
$$= \partial((\partial G/\partial P)_T)/\partial n_i$$
$$= (\partial G/\partial n_i)_{P,T} = \bar{v}_i \quad (1.6.3)$$

where $\bar{v}_i$ is the partial molal volume (pmv) of the component i. The derivatives of these quantities with respect to temperature and pressure are given in Figure 1.1.

1.7 GENERAL PARTIAL MOLAL EQUATIONS.

The following equations are common to all partial molal quantities. They follow from mathematics for any extensive quantity which is a continuous and single-valued function of the composition of the solution. They are thus independent (in their derivation) of thermodynamics, yet are of fundamental importance in its application.

Let X be any extensive property of a given solution (e.g. volume, entropy, free energy, enthalpy or heat capacity). If temperature and pressure are constant, X depends only on the number of moles of each component, n_1, n_2, n_3 ... .

The partial molal values are defined by (cf. $\bar{G}_i$, above)

$$\bar{X}_1 = (\partial X/\partial n_1)_{P,T,n_2,n_3,\ldots} ,$$

$$\bar{X}_2 = (\partial X/\partial n_2)_{P,T,n_1,n_3\ldots} ,$$

$$\bar{X}_3 = (\partial X/\partial n_3)_{P,T,n_1,n_2,n_4\ldots},$$

and so on.

The total differential of X is thus (see Chapter 2)

$$dX = \bar{X}_1 dn_1 + \bar{X}_2 dn_2 + \bar{X}_3 dn_3 + \ldots \tag{1.7.1}$$

Since these partial molal quantities depend on the relative amounts of the several components, they are (like density) INTENSIVE properties of the solution. For example, the partial molal volume of $MgSO_4$, say, is the same whether we take 1 kg of seawater from the Pacific or 2 kg of seawater from the Atlantic, provided that the relative number of moles of the components is the same in both samples, and the temperatures and pressures are the same.

Thus, if the several components are added simultaneously at constant T and P, and their ratios are constant, the partial molal quantities will remain constant. Integrating eq 1.7.1 under these conditions gives

$$dX = (F_1\bar{X}_1 + F_2\bar{X}_2 + \ldots)dn$$

$$X = (F_1\bar{X}_1 + F_2\bar{X}_2 + \ldots)\, n$$

$$\text{or} \quad X = n_1\bar{X}_1 + n_2\bar{X}_2 + \ldots \tag{1.7.2}$$

where $F_1 = n_1/(n_1+n_2+n_3+\ldots)$, $F_2 = n_2/(n_1+n_2+n_3+\ldots)$, etc.

and $n = n_1+n_2+n_3+\ldots$, F_i representing the mole fraction of the component, i, in the solution.

Since n_1, n_2 may have any value, eq 1.7.2 may be differentiated with respect to any change in composition, no matter how this is produced. The change may be by addition or subtraction of infinitesimal amounts of any or all of the components. The general differentiation of eq 1.7.2 gives

$$dX = n_1 d\bar{X}_1 + \bar{X}_1 dn_1 + n_2 d\bar{X}_2 + \bar{X}_2 dn_2 + \cdots \quad (1.7.3)$$

Combination of eq 1.7.3 and eq 1.7.1 gives, for constant T and P,

$$n_1 d\bar{X}_1 + n_2 d\bar{X}_2 + \cdots = 0 \quad (1.7.4)$$

which gives the relation between the change in $\bar{X}_i$, say, and the change in all the other such quantities, when the composition changes by an infinitesimal amount.

A special form of this equation, which is often useful, is obtained by taking the number of moles of one of the components, say n_3, as the main variable, then

$$n_1(\partial\bar{X}_1/\partial n_3)_{P,T} + n_2(\partial\bar{X}_2/\partial n_3)_{P,T} + n_3(\partial\bar{X}_3/\partial n_3)_{P,T} + \cdots = 0 \quad (1.7.4\underline{a})$$

This equation may be derived directly, and is easily verified if it is remembered that the partial molal quantities are already partial derivatives, so that

$$\partial\bar{X}_1/\partial n_3 = \partial((\partial X/\partial n_1))/\partial n_3 = \partial((\partial X/\partial n_3))/\partial n_1 = \partial\bar{X}_3/\partial n_1$$

since the order of differentiation does not matter.

Equation (1.7.4a) becomes

$$n_1 (\partial\bar{X}_3/\partial n_1)_{P,T} + n_2(\partial\bar{X}_3/\partial n_2)_{P,T} + n_3(\partial\bar{X}_3/\partial n_3)_{P,T} + \cdots = 0 \quad (1.7.5)$$

Equations 1.7.4a and 1.7.5 both give the change in $\bar{X}_3$ per mole addition of the specific component. If the relative composition of the added components is the same as originally in the solution $(n_1, n_2, n_3, \ldots)$ then the composition of the solution has not changed and the value of $\bar{X}_3$ must remain unchanged. Thus the expression in eq 1.7.5 should equal zero.

If $\bar{X}$ is the partial molal free energy, then eq 1.7.4a becomes

$$n_1(\partial\bar{G}_1/\partial n_3)_{P,T} + n_2(\partial\bar{G}_2/\partial n_3)_{P,T} + n_3(\partial\bar{G}_3/\partial n_3)_{P,T} + \ldots = 0 \quad (1.7.6)$$

This equation was first obtained by Gibbs in 1875 and has proved very useful in solution studies. In a binary solution, NaCl-water, say, we have, with n_{NaCl} as the main variable

$$n_w(\partial\bar{G}_w/\partial n_{NaCl})_{P,T} = -n_{NaCl}(\partial\bar{G}_{NaCl}/\partial n_{NaCl})_{P,T} \quad (1.7.7)$$

Thus, if the change of the chemical potential of sodium chloride is known, then the corresponding change for the other component, water, can be determined. Conversely, if the change in the chemical potential of the water is known then the corresponding change for sodium chloride can be determined. The isopiestic method depends upon the equilibration of one solution (the standard) with another (the test) until the vapor pressure of the water in equilibrium with both solutions is the same. The chemical potentials of the components of the standard solutions are known, thus the chemical potential of water in the test solution is known and changes can be monitored. The chemical potentials of other components of the test solution can therefore be determined.

1.71 THREE PHASES IN EQUILIBRIUM

Suppose we have a saturated solution of sodium chloride, solid sodium chloride and ice at the eutectic point (where both solids are in equilibrium with the same solution). We thus have three phases (ice, solid sodium chloride, aqueous sodium chloride) in equilibrium. This is analogous to two phases of a pure substance with one degree of freedom. That is, if any one of the variables T, P or composition is specified then the other two variables are fixed. If the pressure, say, is changed then the temperature and composition of the solution must change by a fixed amount if equilibrium is to be maintained.

We can use the above set of equations to calculate the escaping tendency of the salt and of water in the different phases. The chemical potentials or molal free energies of the solids (ice and NaCl) are given by $G_w(s)$ and $G_{NaCl}(s)$. The chemical potentials or pmfe's of water and salt in the solution are given by $\bar{G}_w$ and $\bar{G}_{NaCl}$. Assuming equilibrium is maintained, for water,

$$G_w(s) = \bar{G}_w \tag{1.71.1}$$

$$dG_w(s) = d\bar{G}_w \tag{1.71.2}$$

But
$$d\bar{G}_w = (\partial\bar{G}_w/\partial T)dT + (\partial\bar{G}_w/\partial P)dP + (\partial\bar{G}_w/\partial n_w)dn_w \tag{1.71.3}$$

and for the pure phase, ice

$$dG_w(s) = (\partial G_w(s)/\partial T)dT + (\partial G_w(s)/\partial P)dP \tag{1.71.4}$$

Thus, equating 1.71.3 and 1.71.4 according to 1.71.2,

$$(\underline{S}_w(s) - \bar{s}_w)dT + (\bar{v}_w - \underline{V}_w(s))dP + (\partial\bar{G}_w/\partial n_w)dn_w = 0 \tag{1.71.5}$$

since $(\partial G_w(s)/\partial T) = -\underline{S}_w(s)$; $(\partial\bar{G}_w/\partial T) = -\bar{s}_w$

$(\partial G_w(s)/\partial p) = \underline{V}_w(s)$ and $(\partial\bar{G}_w/\partial P) = \bar{v}_w$

The same result applies to the salt, so that

$$(\underline{S}_{NaCl}(s) - \bar{s}_{NaCl})dT + (\bar{v}_{NaCl} - \underline{V}_{NaCl}(s))dP + (\partial\bar{G}_{NaCl}/\partial n_w)dn_w = 0 \tag{1.71.6}$$

But, from Gibbs eq (1.7.6)

$$n_w(\partial\bar{G}_w/\partial n_w) + n_{NaCl}(\partial\bar{G}_{NaCl}/\partial n_w) = 0$$

so that eqs 1.71.5 and 1.71.6 combined become

$$(n_w\underline{S}_w(s) - n_w\bar{s}_w + n_{NaCl}\underline{S}_{NaCl}(s) - n_{NaCl}\bar{s}_{NaCl})dT = (n_wV_w(s) - n_w\bar{v}_w + n_{NaCl}V_{NaCl}(s) - n_{NaCl}\bar{v}_{NaCl})dP \tag{1.71.7}$$

Since, by eq 1.7.2 the entropy of the solution is

$$\mathcal{S}(\text{soln}) \;=\; n_w\bar{s}_w \;+\; n_{NaCl}\bar{s}_{NaCl} \qquad (1.71.8)$$

and the total entropy of the solids

$$\underline{\mathcal{S}}(\text{solids}) \;=\; n_w\underline{\mathcal{S}}_w(s) \;+\; n_{NaCl}\underline{\mathcal{S}}_{NaCl}(s) \qquad (1.71.9)$$

The L.H.S. of eq 1.71.7 therefore represents $\Delta\underline{\mathcal{S}}dt$ where $\Delta\underline{\mathcal{S}}$ is the increase in entropy when the solution forms from the two solids. Similarly, the R.H.S. represents ΔVdP with ΔV the change in volume during the formation of the solution. Thus,

$$dP/dT \;=\; \Delta\underline{\mathcal{S}}/\Delta V \;=\; \Delta H/T\Delta V \qquad (1.71.10)$$

This is the equation that would have been obtained for a one component, two phase system (e.g. water in equilibrium with ice or with water vapor). So using the first paragraph of this section we could have obtained the result (eq 1.71.10) directly. For example, with water in equilibrium with its vapor,

$$\Delta\mathcal{S} \;=\; q/T \;=\; \Delta H/T \qquad (1.71.11)$$

since this is a reversible process ($\Delta G=0$, see eq 1.2.5) and q is ΔH, the heat of vaporization. If ΔV is the increase in volume when an amount of the water vaporizes, eq 1.71.11 can be written

$$\Delta\mathcal{S}/\Delta V \;=\; \Delta H/T\Delta V \qquad (1.71.12)$$

and, since in this case $\partial\mathcal{S}/\partial V \;=\; \Delta\mathcal{S}/\Delta V$,

and since the vapor pressure does not depend on the volume of the system.

$$dP/dT \;=\; \Delta H/T\Delta V \qquad (1.71.10)^*$$

from the function for the variation of entropy with volume

$$(\partial\mathcal{S}/\partial V)_T \;=\; (\partial P/\partial T)_V$$

(see section 1.9, eq 1.8.4 for the derivation).

The eq 1.71.10 was obtained by Clapeyron in 1834 and represents the first physicochemical application of what is now called the Second Law of Thermodynamics (See section 1.2). The equation is also valid for a solid-solid equilibrium.

A recent study by Platford (1974) illustrates the usefulness of equilibrium measurements involving different phases. Platford made isopiestic measurements on the H_2O-NaCl-Na_2SO_4-$MgSO_4$ system at 25°C. In the H_2O-Na_2SO_4-$MgSO_4$ sub-system, the double salt bloedite, $Na_2SO_4 \cdot MgSO_4 \cdot 4H_2O$ can appear as a solid phase above 18.7°C. This is an interesting system as it has two eutonics at 25°C; each containing bloedite as one solid phase. Another interesting study is that on the freezing point, FP, of seawater. The FP is defined as the temperature at which pure ice and seawater are in thermodynamic equilibrium. See, for example, the recent studies of Doherty and Kester (1974), Fujino, Lewis and Perkin (1974) and Kester (1974).

1.72 EQUILIBRIUM BETWEEN PHASES PLUS AN EQUILIBRIUM IN A CHEMICAL REACTION

As an example, it is of interest to note some points on the solubility of a hydrated salt such as $MgCO_3(H_2O)_3$. We can consider the system as a two component system, i.e. $MgCO_3$ and H_2O and the two phases are $MgCO_3 \cdot 3H_2O$ and a solution. If one of the variables is fixed, say temperature, then the system of two phases has one remaining degree of freedom and either pressure or composition may be arbitrarily varied, but not both. Alternatively, the pressure may be kept constant and either temperature or composition varied.

The temperature versus composition of the solution may be represented graphically by a continuous curve, (with temperature as the ordinate and the mole fraction f_2 of $MgCO_3$ in the solution as the abscissa). At the maximum point of the curve (which has a roughly parabolic shape), where the composition of the solution is the same as that of the solid, any finite addition of either H_2O or $MgCO_3$ will lower the equilibrium temperature. But if in the solution there is any appreciable dissociation of $MgCO_3.3H_2O$ into water and $MgCO_3$, an infinitesimal addition of either component does not change the equilibrium temperature. In other words, the point in question is a true maximum of a continuous curve, and not a cusp, or a point of intersection of two curves. This theorem was developed in 1892 by Lorentz and Stortenbeker and may be demonstrated as follows:

The system of two independent components may be considered as consisting of three components in the solution phase, namely n_1 moles of H_2O, n_2 moles of $MgCO_3$, and n_3 moles of $MgCO_3.3H_2O$, with partial molal free energies $\bar{G}_1$, $\bar{G}_2$, and $\bar{G}_3$. At the melting point of the pure hydrate, its composition is that of the solution, and therefore f_1 (the mole fraction of water) = $3f_2$ (however far dissociation may occur according to the equation $MgCO_3.3H_2O = MgCO_3 + 3H_2O$).

Suppose dn_1 moles of water are added. By eq 1.7.4 at constant temperature and pressure

$$n_1 d\bar{G}_1 + n_2 d\bar{G}_2 + n_3 d\bar{G}_3 = 0 \qquad (1.72.1)$$

or

$$n_2(3d\bar{G}_1 + d\bar{G}_2) + n_3 d\bar{G}_3 = 0 \qquad (1.72.2)$$

But $MgCO_3.3H_2O$ $MgCO_3$, and H_2O are in equilibrium and will remain so, thus, since the state of equilibrium dictates that the change in the total free energy is zero for any infinitesimal process occurring at constant temperature and pressure, i.e. dG = 0,

$$\Delta G = \bar{G}_3 - 3\bar{G}_1 - \bar{G}_2 = 0$$

$$d\bar{G}_3 = 3d\bar{G}_1 + d\bar{G}_2 \qquad (1.72.3)$$

Combining this with eq 1.72.2

$$n_2 d\bar{G}_3 + n_3 d\bar{G}_3 = 0 \qquad d\bar{G}_3 = 0$$

In other words, the escaping tendency of $MgCO_3.3H_2O$ in the liquid phase is not changed by an infinitesimal addition of water, and it will therefore remain in equilibrium with the solid $MgCO_3.3H_2O$ without change in the equilibrium temperature.

Thermodynamics is unable to predict how flat such a solubility curve will be. This must depend upon the extent to which the compound dissociates, for if there were no dissociation at all in the solution, adding $MgCO_3$ or H_2O would be like adding some foreign constituent and the continuous curve would be replaced by two curves intersecting at the melting point.

1.8 SOME USEFUL THERMODYNAMIC RELATIONSHIPS FOR SOLUTIONS AT CONSTANT COMPOSITION - SYSTEMS IN INTERNAL EQUILIBRIUM

For the open ocean, the assumption that seawater is of constant composition is often a good approximation. In any case, it is common for the effects of temperature and pressure on seawater properties to dominate over the changes due to the small compositional changes that occur.

From the 2nd law equations, when dn_i is zero for all the components present, we can derive the following set of equations. These are state functions which have perfect differentials. These equations, unless otherwise stated, only apply to systems in internal equilibrium.

From eq 1.2.4a, with $\Sigma_i \mu_i dn_i$ zero

$$dA = -\underline{S}dT - PdV \qquad (1.8.1)$$

with the volume constant

$$(\partial A/\partial T)_V = -\underline{S} \qquad (1.8.2)$$

with the temperature constant

$$(\partial A/\partial V)_T = -P \qquad (1.8.3)$$

and $(\partial^2 A/\partial V \partial T) = -(\partial \underline{S}/\partial V)_T = \partial^2 A/\partial T \partial V$

$$= -(\partial P/\partial T)_V$$

so that

$$(\partial \underline{S}/\partial V)_T = (\partial P/\partial T)_V \qquad (1.8.4)$$

Also, from eq 1.2.5b at constant composition

$$dG = -\underline{S}dT + VdP \qquad (1.8.5)$$

so that $(\partial G/\partial T)_P = -\underline{S} = (G - H)/T$ (1.8.6)

$(\partial G/\partial P)_T = V$ (1.8.7)

and $(\partial \underline{S}/\partial P)_T = -(\partial V/\partial T)_P$ (1.8.8)

Changes of entropy with volume or pressure can be calculated from these equations using the coefficients of expansion $((1/V)(\partial V/\partial T)_P)$ and compressibility $(-(1/V)(\partial V/\partial P)_T)$ and, for eq 1.8.4 the equation

$$(\partial P/\partial T)_V = -(\partial V/\partial T)_P/(\partial V/\partial P)_T$$

$$= \alpha/\beta \qquad (1.8.9)$$

(see Chapter 2 for the derivation of eq 1.8.9, which follows from the mathematics).

From eq 1.2.2, at constant composition

$$dU = Td S - PdV$$

$$(\partial U/\partial V)_T = T(\partial S/\partial V)_T - P$$

and substituting eq 1.8.4

$$P = T(\partial P/\partial T)_V - (\partial U/\partial V)_T \qquad (1.8.10)$$

From $dH = dU + PdV + VdP$

$$= Td S + VdP$$

which is eq 1.2.3c when the composition is constant, and with eq 1.8.8

$$V = T(\partial V/\partial T)_P + (\partial H/\partial P)_T \qquad (1.8.11)$$

If either $(\partial U/\partial V)_T$ or $(\partial H/\partial P)_T$ have been obtained experimentally, eqs 1.8.10 and 1.8.11 give the thermodynamic requirements for any empirical equation of state.

The last term in eq 1.8.11 viz. $(\partial H/\partial P)_T$ is related to the Joule-Thomson coefficient μ_{JT} (or $(\partial T/\partial P)_H$), viz.

$$(\partial H/\partial P)_T = -\mu_{JT}C_p \qquad (1.8.12)$$

where C_p (or $(\partial H/\partial T)_p$) is the heat capacity at constant pressure. The coefficient μ_{JT} will be referred to in more detail in Chapter 6. If measurements of μ_{JT} for seawater were available, eq 1.8.11 could be used to determine the V-T relationship.

The Joule-Thomson coefficient relates to a process that is adiabatic (i.e. no heat enters or leaves the system being studied) - see the first law above. However, in this case, the process is irreversible.

For an adiabatic process that is reversible (i.e. during the adiabatic compressions or expansions no sensible pressure gradients develop within the system) there is no entropy change, i.e. $\Delta S = 0$. Such a process is called isentropic. The compressibility derived directly from sound velocity measurements is assumed to be the isentropic compressibility. This is converted to the isothermal compressibility via suitable thermodynamic relationships. These compressibilities will be discussed in Chapter 6.

There is normally a temperature change during isentropic compression and the measurement of the ratio between this temperature change and the pressure change gives $(\partial T/\partial P)_S$, and

$$(\partial T/\partial P)_S = -(\partial S/\partial P)_T/(\partial S/\partial T)_P = (T/C_p)(\partial V/\partial T)_P$$

$$= (T/C_p)\alpha V \qquad (1.8.13)$$

from the mathematics, and using eq 1.8.8 and the definition of heat capacity at constant pressure in the form:

$$(\partial S/\partial T)_P = C_p/T \qquad (1.8.14)$$

Similarly, with the heat capacity at constant volume, C_V, given by

$(\partial S/\partial T)_V = C_V/T$, and eq 1.8.4 and 1.8.9

$$(\partial T/\partial V)_S = -(\partial S/\partial V)_T/(\partial S/\partial T)_V$$

$$= -(T/C_V)(\partial p/\partial T)_V = -(T/C_V)\alpha/\beta \qquad (1.8.15)$$

1.9 THE EQUILIBRIUM CONSTANT - STANDARD STATES

The free energy change in a chemical reaction is measured under conditions convenient to the experimenter. One substance may be in solution while another is in the gaseous state at high pressure. In order to make such measurements useful for further calculations we need to obtain from them the corresponding free energy change when each of the participating substances is in its standard state (a conveniently defined reference state). This is achieved by using the very useful function, the activity function, which is given by

$$RT\ln \underline{a} = \bar{G} - \bar{G}^{o} \qquad (1.9.1)$$

As can be seen, the activity $\underline{a}$ is obtained by taking the antilog of the free energy. This eliminates the main disadvantage of the free energy, that is, the large variation its value can have (plus infinity to minus infinity). At a given temperature and pressure, $\bar{G}$ is the chemical potential in any state, and the superscript zero indicates the standard state. Thus, the problem of converting free energies in various states into free energies in standard states reduces to the determination of the activities of the various substances concerned. From eq 1.9.1 it can be seen that the activity of a substance must be unity when the substance is in the standard state.

Consider a chemical reaction with reactants F, B and products, C, D and E in various states (the state and concentration in the solution is shown in brackets). The number of moles of each component participating in the reaction is indicated by the corresponding lower case letters (f,b,c,d.e). The substances are all at atmospheric pressure, except the gas.

$$\text{fF(solid)} + \text{bB(aq, 3}\underline{m}\text{)}$$
$$= \text{cC(solid)} + \text{dD(liq,with B at 3}\underline{m}\text{)} + \text{eE(gas,p atm)} \qquad (1.9.2)$$

Once ΔG, the free energy change, for this reaction is known, the corresponding change ΔG^{o} for the standard state can be calculated from $\bar{G} - \bar{G}^{o}$ for each substance.

For a liquid or solid which may act as a solvent, the pure liquid or solid at 1 atm pressure is usually chosen as the standard state. This allows convenient tabulation of heat capacities, enthalpies, entropies and free energies. The solids in the above reaction are therefore in their standard states and their activities are unity (since $\bar{G}-\bar{G}^o = 0$ for these cases). The activities of B and D can be obtained from suitable measurements such as investigation of the vapor pressure of the solutions or, potentiometric, solubility or freezing point measurements. The value of $\bar{G}-\bar{G}^o$ is then calculated from eq 1.9.1. The activity of the gas could be calculated approximately by assuming it to be a perfect gas, or from volume-pressure data of the gas. If the gas was perfect and at 1-atm pressure, then it would have been in its standard state. This is the state generally chosen for tabulating entropies, and free energies of gases. More explicit descriptions of the various standard states will be given below (see, also Table 1.1).

The value of ΔG^o, the change of free energy for the reaction (1.9.2) when each substance is in its standard state can then be calculated since

$$\Delta G - \Delta G^o = RT\ln(a_C^c a_D^d a_E^e / a_F^f a_B^b) \qquad (1.9.3)$$

or

$$\Delta G - \Delta G^o = RT\ln\ a_D^d a_E^e / a_B^b \qquad (1.9.3\underline{a})$$

since, in this case, a_F and a_C are each unity.

This equation is obtained by combining the individual

$$x(\bar{G} - \bar{G}^o) = xRT\ln\underline{a}$$

equations, where x is the number of moles of the component participating in the reaction (i.e. f,b,c,d, or e in eq 1.9.2).

There is no restriction on the number of components participating (as far as eq 1.9.3 is concerned) there could be more or less reactants and products than those given for eq 1.9.3. The quotient on the RHS of this equation is called the activity quotient of the reaction.

From tabulations of ΔG^o built up in this way and measurements of the activities we can calculate ΔG for the state of interest.

If chemical equilibrium prevails in the reaction then from eq 1.3.9

$$\Delta G = 0$$

Substituting into eq 1.9.3

$$\Delta G^o = -RT\ln \underline{a}_C^c \underline{a}_D^d \underline{a}_E^e / \underline{a}_A^a \underline{a}_B^b \quad (1.9.4)$$

For a given temperature and pressure ΔG^o is a constant. Thus, the activity quotient must also be constant as a condition of equilibrium. This quotient, for equilibrium conditions, is therefore called the equilibrium constant, K, so that

$$\Delta G^o = -RT\ln K \quad (1.9.5)$$

If, in eq 1.9.4 the partial pressures (for gases) or molalities (for components in solution) are substituted for the activities, the quotients will approach the true equilibrium constant (the quotient of activities) only at the approach to zero pressure (gases) or zero concentration of the solutes (solutions). However, provided the activities are used the constancy of the quotient should prevail provided a state of equilibrium exists.

Suppose ΔG^o is taken as zero. Then the activity quotient (or K) must be equal to unity. This is the convention usually adopted for strong electrolytes (which are appreciably dissociated in solution, for example, NaCl, in contrast to weak electrolytes, such as acetic acid). In effect, this convention means that in the chosen standard state there is no difference in the standard free energies of dissociated and undissociated species and hence $\Delta G^o=0$. This is the basis for the definitions of the mean activity, mean activity coefficient and mean molality. When measurable association occurs, as for example, in $MgSO_4$ or HNO_3 aqueous solutions, the $\Delta G^o=0$ convention may still be used. It is possible to give an exact thermodynamic treatment on the assumption of complete dissociation, whether this occurs or not. We can treat the electrolyte solution as consisting of a mixture of undissociated molecules and of ions, or assume complete dissociation, but take into account the interaction between the ions.

As an example of the $\Delta G^o=0$ convention, or choice of standard state, consider an aqueous electrolyte such as KCl, NaCl, $CaCl_2$ or $LaCl_3$, say. If the activity of the solute (undissociated electrolyte, x) is represented by a_x and the activity of the +ve and -ve ions by a_{x+}, a_{x-} and if x dissociated into $\nu(=\nu_+ + \nu_-)$ ions according to

$$x = \nu_+ x^+ + \nu_- x^- \tag{1.9.6}$$

then
$$K = (a_{x+}^{\nu+} a_{x-}^{\nu-})/a_x$$
$$= 1 \tag{1.9.7}$$

so that
$$a_x = a_{x+}^{\nu+} a_{x-}^{\nu-} \tag{1.9.8}$$

or, in terms of the molalities and activity coefficients, γ ,

$$a_x = (m_+^{\nu+} m_-^{\nu-} \gamma_+^{\nu+} \gamma_-^{\nu-}) \tag{1.9.9}$$

(since $\gamma_+ = a_+/m_+$ and $\gamma_- = a_-/m_-$)

In practice, it is more generally useful to take the geometric mean of the ion activities and activity coefficients, viz.

$$a_{\pm x} = a_x^{1/\nu} = (a_{x+}^{\nu+} a_{x-}^{\nu-})^{1/\nu} \tag{1.9.10}$$

and
$$\gamma_\pm = a_\pm/(m_+^{\nu+} m_-^{\nu-})^{1/\nu} = a_\pm/m_\pm \tag{1.9.11}$$

$$\gamma_\pm = (\gamma_+^{\nu+} \gamma_-^{\nu-})^{1/\nu} \tag{1.9.12}$$

which is applicable to any solution, and thus to a mixed electrolyte solution.

The solute standard state is usually adopted for solutions. Apart from the convention of $\Delta G^o = 0$ that can be used, other specifications (related to the properties of a perfect solution) apply and these will be discussed in detail below.

With the convention $\Delta G^o=0$ or $K = 1$ we have, in effect, defined the solute in terms of the ions. In other words, we have taken the free energies of the ions and undissociated molecule as equivalent in the standard state. This convention is therefore compatible with the properties of a strong electrolyte which has "ideal" behavior at infinite dilution only if complete ionization is assumed.

For <u>weak</u> electrolytes (e.g. acetic acid) on the other hand, it is necessary to make two types of extrapolations to infinite dilution in order to establish the solute standard state for the ions as well as the undissociated molecule. Thus the free energies of the ions are not equivalent to the free energy of the undissociated molecule in the standard state and $\Delta G^o \neq 0$. For example, for the equilibrium

$CH_3COOH = H^+ + CH_3COO^-$

$$\Delta G^o_{298} = -RT\ln K = 6{,}487 \text{ cal}$$

and

$$H_2CO_3(aq) = H^+ + HCO_3^- \qquad \Delta G^o_{298} = 8{,}683 \text{ cal}$$
$$HCO_3^- = H^+ + CO_3^{2-} \qquad \Delta G^o_{298} = 14{,}091 \text{ cal}$$
$$H_2CO_3(aq) = 2H^+ + CO_3^{2-} \qquad \Delta G^o_{298} = 22{,}776 \text{ cal}$$

(data from Pitzer and Brewer 1961).

These values are for a temperature of 298°K and 1 atm pressure. Generally, the term <u>standard</u> state is reserved for 1 atm pressure for a given temperature. At any other pressure the term <u>reference</u> state is used.

Some strong electrolytes may dissociate to give an ion which is, in effect, a weak electrolyte. For example, $KH_2PO_4 = K^+ + H_2PO_4^-$ $\Delta G^o = 0 \quad K = 1$.

But $H_2PO_4^-$ is a weak electrolyte and HPO_4^{2-} is weaker still, so that ΔG^o for dissociation of these ions is not zero, e.g.

$$H_2PO_4^- = H^+ + HPO_4^{2-} \qquad \Delta G^o_{298} = 9{,}830 \text{ cal}$$
$$HPO_4^{2-} = H^+ + PO_4^{3-} \qquad \Delta G^o_{298} = 16{,}370 \text{ cal}$$

In some cases, it may be desirable to distinguish between the ions and the undissociated molecule for a strong electrolyte (e.g. in order to treat ionpairing and complex formation). The separate solute standard states are established for the ions and undissociated species and ΔG^o has a finite value.

For example

$MgSO_4^o$	=	Mg^{2+}	+	SO_4^{2-}	ΔG^o_{298}	= 3,231 cal
$NaOH^o$	=	Na^+	+	OH^-	ΔG^o_{298}	= -950 cal
$CdCl^+$	=	Cd^{2+}	+	Cl^-	ΔG^o_{298}	= 2,680 cal

(Data from Nancollas 1966)

STANDARD STATES

General

We have already noted some of the essentials of the standard states for a gas, for a liquid or solid which may act as a solvent, and for a solute. These essentials (or rules) lessen the arbitrariness in the choice of a suitable standard state without undue loss of flexibility.

The common reference taken for the standard state is the state of the substance where its behavior is simplest (or most ideal). A gas approaches ideal behavior as the pressure approaches zero, while a solution becomes more ideal as the concentration of the solute approaches zero. Hence, the behavior of the substances in their standard states conform to their behavior in these "ideal" states. However, the actual standard state represents a hypothetical gas or solution behaving <u>ideally</u>, but where the activity is unity <u>to conform to eq 1.9.1</u>.

The standard state is the state that real substances would have at 1 atm (for gases) or 1 molal (for solutes) if they behaved in an ideal way. For example, a solution would have to obey Henry's Law (which infinitely dilute solutions obey). This law specifies that the activity of the solute is proportional to its mole fraction. In infinitely dilute solutions the mole fraction is proportional to the molality so that Henry's Law specifies that a/m = 1.

For strong electrolytes, where the products of dissociation are considered as the solute $a_{\pm}/m_{\pm} = 1$, or $\gamma_{\pm} = 1$.

The experimental evidence indicates that Henry's Law will be approached in the limit of zero molality only if complete ionization is assumed for strong electrolytes. This is why the products of dissociation are considered as the solute, in this case. The standard state for a solution is represented in Figure 1.2. The behavior of the real solution is illustrated by the curved line. At infinite dilution the real solution approaches ideal behavior. Thus if the section of the curve at very low concentration (which is linear, according to Henry's Law) is extrapolated to unit activity (or unit molality, since they are proportional for the ideal solution), then this is the standard state. This is often referred to as the "infinite-dilution standard state" for obvious reasons, as for the "ideal gas standard state". We shall now give a little more detail for each specific case; the essential points are then summarised in Table 1.1 at the end of this Section.

Gas. a = f, or a/p = 1 when p = 0, where f is the fugacity (the "ideal" or "corrected" vapor pressure), a is the activity and p the pressure.

It is convenient in the case of a gas to make the activity equal to the fugacity. Thus, at each temperature the standard state is chosen so that the fugacity is unity. For a perfect gas this means the activity is unity at unit pressure. This explains the description "a 1-atm perfect-gas standard state". If the fugacity of a real gas is plotted against the pressure, the "ideal" behavior becomes apparent as zero pressure is approached. Extrapolation of this "ideal" part of the curve to 1-atm pressure will give the standard state (cf. the straight line in Figure 1.2 for solutions).

Liquid or solid. (solvent standard state). The standard state is taken as the pure liquid or pure solid at 1-atm pressure at each temperature, and activity/mole fraction = 1. This standard state is nearly always used for the component in largest amount, for solutions, i.e. the solvent, but may be used for all components. If a_s is the activity of a solvent, say water, in a solution, then $a_s/x_s = 1$ (where x represents the mole fraction) for the infinitely dilute solution.

Solute. This standard state will be used for the components of seawater, with the exception of water for which the previously given standard state will be used.

For aqueous solutions, this state may be referred to as the "infinite-dilution-in-pure-water", IDW, standard state. An important feature of this state is that the activity coefficient equals unity (irrespective of the concentration scale used) at infinite dilution of solute at all temperatures and pressures. For a typical solution at infinite dilution, by Henry's Law, the activity of the solute is proportional to its mole fraction. Thus, it is convenient to choose the standard state so that at infinite dilution the ratio of the activity to the measure of concentration is unity. Thus, a characteristic of the solute standard state is that $a_x/m = 1$ at 1-atm pressure, (a characteristic of an ideal solution), when m = 0; where x is the solute. As outlined above, such a definition gives a hypothetical state as the standard state, and this can be reached by extrapolating to m=1 along the "ideal" solution a versus m line of Figure 1.2.

In this hypothetical state, which we have conveniently taken as our standard state, the partial molal enthalpy and heat capacity of the solute have the values of the infinitely dilute solution but the free energy and entropy correspond to unit activity (in this text the unit of activity is based on molality, moles per kg of water). As the solution becomes infinitely dilute $\bar{G} = -\infty$, thus unit activity is specified. In mathematical terms the standard chemical potential is defined.

$$\mu^o = \bar{G}^o = \lim_{m \longrightarrow 0} (\bar{G}_m - RT\ln(m/1))$$

(m/1 signifies that the dimensions are moles per kg solvent) $\bar{G}_m$ is the chemical potential at concentration m.

An important point to note for the investigation of polyphase systems is that we have defined different standard states for a given substance in different phases.

Thus the activities in the different phases at equilibrium will not be equal. However, they will be proportional to one another as long as equilibrium persists. The factor of proportionality depending upon the choice of standard states. In other words, the activity of a component at constant temperature pressure and charge is the same in every phase, provided that it is defined in each phase in reference to the same standard state.

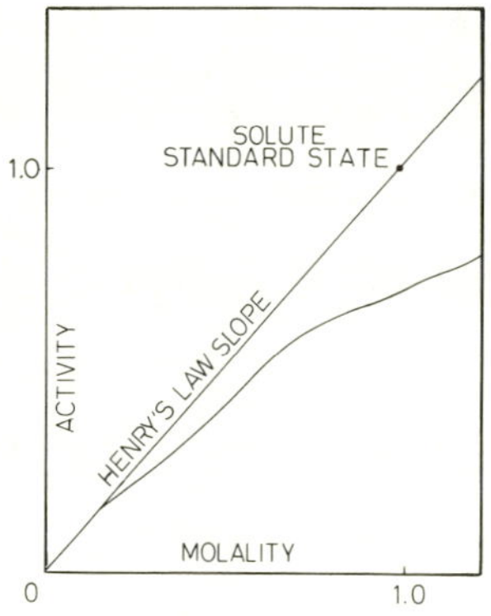

FIGURE 1.2

Straight line - dilute solution function extrapolated to unit activity or 1 molal (hypothetical solution behavior).

Curved line - real solution behavior.

TABLE 1.1 CHARACTERISTICS OF STANDARD STATES

Phase	Standard State or Hypothetical State of Unit Activity
gas	1-atm perfect gas standard state. Activity = fugacity. Fugacity is unity for a given temperature. a/p = 1 when p = 0. For a perfect gas a = 1 when p = 1
Liquid or solid (that can be a solvent)	Pure liquid or pure solid at 1-atm pressure at each temperature
Solute	Hypothetical ideal solution in which the partial molal enthalpy and heat capacity of the solute have the values of the infinitely dilute solution, and the free energy and entropy correspond to unit activity.

1.10 SEAWATER SOLUTIONS.

As noted in Section 1.4, seawater will be treated as an aqueous mixture of electrolytes. The concentrations of ions in seawater are not independent variables, i.e. the number of moles of an individual ionic species cannot be varied independently. For this reason, the term component will be reserved for the electrolytes and water; the number of moles of a component can be varied independently. The ionic species will be referred to as constituents of the seawater solution. Thermodynamics does not permit the evaluation of the chemical potentials, free energies, partial molal volumes, etc., of these constituents. However, it is the ions that are of fundamental importance in the determination of the behavior and properties of the electrolyte solution. Generally, the problem is overcome by using ionic products or ratios or adopting a suitable convention. The more useful conventions are based on theoretical models, or at least, use the physical properties of the ions as a guideline. In this regard, the emphasis in this text will be on the components.

TABLE 1.2 A mixture of salts and water representing average seawater[a]

Seawater Solution A[b]

Salt	m_i (mol/kg water)	f_i	ν_i
NaCl	0.424310	0.796440	2
Na_2SO_4	0.029245	0.054895	3
$NaHCO_3$	0.002418	0.004539	2
NaF	0.000074	0.000139	2
KCl	0.009412	0.017666	2
KBr	0.000854	0.001602	2
$MgCl_2$	0.055211	0.103630	3
$CaCl_2$	0.010707	0.020098	3
$SrCl_2$	0.000093	0.000173	3
H_3BO_3	0.000436	0.000818	2

a Based on the artificial seawater recipe of Kester et al. (1967)
b See Table 4.16. S = 35‰

Usually, seawater will be considered as being made up of salts as shown in Table 1.2. The concentrations given are for seawater according to the practical recipe of Kester et al. (1967) and a salinity of 35‰. The relative concentrations are considered to remain constant. Following the usual practice for binary solutions, a solution of seawater containing 1 kg of water is chosen as a basis for developing the various equations. This immediately defines the number of moles of water present, and the number of moles of each electrolyte component will then be equivalent to the familiar quantity, the molality, m, (moles per kg of water).

Using X to represent any extensive property (volume, free energy, entropy, etc.) of the seawater solution, the following general equations can be developed. For a given temperature and pressure, X depends only on the number of moles of each component, i.e. from eq 1.7.2, with the subscripts w and i representing the components water and electrolytes, respectively,

$$X = n_w\bar{X}_w + \Sigma_i n_i\bar{X}_i \qquad (1.10.1)$$

with n representing the number of moles and $\bar{X}$ the partial molal X.

The second term on the R.H.S. of eq 1.10.1 represents the solute contribution to the extensive property of seawater. For simplicity, this contribution can be referred to as the "seasalt" contribution. For a solution containing 1 kg of water,

$$\Sigma_i n_i\bar{X}_i = \Sigma_i m_i\bar{X}_i$$

$$= m_s\Sigma_i f_i\bar{X}_i \qquad (1.10.2)$$

where $m_s = \Sigma_i m_i$ and $f_i = m_i/m_s$, m_i is the molality of the electrolyte component, i.

Again for simplicity, we can let

$$\Sigma_i f_i\bar{X}_i = \bar{X}_s \qquad (1.10.3)$$

where $\bar{X}_s$ is referred to as the mean partial molal quantity (volume, entropy, etc.) of "seasalt", or the solute in seawater. This quantity is the salt mean since

$$\bar{X}_s = \Sigma_i m_i \bar{X}_i / \Sigma_i m_i \tag{1.10.4}$$

Since $\Sigma_i m_i$, the total number of moles of salts, depends on the recipe chosen, $\bar{X}_s$ will also depend on the recipe. However, as long as the same recipe is used no problems arise and this quantity can be used as a very convenient guide to the effect of the solute on seawater properties. In addition, alternative seawater recipes can be used to calculate seawater properties from binary solution data. Comparison with actual seawater data can then provide information on the short-range ionic interactions (see, for example, Section 4.61 and Figure 4.8).

Alternatively, the total solute contribution to the seawater property X can be averaged per mole of ions, or, per gram of seawater.

Per mole of ions. The concentration of each salt, i, is related to the concentrations of its ions, m_+, m_-, by:

$$m_+ + m_- = \nu_+ m_i + \nu_- m_i$$

$$= \nu_i m_i \tag{1.10.5}$$

where ν_+, ν_- represent the number of moles of ions dissociating from 1 mole of salt, and

$$\nu_i = \nu_+ + \nu_-$$

Thus, $\Sigma_{ion} m_{ion} = \Sigma_i \nu_i m_i = m_s \Sigma_i \nu_i f_i$

and $(m_s \Sigma_i f_i \bar{X}_i)/(m_s \Sigma_i \nu_i f_i) = (\Sigma_i f_i \bar{X}_i)/(\Sigma_i \nu_i f_i)$

$$= \bar{X}_s/(\Sigma_i \nu_i f_i) \tag{1.10.6}$$

For the recipe A given in Table 1.2,

$$\Sigma_i \nu_i f_i = 2.178796$$

so that the mean solute contribution to X (per mole of ions) will be $\bar{X}_s^A/2.178796$. The superscript A indicates that recipe A has been used to calculate $\bar{X}_s$.

Since $\Sigma_i \nu_i f_i$ will be constant for a given recipe, it can be seen that $\bar{X}_s^A$ is proportional to $\bar{X}_s^B$, where B represents any other suitable recipe. (See, for example, Table 4.16), i.e. with $\Sigma_i \nu_i f_i = k$, say,

$$\bar{X}_s^A = (k_A/k_B)\bar{X}_s^B \tag{1.10.7}$$

since the mean partial molal X (per mole of ions), $\bar{X}_s/k$, is independent of any recipe (see Tables 4.16 and 4.17).

It can be seen that interconversion between different recipes, using $\bar{X}_s$, is analogous to a unit interconversion.

Another mean that could be used is the partial molal X of seasalt per gram of seawater. This is simply obtained by dividing the solute contribution by the total mass of the seawater solution.

Per gram of seawater. The total mass of a seawater solution containing 1 kg of water and $\Sigma_i m_i$ moles of salts is $(1000 + \Sigma_i m_i M_i)$ grams, where M_i is the molecular weight of the salt. The solute mass may be given in alternative forms, viz.

$$\Sigma_i m_i M_i = m_s \Sigma_i f_i M_i = m_s M_t \tag{1.10.8}$$

where M_t is the mass of seasalt per mole of salts. OR,

$$\Sigma_i m_i M_i = \Sigma_i m_i (\nu_+ M_+ + \nu_- M_-)$$

$$= \Sigma_{ion} m_{ion} M_{ion}$$

In this text, $m_s M_t$ is commonly used, so that the mean partial molal quantity (per gram of seawater) is given by

$$\bar{X}_s^{sw} = m_s \bar{X}_s / (10^3 + m_s M_t) \tag{1.10.9}$$

where the superscript sw distinguishes it from the corresponding mean (per mole of salts) and from $\bar{X}_s^I$ which is used for the mean (per mole of ions).

In the application to the mean partial molal volume of seasalt, for example,

$$\bar{V}_s^{sw} = m_s\bar{V}_s/(10^3+m_sM_t)$$

(see Table 4.17 for the numerical values of the various weights for different recipes).

This latter mean has been used in many of the figures that illustrate how the contribution of the solute changes with temperature and pressure. A specific outline of these principles is given in Chapter 4, Section 4.6, where they are applied to the volume of seawater (i.e. X = V in this case).

Relationship of m_s and salinity. It is more convenient to use salinity, S‰, than m_s in dealing with the properties of seawater solutions; m_s provides the link with binary solution data. Since a solution containing 1 kg of water has a mass of $(10^3+m_sM_t)$ grams, and contains m_sM_t grams of solute, there is

$$10^3m_sM_t/(10^3+m_sM_t) \text{ g of solute in 1 kg of seawater.}$$

If the mass of solute per kg of seawater is denoted by S*, then

$$S^* = 10^3m_sM_t/(10^3+m_sM_t) \tag{1.10.10}$$

so that
$$m_s = S^*/(M_t(1-c^*)) \tag{1.10.11}$$

where $c^* = 10^{-3}S^*$.

The total solute mass S* is proportional to the salinity, S (as defined by Unesco (1966), S‰ = 1.80655Cl‰, assuming this salinity to be the same as that obtained by the Knudsen definition (Knudsen et al., 1902)), i.e. $S^* = k'S$, at least for average seawater. Thus, with $M_t' = M_t/k'$,

$$m_s = S/(M_t'(1-c)) \tag{1.10.12}$$

neglecting the difference between c* and c $(10^{-3}S)$. The quantity M_t' may be thought of as the "corrected" molecular mass of seasalt. The value of k' (which is independent of the recipe) has been estimated as 1.0049 (Leyendekkers 1973, using the data of Morcos (1970)).

Values of M_t, M_t' for two different recipes are given in Table 4.16.

Thus, m_s in all the above equations may be replaced by $S/(M_t'(1-c))$.

<u>m_s and the ionic strength, I.</u> The ionic strength contribution of each salt, I_i, is given by

$$I_i = \tfrac{1}{2}(z_+z_-\nu)_i m_i \qquad (1.10.13)$$

so that, the total ionic strength, I, is given by

$$I = \Sigma_i I_i = m_s \tfrac{1}{2}\Sigma_i (z_+z_-\nu)_i f_i$$

$$= m_s \underline{W} \qquad (1.10.14)$$

where $\underline{W} = \tfrac{1}{2}\Sigma_i (z_+z_-\nu)_i f_i$, (for solution A, $\underline{W} = 1.3576$)

Substituting into eq 1.10.12, gives

$$I = (\underline{W}/M_t')S/(1-c)$$

The quantity $(\underline{W}/M_t')$ is independent of the recipe and has the value <u>0.01994</u>, so that

$$I = 0.01994S/(1-c) \qquad (1.10.15)$$

Since comparisons with the binary solutions and seawater are made on the basis of the same ionic strength, eq 1.10.15 will be used frequently.

Before going on to discuss the partial molal quantities in detail, a brief outline is given of the basic mathematics to be used (Chapter 2). In <u>Part I</u> of this monograph, the partial molal entropies, volume, expansivity (expansibility) and compressibilities will be dealt with. In <u>Part II</u>, the partial molal free energies, enthalpies and heat capacities will be covered. It has been considered necessary to include quite a lot of data on binary solutions, since these can form a basis of prediction for the seawater properties. In addition, the principles are more simply illustrated with reference to a binary solution than a multicomponent solution. Extension to the more complicated system is then fairly straightforward.

The reader can choose his own starting point, and generally will go first to the Section which interests him most. Therefore, the sequence of the Chapters 3-6 was not considered very critical; except perhaps, that the partial molal volume should precede its temperature and pressure derivatives. Entropy begins the sequence, and its basics are dealt with in somewhat more detail than those of the other quantities which are considered to be more well known.

REFERENCES — Chapter 1

Doherty B.T. and D.R. Kester 1974.

Freezing point of seawater.
J. Mar. Res. 32(2): 285-300.

Fujino K. E.L. Lewis, and R.G. Perkin 1974.

The freezing point of seawater at pressures up to 100 bars.
J. Geophys. Res. 79(12): 1792-1797.

Kester D.R. 1974.

Comparison of recent seawater freezing point data.
J. Geophys. Res. 79(30): 4555-4556.

Kester D.R., I.W. Duedall, D.N. Connors, and R.M. Pytkowicz. 1967.

Preparation of artificial seawater.
Limnol.Oceanogr., 12(1): 176-179.

Knudsen M., C. Forsch, and S.P.L. Sorensen. 1902.

Bericht uber die chemische und physikalische Untersuchungen des Seewassers und die Aufstellung der neuen hydrographischen Tabellen.
Wiss. Meeresunters. Abt. Kiel, 6: 123-184.

Chapter 1.

Larson J.W. 1970.

Entropies of sodium and hydroxide ions and the thermodynamics of ionization of water.
J.Phys.Chem. 74(3): 685-686.

Leyendekkers J.V. 1973.

The chemical potentials of seawater components.
Mar.Chem. 1: 75-88.

McGlashan M.L. 1966.

The use and misuse of the laws of thermodynamics.
J.Chem.Ed. 43(5): 226-232.

Morcos S.A. 1970.

Chemical composition of seawater and the variation of calcium and alkalinity.
J.Conseil.Int.Explor.Mer. 33(2): 126-33.

Nancollas G.H. 1966.

"Interactions in electrolyte solutions", Elsevier, New York, N.Y., 400pp.

Pitzer K.S. and L. Brewer 1961.

Revised edition of "Thermodynamics" by G.N. Lewis and M. Randall. McGraw-Hill, New York, N.Y. 723pp.

Platford R.F. 1974.

Thermodynamics of the system H_2O-NaCl-Na_2SO_4-$MgSO_4$ to the saturation limit of NaCl at 25°C.
J.Sol.Chem. 3(10): 771-779.

UNESCO 1966.

Second Report of the Joint Panel on Oceanographic Tables and Standards. UNESCO Tech.Pap.Mar.Sci.,4: (Mimeographed) 9pp.

FURTHER READING:

Andrews, F.C. 1971.

"Thermodynamics: Principles and Applications", Wiley-Interscience, New York, 288pp.

Harned H.S. and B.B. Owen 1958.

"The physical chemistry of electrolytic solutions", 3rd ed, A.C.S. monograph, Reinhold, New York, 803pp.

Harned H.S. and R.A. Robinson 1968.

"Multicomponent electrolyte solutions", Pergamon Press, 110pp.

Chapter 1.

Klotz I.M. and R.M. Rosenburg 1972.

"Chemical thermodynamics", 3rd ed, Benjamin, Menlo Park, 444pp.

Munster A. 1970.

"Classical thermodynamics", Wiley-Interscience, New York, 387pp.

Pitzer K.S. and L. Brewer 1961.

Revised edition of "Thermodynamics" by G.N. Lewis and M. Randall. McGraw-Hill, New York, N.Y. 723pp.

CHAPTER 2

A FEW MATHEMATICAL NOTES

2.1 INTRODUCTION

The widespread use of computers has greatly facilitated the analysis of data. The useful device of summarising concepts in terms of mathematical statements has accordingly developed in directions not previously taken. Nevertheless, the manipulation of variables, in terms of the typical equations of partial differentiation remains as an essential part of any conceptual analysis. In fact, only a limited number of partial differential equations are needed to characterise the dynamics of nearly all aspects of hydrosystems; similarly, for thermodynamics.

The essentials of the following have been given, in a variety of styles, in a very large number of texts. However, for convenience to the interested reader, it is restated here in a simple outline.

The state of a system is defined by its properties which may be infinite, that is, the number of variables can be infinite. However, for a system that comes to equilibrium with its environment quite readily, all the properties (or variables) are fixed, when only a few properties are fixed. Thus, if we consider a known amount of pure substance and fix two of its properties, then the temperature, pressure and any other of its properties must be compatible with these conditions. If water is chosen, for example, and the dielectric constant and viscosity are fixed at 81.946 and $1.14\text{x}10^{-2}$ poise, respectively, then the temperature will be defined as 15°C, the pressure as 1 atmosphere, the density as 0.999129 gml^{-1} and so on.

Thus, we have arbitrarily chosen the dielectric constant and the viscosity as the independent properties (or variables) and the other properties are dependant on them.

Generally, temperature, T, and pressure, P, are taken as the independent variables, and the volume of a given amount of substance as a function of P, T serves as a basis for defining the state of the system. Therefore, any equation giving the relation between these three variables for a given substance is termed an equation of state. When one of the variables, say T, is kept constant a two-dimensional curve with axes P and V gives the relationship between the two variables. The slope of this curve (or isotherm) at any point is called the partial differential coefficient of V with respect to P, viz.

$$\partial V/\partial P$$

which is read as the rate of change of volume with pressure alone. The notation

$$(\partial V/\partial P)_T$$

denotes explicitly that the temperature is kept constant. Similarly, if P is kept constant the slope of the V-T curve at any point is expressed as $\partial V/\partial T$ or $(\partial V/\partial T)_P$.

These two coefficients are extensive, i.e. they depend on the amount of substance (gram, mole, etc.) however, by dividing them by the volume they become intensive quantities, like density, i.e.

$$(1/V)(\partial V/\partial T) \qquad \text{and} \qquad -(1/V)(\partial V/\partial P)$$

which are the coefficients of thermal expansion and of compressibility, two important quantities in oceanography (Chapters 5 and 6).

2.2 PARTIAL DIFFERENTIATION.

The geometric representation of the P-V-T relationship may be given by means of a surface. Thus, in Figure 2.1, the axes in the horizontal plane indicate the temperature and pressure, and the vertical axis represents the volume. At any point on the surface the height gives the volume corresponding to the temperature and pressure at that point.

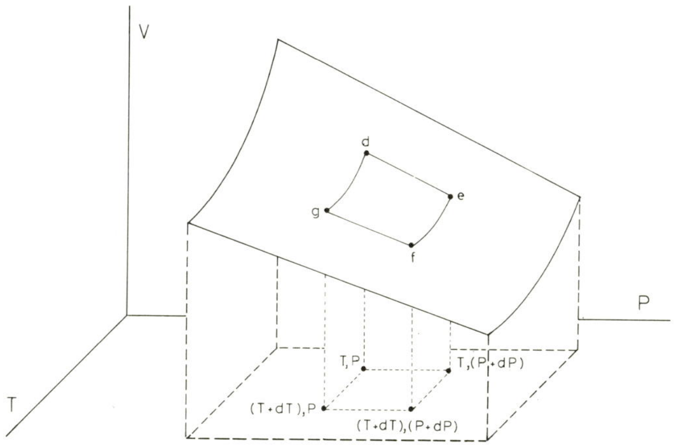

FIGURE 2.1

Consider any two infinitesimally distant points d and f, on the V-T-P surface. On going from d to f the temperature changes from T to T+ dT and the pressure from P to P + dP. The change in V is given by

$$dV = V_d - V_f \qquad (2.2.1)$$

Alternatively, an infinitesimal element of surface defg may be considered, and we can first go to e before going on to f. The volume change in the first step will be $V_d - V_e$, and since the slope of the line <u>de</u> is the V-P coefficient

$$V_d - V_e = (\partial V/\partial P)_T dP \qquad (2.2.2)$$

i.e. the change in V is equal to the rate of change of V with P multiplied by the change in P, T being constant. If we now go to <u>f</u> where T has become T + dT and the pressure is the same as it was at <u>e</u> viz., P + dP, this additional volume change will be

$$V_e - V_f$$

The slope of the line <u>ef</u> is equal to the V-T coefficient $(\partial V/\partial T)_P$ and so the volume change is

$$V_e - V_f = (\partial V/\partial T)_P dT \qquad (2.2.3)$$

The total volume change $(V_d - V_f)$ thus equals

$$(V_d - V_e) + (V_e - V_f), \text{ so that}$$

$$dV = (\partial V/\partial P)_T dP + (\partial V/\partial T)_P dT \qquad (2.2.4)$$

which is the basic equation of partial differentiation.
The path could equally well have gone from <u>d</u> to <u>g</u> and then <u>f</u>, with the same result.

If the height above the T-P plane remains constant in going from one point to another on the surface, we are moving along a contour line, i.e. the volume is constant. In this special case eq 2.2.4 becomes

$$(\partial V/\partial P)_T dP + (\partial V/\partial T)_P dT = 0 \qquad (2.2.5)$$

or, denoting the constancy of V more explicitly,

$$(\partial P/\partial T)_V = -(\partial V/\partial T)_P/(\partial V/\partial P)_T$$

$$= -(\partial V/\partial T)_P(\partial P/\partial V)_T \qquad (2.2.6)$$

since $(\partial V/\partial P)_T = 1/(\partial P/\partial V)_T$

Suppose we consider, in addition to V, some other variable (say E) dependent only on the temperature and pressure. If this variable is kept constant, then in going from <u>d</u> we follow along a line of constant energy to the infinitesimally distant point $\underline{f}'$, say. The change in V corresponding to the infinitesimal change in T is the sum of two terms. Firstly, $(\partial V/\partial T)_P$, the change in V which would occur for the same change in T alone, and, secondly, the change in V caused by the change in P necessary to keep the other variable E constant, which is given by $(\partial V/\partial P)_T(\partial P/\partial T)_E$, thus

$$(\partial V/\partial T)_E = (\partial V/\partial T)_P + (\partial V/\partial P)_T(\partial P/\partial T)_E \tag{2.2.7}$$

When changes in E, or some other dependent variable, are to be considered, equations of the form

$$(\partial V/\partial P)_T = (\partial V/\partial E)_T(\partial E/\partial P)_T \tag{2.2.8}$$

are found convenient.

A property may depend on more than two independent variables. A familiar example is the volume of a given quantity of seawater, which depends not only on temperature and pressure but also upon the <u>salinity</u>, $\underline{S}$. The equation for the infinitesimal change in the volume is just an extension of eq 2.2.4.

$$dV = \partial V/\partial T\, dT + \partial V/\partial P\, dP + \partial V/\partial \underline{S}\, d\underline{S} \tag{2.2.9}$$

If the volume also depends on other variables as well, the corresponding infinitesimal change in each variable, times the corresponding differential, is simply added to the sum. The more explicit presentation, employing the appropriate subscripts (to indicate the independent variables which remain constant during the differentiation) is given by

$$dV = (\partial V/\partial T)_{P,\underline{S}}dT + (\partial V/\partial P)_{T,\underline{S}}dP + (\partial V/\partial \underline{S})_{P,T}d\underline{S} + \ldots \tag{2.2.10}$$

Quite often, it is necessary to consider the second derivatives of the P-V-T relationships and a useful equation is

$$\frac{\partial}{\partial P}(\partial V/\partial T)_P = \partial^2 V/\partial P\partial T = \frac{\partial}{\partial T}(\partial V/\partial P)_T \tag{2.2.11}$$

In terms of the coefficients of thermal expansion and isothermal compressibility this equation gives

$$(\partial\alpha/\partial P)_T = -(\partial\beta_T/\partial T)_P$$

since $\beta_T = -(1/V)(\partial V/\partial P)_T$ (Chapter 6)

and $\alpha = (1/V)(\partial V/\partial T)_P$ (Chapter 5)

Equation 2.2.11 naturally follows from the character of V which is well defined - being a property of the system, and dV is a perfect differential.

2.3 PERFECT DIFFERENTIALS.

Any system in a given state is characterised by a definite amount of energy. If the system undergoes change, the energy change (gain or loss) internal to it is equivalent to the opposite energy change (loss or gain) in the surrounding systems. This follows from the first law of thermodynamics. If no transfer of matter occurs (in any physical or chemical process) the infinitesimal change of energy will be given by the sum of the corresponding changes in heat and work, viz.

$$dE = \delta q + \delta w \tag{2.3.1}$$

The lower case delta, δ, is used to indicate that the infinitesimal quantities of q and w are not perfect differentials. The quantities q and w are not properties of the system but are defined only for a given path of change.

In general, an infinitesimal quantity, say δQ may be given by

$$\delta Q = L(y,z)dy + M(y,z)dz \tag{2.3.2}$$

If this expression is the total differential of a function of the type of eq 2.2.4 then

$$L(y,z) = (\partial Q/\partial y)_z$$

and $$M(y,z) = (\partial Q/\partial z)_y$$

and $$\partial L/\partial z = \partial^2 Q/\partial y\partial z = \partial M/\partial y \tag{2.3.3}$$

The equality represented by eq 2.3.3 (of which eq 2.2.11 is an example) is a necessary and sufficient condition that eq 2.3.2 is a perfect differential of a function Q. This differential may be integrated from one state y_1 z_1 to another state y_2 z_2 and

$$\Delta Q = Q(y_2 z_2) - Q(y_1 z_1) \qquad (2.3.4)$$

However, if eq 2.3.3 does not hold, the integration is impossible unless the path from the one state to the other is specified. Since the result will depend on the path chosen, it is not possible to regard Q as a property of the system.

2.4 Solution of differential equations by digital-analog computer systems.

Finally, in regard to computer applications, we note a very interesting article by V. Vemuri (1973). This includes a section on the solution of differential equations using a computer, and gives the related references. The problems of the computation of partial differential equations is discussed. The conventional procedure for solving a partial differential equation on a digital computer (and the extension to the hybrid system) is outlined by considering a one-dimensional diffusion equation with defined initial, and boundary conditions.

Since only algebraic expressions can be solved via a digital computer, it is necessary to discretize all independent variables. To accomplish this, the whole of the space-time "field" over which the diffusion equation is defined is discretized by superimposing a grid. Only those finite set of points characterised by this grid are considered. This enables the space derivatives of the diffusion equation to be replaced by a central difference approximation, and a quantity, ϕ, say, becomes the discrete analog to the continuous variable of the diffusion equation.

The more important computational schemes for producing the final solution of the equation, using a hybrid computer, are then outlined. A hybrid computer uses both analog (either active or passive) and digital computing elements in an attempt to utilize the better features of both. It is conceivable to link electrical, mechanical and optical data processing elements in the design of a hybrid computer. The computational schemes discussed are:

1. The DSDT (discrete-space and discrete-time) methods, in which all derivatives in the equation are replaced by suitably chosen finite-difference approximations.
2. The DSCT (discrete-space and continuous-time) methods. These retain the time variable in its original continuous form and the space variable is approximated by finite difference approximations.
3. The CSDT (continuous-space and discrete-time) methods. While these methods have the advantage of being able to handle problems involving moving boundaries, there are difficulties regarding its implementation. The application of this method to the study of pollution in rivers and estuaries is given as an example.

   Other methods, not involving finite-difference approximations, are also briefly referred to. For example, a method found to be successful in solving a nonlinear flow problem.

   The solutions of integral equations are also considered.

   It appears that the hybrid computer is ideal for the study of large scale physical problems, such as occur in oceanography.

2.5 UNITS.

Usually, the units of most common usage (up to 1974) have been adhered to. Where experimental data are cited, the units of the reference are given, together with the more common units. For example, the pressure unit, bar, is the most commonly cited one in oceanographic work, however, $kgcm^{-2}$ is also used. In tabulations of data measured at atmospheric pressure

1 atm and 1 bar are used interchangeably. The ml and cm^3 are both used but are simply interconverted when this is necessary (for example, for specific volume data).

The general equations given in Chapter 1 must be used with some self-consistent set of units. See, for example, eq 6.61.8. Since the practical units commonly used are not always mutually consistent it is necessary to know the exact numerical relationship between these units. The interconversion factors for a number of units, likely to be more commonly used, are given below, together with some important physical constants that are used throughout the text.

<u>Physical constants</u>: (Pitzer and Brewer 1961)

The Gas Constant, R	$82.0597 cm^3 atm/degmol$
	1.98726 cal/degmol
	8.31470 joules/degmol
Planck's constant, h	$6.6252 x 10^{27}$ erg sec
Avogadro's number, N_o	$6.0232 x 10^{23}$ per mol
Boltzmann's constant, $k = R/N_o$	$1.38045 x 10^{-16}$ erg/deg
Electron charge e	$4.8029 x 10^{-10}$ esu
	$1.60206 x 10^{-19}$ coulomb
Absolute temperature of ice point (0°C)	273.150°K

(mole is based on 16 g of natural oxygen)

<u>Pressure</u>:

		atm	$dynes/cm^2$	kg/cm^2	bars
1 atm	=	1	1,013,250	1.03323	1.01325
1 kg/cm^2	=	0.967751	980,663	1	0.980663
1 bar	=	0.986923	10^6	1.01972	1

Energy:

	cal/mol	joule/mol	cm^3 atm/mol
1 cal/mol =	1	4.1840	41.292
1 Joule/mol* =	0.23901	1	9.8692
1 cm^3 atm/mol =	0.024218	0.10133	1

* 1 joule = 10^7cm-dyne = 10cm^3 bar

Others:

1 ml = 1.000028cm^3

1 phys. mole = 1.000275 chem.mole

Ionic strength, I, mol/kg of water

$= 0.01994\ S(‰)/(1-10^{-3}(S‰))$

(See Section 1.10)

REFERENCES Chapter 2.

Pitzer K.S. and L. Brewer 1961.

revised edition of "Thermodynamics" by
G.N. Lewis and M. Randall.
McGraw-Hill, New York, N.Y., 723pp.

Vemuri V. 1973.

Hybrid computation in hydrosystems, in:
Advances in Hydroscience, Vol 9, edited by
Ven Te Chow, Academic Press, N.Y. pp1-48.

CHAPTER 3

ENTROPY

3.1 INTRODUCTION

As pointed out in Chapter 1 the concept of entropy is needed in thermodynamics because irreversible processes exist. However, for systems in equilibrium entropy changes occur in different parts of the system (with $d\$ = dq_{rev}/T$ from eq 1.3.1) even though the overall entropy change (for the whole system) is zero. Thus, equations involving entropy find useful application in studies involving reversible processes, see for example sections 1.71 and 1.8. As in the previous chapters, the symbol for entropy will be $\$$, in order to avoid any confusion with salinity. The partial molal entropies are represented by $\bar{s}$.

3.2 Entropy exchanges under equilibrium conditions.

(change of state - solid to liquid, liquid to vapor)

Consider the fusion of 1 mole of ice at its melting point, at atmospheric pressure.

For equilibrium any infinitesimal change in the temperature will either cause the water to freeze (decrease in T) or the solid to melt (increase in T). Similarly, at constant T, infinitesimal changes in pressure will cause corresponding melting or freezing according to the direction of change. In other words, the process is reversible. The change in entropy during fusion is equal to the heat absorbed (δq) divided by the absolute temperature, T (see Section 1.3).

$$d\text{\$} = \delta q_{rev}/T \qquad (3.2.1)$$

Since water is a pure substance the temperature is constant during fusion and

$$\Delta\text{\$} = \Delta H/T \qquad (3.2.2)$$

where ΔH is the heat of fusion of 1 mole, and equals 1436 cals for ice. Thus,

$$H_2O(\text{solid}) = H_2O(\text{liq}) \qquad \text{\$} = 1436/273.15 = 5.257 \text{ caldeg}^{-1}$$

The entropy change in vaporisation is obtained in the same way with ΔH being the heat of vaporisation.

3.3 Change of Entropy with Temperature

From eq (1.3.5a)

$$(\partial\text{\$}/\partial T)_p = c_p/T \qquad (3.3.1)$$

Thus, at constant pressure the change in entropy of a substance due to a temperature change is given by

$$\int d\text{\$} = 2.303 \int c_p d\log T \qquad (3.3.2)$$

In the simple case, where c_p is constant, for the temperature range T_1 to T_2

$$\text{\$}_1 - \text{\$}_2 = c_p \ln(T_1/T_2) = 2.303 c_p \log(T_1/T_2) \qquad (3.3.3)$$

For example, if ice at 0°C is cooled to -10°C, with $c_p(\text{ice})=9$ $\text{caldeg}^{-1}\text{mol}^{-1}$, then

$$\text{\$}_{0°} - \text{\$}_{-10°} = 2.303\text{x}9(\log(273.15/263.15))$$

and if water is heated to 0°C from -10°C with,

$$c_p(\text{water}) = 18,$$

$$\text{\$}_{-10°} - \text{\$}_{0°} = 2.303\text{x}18(\log(263.15/273.15))$$

The sum of these two processes gives

$$2.303x9(\log 263.15/273.15) = -0.336 \text{ caldeg}^{-1}$$

These results can be used to calculate the entropy change for the irreversible isothermal process of fusion of ice at -10°C from supercooled water at -10°C.

$$H_2O(\text{liq}) \quad = \quad H_2O(\text{solid}) \text{ irrev.}$$

The process can be carried out via the following steps with the entropy changes as shown.

$$H_2O(\ell, -10°C) = H_2O(\ell, 0°C) \qquad \Delta\underline{S} = 2.303x18\log\frac{(263.15)}{(273.15)}$$

$$H_2O(\ell, 0°C) = H_2O(\text{solid}, 0°C) \qquad \Delta\underline{S} = \Delta H/T = -1436/273.15$$

$$H_2O(\text{solid}, 0°C) = H_2O(\text{solid}, -10°C) \quad \Delta\underline{S} = 2.303x9\log\frac{(273.15)}{(263.15)}$$

Summing these entropy changes gives -5.593 caldeg^{-1} which is the entropy change for the process. Thus, ice at -10°C has a molal entropy less than water by -5.59caldeg^{-1}.

Note that the irreversible step is replaced by a reversible path between initial and final states.

3.4 Absolute entropies of pure substances

Chemical reactions in solution are determined by the ΔG, free energy change of reaction. The value of ΔG is related to the corresponding entropy and enthalpy change via

$$\Delta G = \Delta H - T\Delta\underline{S} \qquad (3.4.1)$$

from eq1.2.5 (with temperature constant).

For equilibrium $\Delta G=0$ and $T\Delta\underline{S}$ is the reversible heat change.

The $\Delta\underline{S}$ for a chemical reaction is obtained from the absolute entropies of the reactants and products (as well as the absolute entropies of the components of the solution).

The entropy of each element in its solid state is taken as zero at the absolute zero of temperature. This, in effect, is an arbitrary fixing of the entropy in a particular state. However, it is in accord with the Third Law of Thermodynamics and applies exactly to perfect crystalline substances (regardless of the pressure or volume). Most pure substances have perfect crystalline behavior at 0°K, some exceptions are, CO, N_2O, NO (or N_2O_2 in the solid), H_2O and D_2O and $Na_2SO_4.10H_2O$ and the 1-olefins with more than 10 carbon atoms. The residual entropy for ice is Rln(3/2) or 0.806 $caldeg^{-1}mol^{-1}$, and for $Na_2SO_4.10H_2O$ it is Rln2 (1.378 $caldeg^{-1}mol^{-1}$).

The standard entropy at temperature T_1 of a solid which is perfectly crystalline at 0°K is given by

$$\underline{S}^o_{T_1} = \int_0^{T_1} (c^o_p/T)dT \qquad (3.4.2)$$

Heat capacities, c_p, have been measured to very low temperatures for most chemical elements and many compounds. At extremely low temperatures

$$c_p = aT^3 + bT \qquad (3.4.3)$$

where a and b are constants.

Deviations from this equation(which is derived from the Debye function) occur, for example, at around 10°K for KCl and 17°K for Na_2SO_4. By plotting c_p/T versus T^2 and extrapolating to 0°K the low temperature values of c_p are obtained. These can then be transferred to a plot of c_p/T versus T for the calculation of entropy by integration.

With $\underline{S}$ = 0 at 0°K the entropy at T°K is, from eqs 3.4.2 and 3.4.3

$$\underline{S} = (a/3)T^3 + bT \qquad (3.4.4)$$

The absolute entropy of a liquid, such as water or mercury, say, at 25°C is thus calculated from the entropy of the solid(obtained via graphical integration of the c_p/T versus T curve), from the changes in entropy when the solid melts, and when the temperature of the liquid is raised to 25°C. Absolute entropies have been obtained in this way and are tabulated in a number of texts (Kelley and King 1961, Pitzer and Brewer 1961).

TABLE 3.1 Entropies of pure substances $cal\,deg^{-1}mol^{-1}$ at 25°C

Substance	$\underline{S}^o_{298}$	Substance	$\underline{S}^o_{298}$
$Cl_2(g)$	45.93	$CaSO_4$	25.5
$CO_2(g)$	51.08	$FeCl_2$	28.7
KF(g)	54.19	KBr	22.96
$N_2(g)$	45.77	KCl	19.72
$O_2(g)$	49.01	KF	15.91
$H_2O(l)$	16.73	KI	25.2
<u>solids</u>		$KMg_3AlSi_3O_{10}F_2$	75.9
Al	6.77	LiCl	14.17
C	1.37	LiF	8.52
CaO	9.5	$MgCl_2$	21.40
Cu	7.97	MgF_2	13.68
Fe	6.49	$Mg(OH)_2$	15.09
AgCl	23.00	Mg_2SiO_4	22.75
AgBr	25.6	NaCl	17.33
$CaCl_2$	27.2	Na_2CO_3	32.5
$CaCO_3$	22.2	NaF	12.26
CaF_2	16.46	NaI	23.55
$Ca(NO_3)_2$	46.2	$NaNO_3$	27.85
$Ca(OH)_2$	19.93	Na_2SO_4	35.73
		$PbCl_2$	32.5
		UO_3	23.58

Pitzer and Brewer (1961)

Some examples are given in Table 3.1 for 25°C. While the same methods as above can be applied to gases (using the entropy changes - solid, solid to liquid, liquid at T_1 to T_2, liquid to gas) more precise methods are available using spectroscopic and other molecular data (see Pitzer and Brewer 1961, Chapter 27).

For liquids and solids the "solvent" standard state (see Section 1.9) refers to the pure substance at 1-atm pressure. Thus, the absolute entropies are the standard entropies.

For gases, the standard state is the hypothetical perfect gas at a designated pressure of 1-atm. For a perfect gas, the entropy change on isothermal volume change between some low pressure P and 1 atm (the standard state) is given by

$$\underline{S}^{(1)} - \underline{S}^{(p)} = nR\ln(P/1)$$

$$= R\ln(P/1) \text{ for 1 mole} \qquad (3.4.5)$$

(eq 3.4.5 is obtained from the 2nd law and using the properties of a perfect gas - Pitzer and Brewer 1961 p85).

Thus, since $\underline{S}^{o} = \underline{S}^{(1)}$

$$\underline{S}^{o} = \lim_{P \to 0}(\underline{S}^{(p)} + R\ln P) \qquad (3.4.6)$$

The extension to real gases is given in the next section.

3.5 Change of entropy with pressure.

The change in entropy with pressure is given by

$$(\partial \underline{S}/\partial p)_T = -(\partial V/\partial T)_p \quad \text{from eq 1.8.14} \qquad (3.5.1)$$

so that

$$dS = -\int(\partial V/\partial T)_p dp$$

$$= -\int (\alpha V)dp \qquad (3.5.2)$$

For real gases the change of entropy with pressure is, from eq 3.5.2

$$(\underline{S}^{(p)} - \underline{S}^{(0)})_T = -\int_0^P (\partial V/\partial T)_p dp \qquad (3.5.3)$$

The entropy of a gas is infinite at zero pressure and this equation is not useful by itself. However, if the standard state is defined as above (the hypothetical ideal gas at 1 atm pressure), then, with $\underline{v}$ representing the molal volume, and $\underline{S}^{o}$ the standard entropy,

$$(\underline{S}^{(p')} - \underline{S}^{o})_T = -\int_0^{p'} (\partial \underline{v}/\partial T)_p dp + \int_0^1 R/P\, dp \qquad (3.5.4)$$

since $R/P = (\partial v/\partial T)_p$ for a perfect gas. Rearrangement gives

$$\int_0^{p'} (R/P - (\partial v/\partial T)_p)dp - R\ln P' \qquad (3.5.5)$$

for the R.H.S.

Since, the virial equation of state gives

$$Pv = RT + BP + CP^2 + DP^3 + \ldots$$

$$(\underline{\mathcal{S}}^{(p)} - \underline{\mathcal{S}}^{o})_T = -(R\ln P + (dB/dT)P + (dC/dT)P^2/2 + (dD/dT)P^3/3 + \ldots) \qquad (3.5.6)$$

From which the entropy of a real gas may be obtained when the temperature derivatives of the virial coefficients are known.

3.6 Solutions of electrolytes.

For a solution, the total entropy is just the sum of the contributions from the various components, as for the volume. If $\underline{\mathcal{S}}$ is the total entropy, $\bar{s}_w$ the partial molal entropy of the water and $\bar{s}_i$ the partial molal entropy of the component i, then

$$\underline{\mathcal{S}} = n_w\bar{s}_w + \Sigma_i n_i\bar{s}_i \qquad (3.6.1)$$

For a volume of seawater containing 55.51 moles of water and m_i moles of each of the salts of a given recipe (e.g. A in Table 4.16, say) the entropy is given by

$$\underline{\mathcal{S}}_{sw} = 55.51\bar{s}_w + \Sigma_i m_i\bar{s}_i^{sw} \qquad (3.6.2)$$

It will be routine, in all chapters, to look at the results for binary solutions before going on to consider seawater.

3.7 Binary solutions

The partial molal entropy is the derivative of the pmfe with respect to temperature, at constant pressure and composition (moles of components), and is given by

$$(\partial\bar{G}_i/\partial T)_p = \partial((\partial G/\partial n_i)_{P,T})/\partial T$$
$$= \partial((\partial G/\partial T)_p)/\partial n_i$$
$$= -(\partial \underline{S}/\partial n_i)_{p,T} = -\bar{s}_i \quad (3.7.1)$$

Then, since

$$\bar{G}_i - \bar{G}_i^o = RT\ln \underline{a}_i \quad (1.9.1)^*$$

and differentiating with respect to temperature at constant pressure and composition,

$$(\partial\bar{G}_i/\partial T)_p - (\partial\bar{G}_i^o/\partial T)_p = RT(\partial \ln \underline{a}_i/\partial T)_p + R\ln \underline{a}_i \quad (3.7.2)$$

but $$RT(\partial \ln \underline{a}_i/\partial T)_p = -\bar{L}_i/T \quad (3.7.3)$$

where $\bar{L}_i$ is the partial molal enthalpy,

so that, eq 3.7.2 becomes (using eq 3.7.3)

$$-\bar{s}_i + \bar{s}_i^o = -\bar{L}_i/T + R\ln \underline{a}_i$$

or $$\bar{s}_i - \bar{s}_i^o = \bar{L}_i/T - R\ln \underline{a}_i \quad (3.7.4)$$

If $\bar{L}_i$ is constant with temperature, then $R\ln \underline{a}_i = \bar{L}_i/T$ + constant (from eq 3.7.3), so that $\bar{s}_i - \bar{s}_i^o$ = -constant.

However, this is generally not true for aqueous electrolyte solutions.

If the partial molal heat capacities, $\bar{c}_p$, are known, $\bar{L}$ is expressed as a function of temperature by

$$d\bar{L}/dT = \bar{c}_p - c_p^o = \bar{j} \quad (3.7.5)$$

where $\bar{j}$ is the relative partial molal heat capacity in the given state. Assuming $\bar{j}$ to be effectively constant with temperature, $\bar{L}$ may be expressed as a linear function of temperature ($\bar{L}$ = const + $\bar{j}T$), so that integration of eq 3.7.3 now gives

$$R\ln \underline{a}_i = \bar{L}_i/T - \bar{j}\ln T + \text{constant} - \bar{j}$$
$$\bar{s}_i - \bar{s}_i^o = \bar{j}(1 + \ln T) + \text{constant}$$
$$= \bar{j}(1 + 2.303\log T) + \text{constant} \quad (3.7.6)$$

So that, $\Delta\bar{s}_i(t) = \Delta\bar{s}_i(25) + 2.303\bar{j}\log(T/298)$ (3.7.7)

The standard partial molal entropy $\bar{s}_i^o$ refers to the hypothetical ideal solution of unit molality (the solute standard state) for the electrolyte component. For water, in the solution, $\bar{s}^o$ is the molal entropy of the pure liquid (see Section 1.9).

As noted in Section 3.4, in order to obtain Δ$\not{S}$ for a chemical reaction, the absolute entropies of reactants and products are needed, including the absolute entropies of the components of the solution. For the component water (solvent standard state) $\bar{s}^o$ is the entropy of pure water, calculated as outlined above (section 3.4). The absolute entropy of water in the solution is thus calculated from $\bar{s}_w - \bar{s}_w^o$ (eq 3.7.4). For the electrolyte components in the solution the standard entropies refer to the solute standard state. The calculation of the standard entropies, in this case, is naturally not as simple as for a pure solid or liquid and will now be outlined in detail.

3.71 Standard entropies of salts in solution

For the system of a salt in equilibrium with an aqueous solution of the salt, NaCl, say, the standard free energy of solution, $\Delta\bar{G}^o_{solution}$, for the reaction, solid ⟶ solution, is given by

$$\Delta\bar{G}^o_{solution} = -RT\ln \underline{a}_{NaCl}(sat.) \qquad (3.71.1)$$

(this equation is derived along the usual lines as outlined in section 1.9; an alternative derivation is as follows).

In other words, for equilibrium the free energy of the solid equals the partial molal free energy of the solid in solution (NaCl) i.e.

$$G^o(solid) = \bar{G}^o_{NaCl} + RT\ln \underline{a}_{NaCl}(sat.) \qquad (3.71.2)$$

$$\text{and} \quad \bar{G}^o_{NaCl} - G^o(solid) = \Delta\bar{G}^o \qquad (3.71.3)$$

since the standard free energy of the pure solid is G^o(solid) so that $\Delta\bar{G}^o_{solution} = -RT\ln a_{NaCl}(sat.)$ (3.71.4)

This indicates the relation between the chemical potential of the solute as a solid and in its standard state in solution.

The corresponding entropy change is obtained, by measurement of the heat of solution of the solid, via

$$\Delta\bar{s}^o_{solution} = (\Delta\bar{H}^o_{solution} - \Delta\bar{G}^o_{solution})/T \quad (3.71.5)$$

where $\Delta\bar{H}^o_{solution}$ is the standard enthalpy of solution.

The solute standard state is so defined that it has the same partial molal enthalpy as the infinitely dilute solution.

That is, the enthalpy change for

NaCl(solid) ⟶ NaCl(in solution) standard state

is the same as NaCl(solid) ⟶ infinitely dilute solution.

This can readily be seen from

$$\bar{H}_i = \bar{H}^o_i - \nu RT^2(\partial\ln\gamma_{\pm_i}/\partial T)_{m,P} \quad (3.71.6)$$

where $\underline{i}$ represents the electrolyte. At infinite dilution $\gamma_\pm = 1$ at all temperatures so that

$$\bar{H}_i(\text{inf.dil.}) = \bar{H}^o_i \quad (3.71.7)$$

Table 3.2 lists values of $\Delta\bar{s}^o_{solution}$ for a number of salts, together with the standard entropy of the salt in the solution, $\bar{s}^o_i$.

Since $\Delta\bar{s}^o_i(\text{solution}) = \bar{s}^o_i(\text{solution}) - s^o_i(\text{solid})$

then $\bar{s}^o_i(\text{solution}) = \Delta\bar{s}^o_i(\text{solution}) + s^o_i(\text{solid})$ (3.71.8)

where s^o_i(solid) is simply the absolute entropy of the pure salt.

As for the partial molal volumes it is convenient to tabulate the $\bar{s}^o_{ion}$ values rather than those for the electrolytes. This will involve choosing a suitable convention as is done for the $\bar{v}^o_{ion}$ values.

TABLE 3.2 Standard partial molal entropies of salts in solution (water as solvent) $caldeg^{-1}mol^{-1}$ at 25°C

Salt	Entropy of pure solid s_i^o(solid) (a)	$\Delta\bar{H}^o$ of solution $calmol^{-1}$	$\Delta\bar{G}^o$ of solution $calmol^{-1}$	$\Delta\bar{s}^o_{sol} = \frac{(\Delta\bar{H}^o - \Delta\bar{G}^o)}{298.15}$ (b)	$\bar{s}_i^o$ (a) + (b)
NaCl	17.33	928	-2134	10.3	27.6
KCl	19.76	4110	-1260	18.0	37.8
LiCl	14.17	-8850	-9575	2.4	16.6
NaF	12.26	218	266	-0.2	12.1
KBr	22.96	4750	-1460	20.8	43.8
$NaNO_3$	27.85	4900	-1525	21.6	49.4

Data from Pitzer and Brewer (1961)

3.72 The Standard Entropy of Ions in Solution (in cal $deg^{-1}mol^{-1}$)

Entropy of Solution.

The usual convention adopted for estimating the standard partial molal entropies of ions is to assign a value of zero to the hydrogen ion. Since $\bar{s}^o_{HCl}$ = 13.16, then $\bar{s}^o_{Cl-}$ is assigned the value 13.16. The value of $\bar{s}^o_{NaCl}$ is 27.6, so that $\bar{s}^o_{Na+}$ becomes 14.4, and so on.

Values of $\bar{s}^o$ for a large number of ions are listed in Appendix 3. (Table 3A.1).

Laidler and Pegis (1957) found that the standard partial molal entropies of simple cations could be fitted to the equation

$$\bar{s}^o_{298} = -5.5z + 3/2R\ln M + 10.2 - 11.6z^2/r_u \quad (3.72.1)$$

The last term is proportional to the temperature derivative of the Born equation, which gives the electrostatic entropy of solvation, viz

$$\Delta\bar{s}_{el} = (z^2e^2N_o/8\Pi\varepsilon_o r_o D^2)(\partial D/\partial T) \qquad (3.72.2)$$

where z is the ionic charge (+ve or −ve) e the charge on the electron, N_o is Avogadro's number, ε_o is the permittivity of free space, D the dielectric constant and r_o is the radius of the spherical boundary between the ion and the dielectric. Introducing the numerical value of $e^2N_o/8\Pi\varepsilon_o$, gives

$$\Delta\bar{s}_{el} = (166\text{x}10^3 z^2/r_o D^2)(\partial D/\partial T) \quad \text{calmol}^{-1}\text{deg}^{-1} \qquad (3.72.3)$$

(in mks units the value of the coefficient is $6.95\text{x}10^{-5}$ joule m/mole) at 25°C, for water from data listed by Harned & Owen (1958)

$$\Delta\bar{s}_{el} = -9.7z^2/r_o \quad \text{caldeg}^{-1}\text{mol}^{-1}$$

Values of the coefficient for z^2/r_o at various temperatures are given in Chapter 4 but are listed in Table 3.3 as well, for convenience.

TABLE 3.3 The Coefficient of the Electrostatic Entropy of Solvation*

Temp. °C	0	5	15	25	35
$\Delta\bar{s}(r_o/z^2)$ Å caldeg^{-1}	8.8	8.9	9.3	9.7	10.1

* eq 3.72.2

The term −5.5z in eq 3.72.1 converts the values to the conventional scale $\bar{s}^o_{H+} = 0$ and r_u is Pauling's univalent ion radius (Pauling 1960). This is the radius the ion would have if subject only to the compressive forces of single charges. Values of r_u for a number of ions are given in Table 3.4. The non-electrostatic quantity (3/2RlnM + 10.2) covers the non-stationary condition of the ion and specific interactions with the solvent, water.

This term is similar in form to the translational entropy of the ion in the ideal gas phase. For monoatomic gases, this is given by the Sackur-Tetrode equation, derived from statistical mechanics, viz.

$$\bar{s}(g) = R(1.5\ln M + 2.5\ln T - \ln P + \ln Q_e) - 2.315 \qquad (3.72.4)$$

with M the atomic weight, P the pressure in atm and Q_e the multiplicity of the ground state. For atoms or ions of the inert gas structure $Q_e = 1$, and for the standard state of 1 atm

$$\bar{s}^o(g) = 1.5R\ln M + 2.5R\ln T - 2.315 \quad \text{caldeg}^{-1}\text{mol}^{-1} \qquad (3.72.5)$$

(Pitzer and Brewer 1961, p639)

This equation is used to calculate the entropy of hydration which will be discussed in the next section.

Another semi-empirical equation that fits the values of $\bar{s}^o_{298}$ as closely as eq (3.72.1) is that derived by Powell and Latimer (1951),

$$\bar{s}^o_{298} = 1.5R\ln M + 37 - 270(|z|/r_e^2) \quad \text{caldeg}^{-1} \qquad (3.72.6)$$

where r_e is given by the crystal radius plus 1.0Å for negative ions and 2.0Å for positive ions.

For the entropies of polyatomic ions, the following equations have been proposed:

For ions of the type $XO_n(OH)_\ell^{-z}$, Couture and Laidler (1957) give,

$$\bar{S}^o_{298} = 5.5|z| + 40.2 + (3/2)R\ln M - 108.8z^2/nr \qquad (3.72.7)$$

where r is the X-O bond distance plus 1.40Å. King (1959) has shown that the effect of the symmetry number should be included.

Cobble (1953) gives a number of empirical equations for complex ions. For example, a large number of inorganic complex ions in solution have entropies that fit the equation

$$\bar{s}^o_{298} = 49 - 99|z|(F/(r_+ + r_-)) + ws^o_{H_2O} \qquad (3.72.8)$$

where F is a structural factor and w is the number of water molecules displaced from the hydration shell of the cation when the complex is

TABLE 3.4 Paulings Crystal and Univalent* Ionic Radii. (Å)

Ion	r_c	r_u	Ion	r_c	r_u
Al(3+)	0.50	0.72	La(3+)	1.15	1.39
Ag(+)	1.26	1.26	Li(+)	0.60	0.60
As(3-)	2.22	2.85	Mg(++)	0.65	0.82
As(5+)	0.47	0.71	Mo(6+)	0.62	0.93
B(3+)	0.20	0.35	N(3-)	1.71	2.47
Ba(++)	1.35	1.53	N(5+)	0.11	0.25
Be(++)	0.31	0.44	O(--)	1.40	1.76
Bi(5+)	0.74	0.98	O(6+)	0.09	0.22
Br(-)	1.95	1.95	P(3-)	2.12	2.79
Br(7+)	0.39	0.62	P(5+)	0.34	0.59
C(4-)	2.60	4.14	Pb(4+)	0.84	1.06
C(4+)	0.15	0.29	Rb(+)	1.48	1.48
Ca(++)	0.99	1.18	S(--)	1.84	2.19
Cd(++)	0.97	1.14	S(6+)	0.29	0.53
Ce(4+)	1.01	1.27	So(3-)	2.45	2.95
Cl(-)	1.81	1.81	Sb(5+)	0.62	0.89
Cl(7+)	0.26	0.49	Sc(3+)	0.81	1.06
Cr(6+)	0.52	0.81	Se(--)	1.98	2.32
Cs(+)	1.69	1.69	Se(6+)	0.42	0.66
Cu(+)	0.96	0.96	Si(4-)	2.71	3.84
F(-)	1.36	1.36	Si(4+)	0.41	0.65
F(7+)	0.07	0.19	Sn(4+)	0.71	0.96
Ga(3+)	0.62	0.81	Sr(++)	1.13	1.32
Ge(4-)	2.72	3.71	Te(--)	2.21	2.50
Ge(4+)	0.53	0.76	Te(6+)	0.56	0.82
He	-	0.93	Ti(4+)	0.68	0.96
H(-)	2.08	2.08	Tl(3+)	0.95	1.15
Hg(++)	1.10	1.25	V(5+)	0.59	0.88
I(-)	2.16	2.16	Y(3+)	0.93	1.20
I(7+)	0.50	0.77	Zn(++)	0.74	0.88
In(3+)	0.81	1.04	Zr(4+)	0.80	1.09
K(+)	1.33	1.33			
Kr	-	1.69			

* radii that multivalent ions would have if (while retaining their electron distributions) their Coulomb interactions were characteristic of univalent ions.

$r_u = r_c z^p$ where p is an integer characteristic of the crystal (Pauling 1960).

formed. F corrects for the "open" structure of the complexes, and has a value of 0.65.

Values of $\bar{s}^o$ for a large range of cations have been plotted against z^2/r_u in Figure 3.1. Values of z^2/r_c, where r_c is the crystal radius, are shown for comparison; r_u values were not available for many ions. With the slope taken as that for $\Delta\bar{s}_{el}$ (-9.7caldeg^{-1}mol^{-1}), the line

$$\bar{s}^o_{298} = -5.5z + (3/2)R\ln M + A - 9.7z^2/r_u \qquad (3.72.9)$$

with A = 22,

gives a good fit to the data. It is considered best to take the theoretical slope (Table 3.3) since changes with temperature are accurately known. The $\bar{s}^o$ values of the anions fell about 22 units below this line. For purposes of summarising the experimental data (Table 3A.1) and for the prediction of $\bar{s}^o$ at other temperatures, it was considered convenient to use eq 3.72.9 as a basis.

The values of $(z^2/r_u)_{calc}$ corresponding to this equation are shown in Table 3.5. It should be noted that z is negative for anions and +ve for cations, as in eq 3.72.1. Values of z^2/r_u with r_u from Table 3.4 are also shown. The average deviation between $(z^2/r_u)_{calc}$ and z^2/r_u is ± 4caldeg^{-1}mol^{-1} (for 17 ions, Be^{2+}, Al^{3+}, Pb^{4+}, Sn^{4+} and Y^{3+} are not included in the average).

The value of the coefficient for z^2/r_u at 0°C, is 8.8 (Table 3.3). Using eq 3.72.5 as a guide, a decrease in the parameter A of around 0.5 [(5/2)Rln(273.15/298.15)] could be expected. A value of −2 for cations and +1 for anions was estimated from values of $\bar{s}^o$(0°C) calculated from $\bar{c}^o_p$ values. The estimated changes in the entropies for the temperature range 0 to 25°C are listed in Table 3.6. The radius r_u was taken as independent of temperature, and changes in $\bar{s}^o_{H+}$ were neglected. If the change in the entropy of the hydrogen ion is taken into account, then

$$\bar{s}^o_{273} - \bar{s}^o_{298} = -1.5z - 0.5 + 0.9(z^2/r_u)_{calc} \qquad (3.72.10)$$

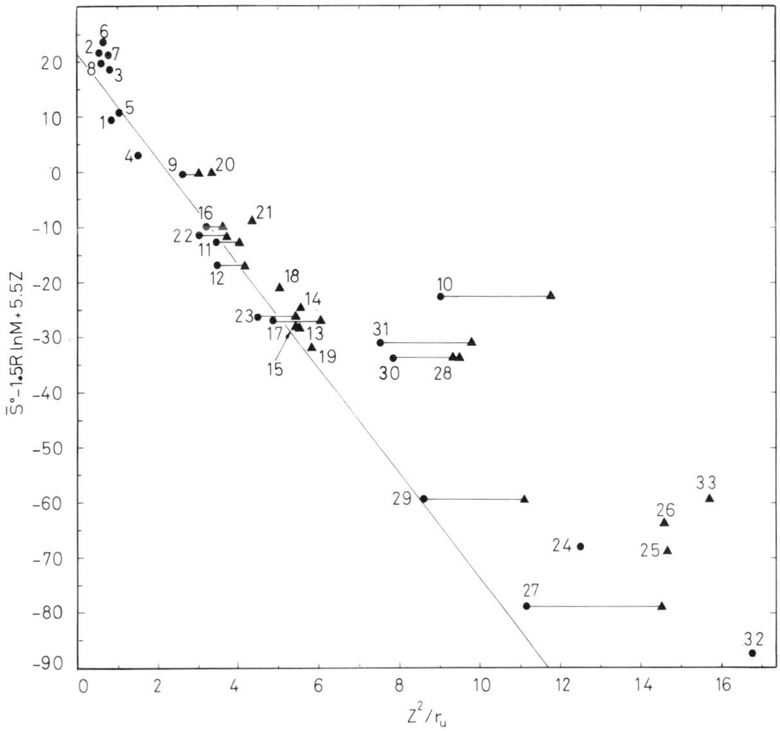

FIGURE 3.1

The standard partial molal entropies (cal/deg mol) of cations (in water) as a function of the electrostatic entropy of solvation (represented by the Born equation, eqs 3.72.2 and 3.72.9, with slope = -9.7Å cal/deg mol) t = 25°C, p = 1-atm.

Charge +1	Code	1	2	3	4	5	6	7	8
	Ion	Ag	Cs	K	Li	Na	NH_4	Rb	Tl
Charge 2+	Code	9	10	11	12	13	14	15	16
	Ion	Ba	Be	Ca	Cd	Co	Cu	Fe	Hg
	Code	17	18	19	20	21	22	23	
	Ion	Mg	Mn	Ni	Pb	Sn	Sr	Zn	
Charge 3+	Code	24	25	26	27	28	29	30	31
	Ion	Al	Cr	Fe	Ga	Gd	In	Tl	Y
Charge 4+	Code	32	33						
	Ion	Sn	Th						

● r_u, the unitary radius of Pauling,

▲ r_c, the crystal radius of Pauling (Table 3.4)

TABLE 3.5 Electrostatic entropy of solvation for various ions*

25°C, 1-atm

$-\Delta\bar{s}_{el}/9.7$ $\mathring{A}^{-1}$

Anion	$(z^2/r_u)_{calc}$	Cation	$(z^2/r_u)_{calc}$	Cation	$(z^2/r_u)_{calc}$
Br^-	2.20	Ag^+	1.32 (0.79)^x	Mn^{2+}	4.43
BrO_3^-	0.36	Al^{3+}	9.30 (12.50)	Na^+	1.18 (1.03)
Cl^-	2.57	Ba^{2+}	2.34 (2.61)	NH_4^+	-0.13 (0.70)
ClO_3^-	0.13	Be^{2+}	4.59 (9.09)	Ni^{2+}	5.58
ClO_4^-	-0.25	Ca^{2+}	3.63 (3.39)	Pb^{2+}	2.25
CN^-	1.26	Cd^{2+}	4.09 (3.51)	Pb^{4+}	10.30 (15.09)
$CO_3^=$	6.00	Co^{2+}	5.17	Rb^+	0.01 (0.68)
$CrO_4^=$	3.77	Cr^{3+}	9.36	Sn^{2+}	3.21
F^-	3.98	Cs^+	-0.08 (0.60)	Sn^{4+}	11.26 (16.67)
$H_2AsO_4^-$	1.47	Cu^{2+}	4.84	Sr^{2+}	3.48 (3.03)
HCO_3^-	1.81	Fe^{2+}	5.16	Th^{4+}	8.38
$HPO_4^=$	5.70	Fe^{3+}	9.03	Tl^+	0.50 (0.68)
$H_2PO_4^-$	2.01	Ga^{3+}	10.44 (11.11)	Tl^{3+}	5.81 (7.83)
HSO_4^-	1.09	Gd^{3+}	6.55	U^{3+}	5.96
I^-	1.63	Hg^{2+}	3.32 (3.20)	U^{4+}	9.72
IO_3^-	1.54	In^{3+}	8.42 (8.65)	V^{3+}	8.48
MnO_4^-	-0.51	K^+	0.30 (0.75)	Y^{3+}	5.45 (7.50)
NO_2^-	0.93	Li^+	1.95 (1.47)	Zn^{2+}	5.04 (4.55)
NO_3^-	0.50	Mg^{2+}	5.02 (4.88)		
OH^-	3.96				
$S^=$	5.13				
$SO_4^=$	4.35				

* calculated from eq 3.72.3

x from univalent ionic radii (see Table 3.4)

TABLE 3.6 Standard partial molal entropy changes with temperature

$\bar{s}^o_{ion}(0°C) - \bar{s}^o_{ion}(25°C)$ caldeg^{-1}mol^{-1}

1-atm

Anion^x	$\Delta\bar{s}^o_-$		Cation^{xx}	$\Delta\bar{s}^o_+$		Cation	$\Delta\bar{s}^o_+$	
Br^-	3.0 (2.7)*		Ag^+	-0.8		Mn^{2+}	2.0	0.5
BrO_3^-	1.3		Al^{3+}	6.4	3.4	Na^+	-1.0 (-0.7)	
Cl^-	3.3 (2.7)		Ba^{2+}	0.1 (1.0)	-1.6	NH_4^+	-2.1	
ClO_3^-	1.1 (1.6)		Be^{2+}	2.1	0.6	Ni^{2+}	3.0	1.5
ClO_4^-	0.9		Ca^{2+}	1.3 (0.8)	-0.2	Pb^{2+}	0.0	-1.5
CN^-	2.1		Cd^{2+}	1.7	0.2	Pb^{4+}	7.3	2.8
$CO_3^=$	6.4	7.9**	Co^{2+}	2.7	1.2	Rb^+	-2.0 (0.8)	
$CrO_4^=$	4.4	5.9	Cr^{3+}	6.4	3.4	Sn^{2+}	0.9	-0.6
F^-	4.6 (-0.35)		Cs^+	-2.1		Sn^{4+}	8.1	3.6
$H_2AsO_4^-$	2.3		Cu^{2+}	2.4	0.9	Sr^{2+}	1.0 (1.0)	-0.5
HCO_3^-	2.6		Fe^{2+}	2.6	1.1	Th^{4+}	5.6	1.1
$HPO_4^=$	6.1	7.6	Fe^{3+}	6.1	3.1	Tl^+	-1.5	
$H_2PO_4^-$	2.8		Ga^{3+}	7.4	4.4	Tl^{3+}	3.2	0.2
HSO_4^-	2.0		Gd^{3+}	3.9	0.9	U^{3+}	3.4	0.4
I^-	2.5 (2.7)		Hg^{2+}	1.0	-0.5	U^{4+}	6.8	2.3
IO_3^-	2.4		In^{3+}	5.6	2.6	V^{3+}	5.6	2.6
MnO_4^-	0.5		K^+	-1.7 (-0.2)		Y^{3+}	2.9	-0.1
NO_2^-	1.8		Li^+	-0.2 (-1.2)		Zn^{2+}	2.5	-1.0
NO_3^-	1.5 (1.6)		Mg^{2+}	2.5 (0.5)	1.0			
OH^-	4.6 (2.8)							
$S^=$	5.6	7.1						
$SO_4^=$	4.9 (5.8)	6.4						

x $\Delta\bar{s}^o_- = 1 + 0.9(z^2/r_u)_{calc}$

xx $\Delta\bar{s}^o_+ = -2 + 0.9(z^2/r_u)_{calc}$

* from $\bar{c}^o_p \ln(273.15/298.15)$, $\bar{c}^o_p$ data from Pitzer and Brewer (1961) and Parsons (1959)

** calculated from eq 3.72.10

on the basis of the experimental values calculated from $\bar{c}_p^o \ln(273.15/298.15)$. The value of $\bar{s}_{H+}^o$ at 0°C was estimated as $-7 \text{cal deg}^{-1} \text{mol}^{-1}$. If eq 3.72.10 is used the values on the right hand side in Table 3.6 are obtained. Only ions with charges greater than unity are affected. This analysis indicates that the changes in the standard entropy, over the range 25°C to 0°C, are only small.

3.73 Entropy of hydration.

The standard entropy of ions in solution may also be related to the entropy of the ions in the ideal gaseous state, via

$$\Delta\bar{s}_{hyd}^o = \bar{s}^o(\text{solution}) - \bar{s}^o(\text{gas}) \qquad (3.73.1)$$

where $\Delta\bar{s}_{hyd}^o$ is the standard entropy change when the ion in the ideal gaseous state is transferred to the aqueous solution (see eq 3.72.5). The advantage of this relationship is that there are no contributions due to the physical state of the pure solute. That is, the problem of the varying stability of various crystals is eliminated.

$\Delta\bar{s}_{hyd}^o$ has been calculated by Eley and Evans (1938) via statistical mechanics using the following steps

ION (gas) —— $-\bar{s}^o(\text{gas})$ ——→ ION(in stationary state in solution)

↓ $\bar{s}_{SR}^o$(solvent reorganization)

ION(associated with reorganised water molecules)

←—— $\Delta\bar{s}_{el}^o$ ——

ION (exerting long-range electrostatic effect)

↓ $\bar{s}_{trans}^o$(solution)

ION (in normal non-stationary state in solution)

So that, summing all these steps, gives

$$\Delta\bar{s}^{o}_{hyd} = -\bar{s}^{o}(g) + \bar{s}^{o}_{SR} + \Delta\bar{s}^{o}_{el} + \bar{s}^{o}_{trans}(\text{solution}) \quad (3.73.2)$$

This equation is to be compared with

$$\Delta\bar{s}^{o}_{hyd(298)} = -5.5z - 4.0 - 9.7z^{2}/r_{u} \quad (3.73.3)$$

obtained from eqs 3.72.5, 3.72.9 and 3.73.1.

Eq 3.73.3 predicts a negative entropy of hydration. The entropy of hydration of ions pairs is also negative. Conversely, the entropy of association, for reactions where coordinated water molecules are released from the ion can have large positive values (see Section 3.75, Table 3.A.7, Nancollas 1966).

3.74 Comparison of the changes in $\bar{s}^{o}_{i}$ and in $\Delta\bar{s}^{o}_{el}$.

The standard partial molal entropy pme of the ion in solution is made up of

$$\bar{s}^{o}(\text{solution}) = \bar{s}^{o}_{SR} + \Delta\bar{s}^{o}_{el} + \bar{s}^{o}_{trans}(\text{solution}) \quad (3.74.1)$$

from eqs 3.73.1 and 3.73.2.

It might be expected that, for moderate concentration, (up to ca. 1 m) the changes in the pme would be related to the change in the electrostatic entropy of solvation, $\Delta\bar{s}_{el}$. The effects of various electrolytes on the dielectric constant of water have been measured (Table 3.7). If the "corrected" value of D at an ionic strength of 0.7 (molkg^{-1}) is used in eq 3.72.3 (neglecting changes in $\partial D/\partial T$), the change in $\Delta\bar{s}^{o}_{el}$ is as shown in Table 3.8.

The electrolyte values were obtained by summing the ionic contributions. Values of $\bar{s}_{i}-\bar{s}^{o}_{i}$ for the electrolytes at the same ionic strength are compared with the summed ionic $\Delta\bar{s}^{o}_{el}$ values, i.e. with

$$\Delta\bar{s}^{o}_{el} = 9.7(\nu_{1}z_{1}^{2}/r_{1} + \nu_{2}z_{2}^{2}/r_{2})((D_{w}-D_{i})(D_{w}+D_{i})/D_{i}^{2}) \quad (3.74.2)$$

where 1 and 2 represent the ions of the electrolyte i and D_{w}, D_{i}

TABLE 3.7 Effect of Salts on the Dielectric Constant of Water

$D = 78.54 - 2A\underline{c} \qquad 0.5 \leq c \leq 2.0 \text{ mol/}\ell$

Salt	Coefficient A ℓ/mol	Salt	Coefficient A ℓ/mol
$BaCl_2$	14	$MgCl_2$	15
HCl	10	NaCl	5.5
KCl	5	NaF	6
KF	6.5	NaI	7.5
KI	8	NaOH	10.5
$LaCl_3$	22	Na_2SO_4	11
LiCl	7	RbCl	5

Parsons (1959)

represent the dielectric constant of water and the aqueous electrolyte solution, respectively. These comparisons are also shown in Figure 3.2.

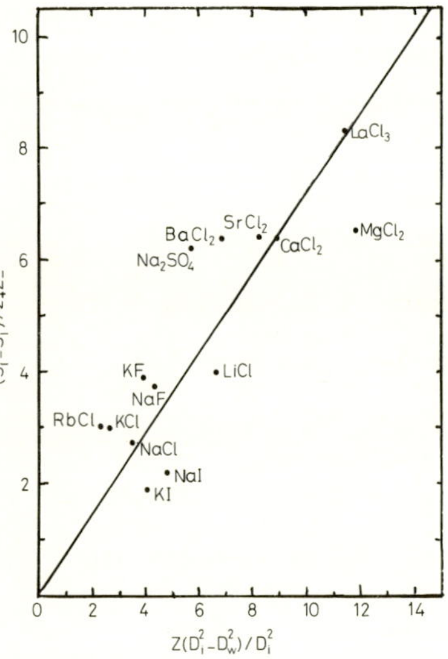

FIGURE 3.2

The relative partial molal entropy, versus the relative electrostatic entropy of solvation, for a number of electrolytes.

$Z = -9.7(\nu_+ z_+^2/r_+ + \nu_- z_-^2/r_-)$ caldeg^{-1}mol^{-1}.

D_i is the dielectric constant of the salt solution at I = 0.723 mol/kg water,

D_w the dielectric constant of water t = 25°C and p = 1-atm.

TABLE 3.8 Comparison of the relative partial molal entropies and "corrected" electrostatic entropies of solvation

I = 0.7 mol/kg water

t = 25°C, p = 1-atm

Salt i	$9.7(D_w^2-D_i^2)/D_i^2$ = Y Åcal/degmol	$\left(\frac{\nu_+ z_+^2}{r_+}+\frac{\nu_- z_-^2}{r_-}\right)Y$ cal/degmol	$(\bar{s}_i - \bar{s}_i^o)$cal/degmol Exp	From line Fig.3.2	From line Fig.3.3
$BaCl_2$	1.84	6.84	12.82	8.4	13.2
$CaCl_2$	1.99	8.95	12.85*	12.9	13.8
KCl	1.99	2.60	2.99*	2.0	2.0
KF	2.63	4.03	3.83*	3.0	3.8
KI	3.50	4.25	1.84	2.9	5.8
$LaCl_3$	1.40	11.42	24.9	24.9	23.2
LiCl	2.96	6.60	3.96*	4.7	4.5
$MgCl_2$	1.99	11.91	13.01	17.0	14.4
NaCl	2.22	3.52	2.74*	2.7	2.7
NaF	2.46	4.35	3.77	3.3	3.5
NaI	3.13	4.82	2.2	3.5	5.2
Na_2SO_4	1.38	5.75	12.41	8.6	10.0
RbCl	1.99	2.45	3.04	1.8	2.0
$SrCl_2$	1.99	8.20	12.85	12.0	13.8^x
HCl			3.16		8.3

D_w, D_i from Table 3.7, radii, r from Table 3.4 ($r_{SO_4^{2-}}$ was taken as 2.01Å, r_{Na^+} was taken as 0.97Å)

Exp* from eq 3.8.1, otherwise eq 3.8.5 and Tables 3.12, 3.13

x Assuming A (Table 3.7) is ca. 14.5 $l mol^{-1}$.

For a number of 1:1 electrolytes the two quantities are almost equivalent, deviations being due to changes in the other two terms of eq 3.74.1. Probably, the solvent re-orientation effect is dominant. These latter changes must be larger for the higher charge systems. In addition, the correct value for the radius is uncertain, especially for the sulfate ion.

A better overall correlation is obtained if the parameter A (the dielectric decrement of the salt, Table 3.8a) is plotted against the corresponding $\bar{s}_i - \bar{s}_i^o$ values (see Table 3.8, Figure 3.3). The iodide salts and HCl are marked exceptions (Table 3.8). The hydrated radii of

TABLE 3.8a Ionic dielectric decrements for eq

$$D_i = D_w - \sigma_+ c_+ - \sigma_- c_-$$

$$(\nu_+ \sigma_+ + \nu_- \sigma_-) = 2A \text{ of Table 3.7.}$$

Ion	σ 1 mol^{-1}	Ion	σ 1mol^{-1}
Ba^{2+}	22	Li^+	11
Cl^-	3	Mg^{2+}	24
F^-	5	Na^+	8
H^+	17	OH^-	13
I^-	7	Rb^+	7
K^+	8	SO_4^{2-}	7
La^{3+}	35		

Hasted et al (1948) as given by Erlander (1968a)

i represents electrolyte, w water

D is the macroscopic dielectric constant

H_3O^+ and I^- calculated by Erlander (1968a), in his study of the effective dielectric constant of ions, had infinitely large values. The ions Li^+, Na^+, K^+, Rb^+, OH^-, F^-, Cl^-, and Br^- on the other hand, had finite values (2-14Å). Thus, the deviations shown in Table 3.8 probably indicate some peculiarity in the behavior of the hydronium and iodide ions.

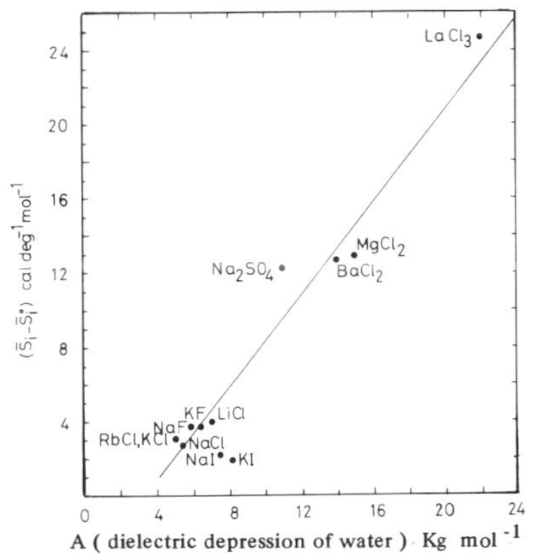

FIGURE 3.3

The relative partial molal entropy of electrolytes versus the change in the dielectric constant of water ($x\frac{1}{2}$) per mole of electrolyte, i.e. $A = (D_w - D_i)/2c$, see Table 3.7.

Ionic strength is 0.723 mol/kg water

(≡ S = 35‰), t = 25°C, p = 1-atm.

For a given concentration, A is directly proportional to the "dielectric depression" of water due to the presence of the solute (i.e. $A = (D_w-D_i)/2c$). Thus, the entropy change and dielectric change are closely related. As indicated by the first column of Table 3.8 (and also Table 3.7), the main contribution to the dielectric depression is from the anion. The value of the parameter A for seawater is estimated as 7 ℓmol^{-1}, using the values for NaCl, $MgCl_2$ and Na_2SO_4, weighted according to their relative concentrations in seawater. From Figure 3.3 the value of $\bar{S}_s-\bar{S}_s^o$ is predicted as 4.5 cal/deg mol at 25°C, 1 atm pressure, S = 35‰. $\bar{S}_s$ represents the seasalt entropy per mole of salts; the same analysis could be done using the corresponding values per mole of ions. In Section 3.11, the values of $\bar{S}_s$ and $\bar{S}_s^o$ have been estimated from the binary solution values. At 25°C, I = 0.723 mol/kgwater, the value of $\bar{S}_s-\bar{S}_s^o$ is calculated as 4.4 cal/deg mol salts (solution A) (Table 3.16a), which is consistent with the value predicted from Figure 3.3.

The dielectric constant of water is a function of the polarization of the water molecules. The polarization consists of (Erlander 1968a) (a) distortion of the electronic distributions of atoms and molecules (induced dipoles, e.g. ion-water interactions) and (b) orientation polarization - the partial mutual alignment of molecules with permanent molecular dipoles. The "effective" (or microscopic) dielectric constant can be explained in terms of (a) whilst the macroscopic dielectric constant reflects the magnitude of (b). The effective dielectric constant gives a measure of dipole-dipole and ion-dipole interactions between water molecules and between ions and associated water molecules. It differs from the macroscopic constant which is greatly increased by large structures such as "clusters" of hydrogen-bonded water. An ionic solute exerts an effective pressure (Tammann's principle - Chapters 4, 5, 6) and since pressure breaks up the larger structures (Section 5.73) the macroscopic dielectric constant would be expected to decrease, as observed (Table 3.7). The "effective" dielectric constant, however, decreases or changes little when large structures form.

The effective dielectric constant of the water surrounding an ion or polar group and water-structure itself, are related to the denaturing or solubilizing of proteins. Investigation of the effect of "seasalt" on the dielectric constant of water and the effect of changing salinity on the denaturing of proteins would probably be of especial interest. The maximum in dispersing power of lithium salts on certain proteins is interesting (Erlander 1968a) because seasalt appears similar to LiCl (quantitatively) in regard to its concentration effect on the properties of water (Figure 3.3, Table 5.4, Leyendekkers 1973). The temperature dependence of these effects may not be the same, however.
Investigations of the changes in the dielectric constant of water due to solutes is also of interest for the study of solubilities of non-electrolytes (e.g. gases). The solubility, s, of a non-polar molecule in a salt

solution, at a given temperature, is related to its solubility in pure water, $\underline{s}_o$, via the equation $\underline{s}_o/\underline{s} = f_3$ where f_3 is the activity coefficient of the non-electrolyte. In addition, if log f_3 is plotted against the salt concentration, then the limiting slope gives a constant, k_s (the Setchenow constant), characteristic for the salt. In turn, the value of k_s is correlated with the dielectric decrement of the ions (Table 3.8a). For the background theory, see Harned and Owen (1958 pp 531-6).

Erlander (1968a) has developed an equation for predicting k_s and compared the results with measured values of k_s. He shows that the discrepancies between the two sets of values are due to differences between the macroscopic and effective dielectric decrements of ions (other than Na, F and possibly Cl). Further discussion is outside the scope of the present chapter and the interested reader is referred to the original papers (Erlander 1968a, b). See also, the interesting papers of Krishnan and Friedman (1974) and of Masterton (1975).

3.75 Entropy of Association - Changes of $\bar{s}^o_{ion}$ with T and P.

The most important factor determining the entropy of association is the release of coordinated water molecules from the ions concerned. In addition, complex-ion formation reduces the amount of charged species in solution and may proceed even if substantially endothermic because of the positive $\Delta\bar{s}^o_{ass}$. The entropy of association may be considered as made up of the following steps

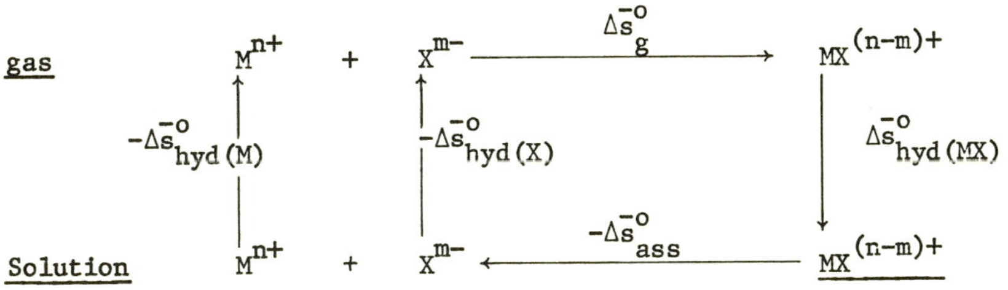

where $\Delta\bar{s}^o_g$ is the standard association entropy change for the hypothetical reaction in the gas phase. The standard entropy of association in solution is therefore given by:

$$\Delta\bar{s}^o_{ass} = \Delta\bar{s}^o_g + \Delta\bar{s}^o_{hyd(MX)} - \Delta\bar{s}^o_{hyd(M)} - \Delta\bar{s}^o_{hyd(X)} \qquad (3.75.1)$$

or, from eq 3.73.1

$$\Delta\bar{s}^o_{ass} = \bar{s}^o_{MX} - \bar{s}^o_M - \bar{s}^o_X \quad \text{(in solution)} \qquad (3.75.2)$$

Thus, if $\Delta\bar{s}^o_{ass}$ is known, $\bar{s}^o_{MX}$ can usually be estimated since the pme's of the ions are known (Table 3A.1). On the other hand, $\bar{s}^o_{ion}$ may be estimated, e.g. for the ionization of water at 298°K (data from Pitzer and Brewer 1961)

$$\Delta\bar{G}^o_{diss} = -RT\ln K_w = 19{,}089 \text{ calmol}^{-1}$$

and $\Delta\bar{H}^o_{diss} = 13{,}500 \text{ calmol}^{-1}$, so that

$$\Delta\bar{s}^o_{diss} = (\Delta\bar{H}^o - \Delta\bar{G}^o)/T = -18.7 \text{ caldeg}^{-1}\text{mol}^{-1}$$

or $\Delta\bar{s}^o_{ass} = 18.7 \text{ caldeg}^{-1}\text{mol}^{-1}$

The entropy of 1 mole of liquid water at 25°C is 16.7 caldeg^{-1} so that, from eq 3.75.2

$$\bar{s}^o_{H+} + \bar{s}^o_{OH-} = \bar{s}^o_{H_2O} - \Delta\bar{s}^o_{ass} = 16.7 - 18.7 = -2.0 \text{ caldeg}^{-1}\text{mol}^{-1}$$

The calorimetric experimental value is -2.52 caldeg^{-1}mol^{-1} (see Larson (1970) concerning this discrepancy).

Values at some other temperature T°K can be calculated from

$$\bar{s}^o_{H_2O}(T) = 16.7 - (c^o_p)_w \ln(T/298.15) \qquad (3.75.3)$$

The changes with pressure are given in Section 3.12.

The entropy change $\Delta\bar{s}^o_{diss}$ for the ionization of water is given by

$$\Delta\bar{s}^o_{diss} \doteq 27.8506 - 0.15610T \qquad (3.75.3\underline{a})$$

(Harned and Owen 1958, Table 15-6-4A).

For $\underline{T}$ = 273.15°K (0°C) this gives

$$\Delta\bar{s}^{o}_{diss} = -14.79$$

and $$\bar{s}^{o}_{H_2O} = 15.12$$

so that $$\bar{s}^{o}_{H+} + \bar{s}^{o}_{OH-} = 15.12 - 14.79 = 0.33$$

From Table 3.6

$$\bar{s}^{o}_{OH-} = 2.8 - 2.5 = 0.3 \text{ from } c^{o}_{p} \text{ data}$$

or $$\bar{s}^{o}_{OH-} = 4.6 - 2.5 = 2.1 \text{ from eq given in footnote.}$$

All entropy units used are $\text{caldeg}^{-1}\text{mol}^{-1}$.

This sort of analysis should be useful for estimating the pressure effects on $\bar{s}^{o}_{ion}$ values. For a pure solution such as water, for example, the entropy change with pressure is given by

$$d\not{s} = -\int(\alpha v)\ dp \qquad (3.5.2)^*$$

with αv given by eq (5.71.6), viz

$$(\alpha v)^{(p)} = (\alpha v)^{(1)} + 0.4343\text{x}0.315\ \frac{dB}{dt}\left[1/(B+1) - 1/(B+p)\right] \qquad (5.71.6)^*$$

with $B = 2671.8 + 19.454t - 0.1028t^2 + 9.798\text{x}10^{-4}t^3$

The change of $\Delta\bar{G}^{o}_{diss}$ with pressure is well characterised for water (Kearns 1966, Whitfield 1972, Millero <u>et. al</u> 1972).

$$\text{Since} \quad \partial\Delta\bar{G}^{o}_{diss}/\partial p = -RT\partial\ln K_w/\partial p = \Delta\bar{v}^{o}_{diss} \qquad (3.75.4)$$

$$\text{and} \quad \partial(\Delta\bar{H}^{o}_{diss}/T)/\partial p = \partial(RT\partial\ln K_w/\partial T)/\partial p$$

$$= \partial(RT\partial\ln K_w/\partial p)/\partial T$$

$$= -\partial\Delta\bar{v}^{o}_{diss}/\partial T \qquad (3.75.5)$$

$$\text{so that} \quad (\partial\Delta\bar{s}^{o}_{diss}/\partial p)_T = (1/T)(\partial\Delta H^{o}_{diss}/\partial p - \partial\Delta G^{o}_{diss}/\partial p)$$

$$= -\partial\Delta\bar{v}^{o}_{diss}/\partial T - \Delta\bar{v}^{o}_{diss}/T \qquad (3.75.6)$$

$$\text{where} \quad \Delta\bar{v}^{o}_{diss} = -23.2 + 0.1108t + (69.28-0.088t)(p-1)\text{x}10^{-4}\ \text{mlmol}^{-1} \qquad (3.75.7)$$

from Kearns (1966), with (t + 273.15)= T, as usual; p is in bars.

For 1000 bars and at 0°C (Table 3.17, eqs 3.75.3a, 3.75.6, 3.75.7),

$$\bar{s}^{o(1000)}_{H_2O} = \bar{s}^{o(1)}_{H_2O} - 0.04 = 15.12 - 0.04 = 15.08$$

$$\Delta\bar{s}^{o(1000)}_{ass} = \Delta\bar{s}^{o(1)}_{ass} - 10.85 = 14.8-10.85 = 3.95 \text{ caldeg}^{-1}\text{mol}^{-1}$$

so that $(\bar{s}^{o}_{H+} + \bar{s}^{o}_{OH-})^{(1000)}_{0°C} = 15.08 - 3.95$

$$= 11.13 \text{ caldeg}^{-1}\text{mol}^{-1}$$

3.8 Calculation of the partial molal entropy of electrolytes.

For an electrolyte i, dissociating into cation 1 and anion 2, values of $\bar{s}_i$ can be calculated directly from

$$\bar{s}_i - \bar{s}^o_i = \bar{L}_i/T - R\ln a_i \qquad (3.7.4)^*$$

which was derived in Section 3.7, with

$$a_i = a_1^{\nu_1} a_2^{\nu_2} = \gamma_1^{\nu_1}\gamma_2^{\nu_2}(m_1^{\nu_1} m_2^{\nu_2})$$

$$= \gamma^{\nu}_{\pm_i}(\nu_1^{\nu_1}\nu_2^{\nu_2})m^{\nu}$$

See Section 1.9

so that

$$\bar{s}_i - \bar{s}^o_i = \bar{L}_i/T - \nu R\ln\gamma_{\pm_i} m$$

$$- R\ln(\nu_1^{\nu_1}\nu_2^{\nu_2}) \qquad (3.8.1)$$

Values of $\bar{s}_i - \bar{s}^o_i$ for a number of electrolytes were calculated from this equation (Table 3.9). The change with temperature is given by

$$(\bar{s}_i - \bar{s}^o_i) = \bar{j}(1 + 2.303\log T) \qquad (3.7.6)^*$$

$$+ \text{ constant}$$

Values of $\bar{j}$ (the relative partial molal heat capacity) and the constant are also listed in Table 3.9. The constant is eliminated via eq 3.7.7 (Section 3.7).

TABLE 3.9 The partial molal entropy of electrolytes at 25°C, 1-atm
$I = 0.7$ molkg^{-1}
($\bar{j}$, $\bar{s}$ in caldeg^{-1}mol^{-1})

Salt	$\bar{s}_i - \bar{s}^o$ eq 3.8.1	$\bar{s}_i$	$\bar{j}_i$	constant eq 3.7.6
$CaCl_2$	12.85	26.05 (25.94)	18	-107.721
HCl	3.16	16.36	6.7	-41.717
KBr	2.61	46.40	13	-84.469
KCl	3.00	40.70 (40.57)	12.3	-79.398
KF	3.83	26.03		
$LaCl_3$	24.25	-2.85		
LiCl	3.96	20.56	6.7	-40.924
NaCl	2.74	30.34 (30.18)	16	-104.435
NaOH	3.02	14.92	14.2	-92.105

bracketed values from eq 3.8.5 and data in Tables 3.12 and 3.13

$\bar{s}^o$ from Table 3A.1

For aqueous solutions the relative partial molal enthalpies and heat capacities generally have large values and change fairly rapidly with molality. This aggravates the problem of an insufficiency in the thermal data. It is therefore convenient to have a form of equation that facilitates estimates of the entropy when the thermal or activity coefficient data are lacking. This is achieved by substituting the equations (Pitzer and Brewer 1961, Appendix 4)

$$\log\gamma_{\pm} = -A_{\gamma}(z_1z_2)\{I^{\frac{1}{2}}/(1+I^{\frac{1}{2}})\} + (2\nu_1\nu_2/\nu)B^{\cdot}m \qquad (3.8.2)$$

$$\bar{L}_i = (\nu/2)A_H(z_1z_2)\{I^{\frac{1}{2}}/(1+I^{\frac{1}{2}})\} - 2\mathrm{x}2.303RT^2\frac{dB^{\cdot}}{dt}m\nu_1\nu_2 \qquad (3.8.3)$$

and $I = \frac{1}{2}m\nu z_1z_2$, into eq 3.8.1. The charges z_1 and z_2 are the absolute values, as usual.

It should be noted that eq 3.8.2, because of its simple form, is used rather than the more recent equation of Pitzer and Mayorga (1973).

This gives,

$$\bar{s}_i - \bar{s}_i^o = \nu z_1 z_2\{I^{\frac{1}{2}}/(1+I^{\frac{1}{2}})\}\{A_H/2T + 2.303RA_\gamma\}$$
$$- 2.303R\left[\nu \log 2I + \log((\nu_1^{\nu_1}\nu_2^{\nu_2})/(\nu z_1 z_2)^{\nu})\right]$$
$$- (9.212R\nu_1\nu_2 I/\nu z_1 z_2)(B_i^{\bullet} + TdB_i^{\bullet}/dT) \qquad (3.8.4)$$

or

$$\bar{s}_i - \bar{s}_i^o = A_{jk} - C_{jk}(B_i^{\bullet} + TdB_i^{\bullet}/dT) \qquad (3.8.5)$$

with $A_{jk} = \nu z_1 z_2(I^{\frac{1}{2}}/(1+I^{\frac{1}{2}}))(A_H/2T + 2.303RA_\gamma)$

$$- 2.303R(\nu \log 2I + \log(\nu_1^{\nu_1}\nu_2^{\nu_2})/(\nu z_1 z_2)^{\nu})$$

and $C_{jk} = (9.212R\nu_1\nu_2/\nu z_1 z_2)I$; j, k represent the charges on the cation and anion, respectively.

A_H and A_γ are the Debye-Huckel limiting constants for water at 1-atm on the $molkg^{-1}$(water) or molal scale, (Table 3.10), R is the Gas Constant ($1.9872 caldeg^{-1}mol^{-1}$).

TABLE 3.10 Debye-Huckel Limiting Constants for water at 1-atm on molal scale (mol/kg water)

t^oC	A_γ	A_H $calmol^{-1}$	A_J $caldeg^{-1}mol^{-1}$
0	0.492	464	7.5
10	0.499	545	8.7
20	0.507	637	9.9
25	0.511	688	10.4
30	0.517	743	11.0

Pitzer and Brewer (1961)

TABLE 3.11 The partial molal entropy function for electrolytes.

Coefficients for the equation

$$\bar{s}_i - \bar{s}_i^o = A_{jk} - C_{jk}(B_i^{\bullet} + T dB_i^{\bullet}/dT) \quad (\mathrm{cal\,deg^{-1}\,mol^{-1}})$$

Charge type*		$I\,\mathrm{mol\,kg^{-1}}$	0.1	0.3	0.5	0.7	0.723	1.0
		S‰	5.0	14.9	24.6	34.1	35.0	48.0
j	k				C_{jk} (independent of temperature)			
1	1		0.9153	2.7459	4.5765	6.4071	6.6177	9.1530
2	1	(or 1,2)	0.6102	1.8306	3.0510	4.2714	4.4118	6.1020
3	1	(or 1,3)	0.4577	1.3730	2.2883	3.2036	3.3088	4.5765
2	2		0.2288	0.6865	1.1441	1.6018	1.6544	2.2883
					A_{jk}			
1	1		10.643	6.981	5.324	4.243	4.139	3.101
2	1	t=0°C	21.995	17.559	15.635	14.398	14.280	13.098
3	1		34.941	30.435	28.619	27.481	27.373	26.300
2	2		20.624	19.076	18.542	18.229	18.200	17.915
1	1		10.713	7.083	5.444	4.375	4.273	3.246
2	1	t=10°C	22.204	17.867	15.996	14.794	14.680	13.534
3	1		35.359	31.051	29.340	28.274	28.173	27.171
2	2		20.903	19.487	19.023	18.758	18.734	18.495
1	1		10.790	7.197	5.578	4.522	4.420	3.407
2	1	t=20°C	22.436	18.208	16.395	15.233	15.123	14.016
3	1		35.822	31.734	30.138	29.153	29.060	28.135
2	2		21.212	19.942	19.555	19.344	19.325	19.138
1	1		10.831	7.258	5.649	4.600	4.499	3.492
2	1	t=25°C	22.559	18.390	16.608	15.467	15.359	14.272
3	1		36.069	32.097	30.564	29.621	29.532	28.649
2	2		21.376	20.184	19.839	19.656	19.639	19.480
1	1		10.879	7.328	5.731	4.690	4.590	3.592
2	1	t=30°C	22.702	18.600	16.854	15.738	15.632	14.570
3	1		36.355	32.518	31.057	30.163	30.079	29.243
2	2		21.567	20.465	20.167	20.017	20.004	19.877

* j, k represent charge on cation and anion, respectively.

Values of A_{jk} and C_{jk} for various charge systems, temperatures and ionic strengths are listed in Table 3.11. Values of $B_i^{\bullet}$ and $dB_i^{\bullet}/dT$ for 25°C and I = 0.723 (S = 35‰) are given in Table 3.12. More extensive data on $B_i^{\bullet}$, dB_i/dT and $\bar{L}_i$ are given in Appendix 3, Table 3A.2. Values of $dB^{\bullet}/dT$ as a function of $m^{\frac{1}{2}}$, over the experimental ranges, are also given.

The second derivative of $B^{\bullet}$ with respect to temperature is calculated as follows: (Pitzer and Brewer 1961, Appendix 4)

When $dB^{\bullet}/dT$ and $\bar{J}_i$ are known,

$$d^2B^{\bullet}/dT^2 = \left[(\tfrac{1}{2}\nu z_1 z_2 A_J I^{\frac{1}{2}}/(1+I^{\frac{1}{2}}))-\bar{J}_i\right]/4.606RT^2 m\nu_1\nu_2 - (2/T)dB^{\bullet}/dT \qquad (3.8.6)$$

Values of A_J are given in Table (3.10).

Or, using a reference electrolyte of the same charge type, denoted by subscript r (e.g. KCl for 1:1 electrolytes and $CaCl_2$ for 2:1 electrolytes - Pitzer and Brewer 1961) and calculating directly from the experimental data,

$$d^2(B_i^{\bullet} - B_r^{\bullet})/dT^2 = -(1/(2x2.303RT^2\nu_1\nu_2 m))\left[(\bar{J}_i-\bar{J}_r)-(2/T)(\bar{L}_i-\bar{L}_r)\right] \qquad (3.8.7)$$

Values of $d^2B^{\bullet}/dT^2$ at 25°C, for a number of electrolytes are listed in Tables 3.12 and 3.13 for the ionic strength equivalent to S = 35‰. Further values are given in Appendix 3.

Neglecting the small temperature variations of $d^2B^{\bullet}/dT^2$, the following equations can be derived by integration.

$$d^2B^{\bullet}/dT^2 = \text{constant}$$

Thus
$$(dB^{\bullet}/dT)_t = (dB^{\bullet}/dT)_{25} + (d^2B^{\bullet}/dT^2)(t-25) \qquad (3.8.8)$$

and
$$B_t^{\bullet} - B_{25}^{\bullet} = (dB^{\bullet}/dT)_{25}(t-25) + (\tfrac{1}{2}d^2B^{\bullet}/dT^2)(t-25)^2 \qquad (3.8.9)$$

Whence,

$$(B^{\bullet}+TdB^{\bullet}/dT)_t = B_{25}^{\bullet} + (dB^{\bullet}/dT)_{25}(2t+248.15) + \tfrac{1}{2}(d^2B^{\bullet}/dT^2)(t-25)(3t+521.3) \qquad (3.8.10)$$

These equations, together with values of $B^{\bullet}$ and its derivatives at 25°C

TABLE 3.12 $B^{\cdot}$ and the temperature derivatives at I = 0.723 (mol/kg water) (equivalent to S = 35‰)

25°C 1-atm

1:1 Charge type Bracketed values estimated from Figures 3.4,3.6 and 3.7

Salt i	$B_i^{\cdot}$ kg mol^{-1}	$dB_i^{\cdot}/dT$ x10^4 kg mol^{-1}deg^{-1}	$d^2B_i^{\cdot}/dT^2$ x10^5 kg mol^{-1}deg^{-2}	$\bar{s}_i^o$ caldeg^{-1}mol^{-1}
$AgNO_3$	-0.109	(13)	(-2.7)	52.7
CsBr	-0.016	12.4	-2.6	51.1
CsCl	-0.0122	(10)	(-1.8)	45.0
CsF	(0.11)	1.3	-0.9	29.5
CsI	-0.0222	16.1	-3.1	57.9
HBr*2	0.2048	(-1.6)	-0.18	19.3
HCl	0.1713	-3.6	-0.13	13.2
$HClO_4$*1	0.1798	7.3	(-2.2)	43.6
HF	(-0.06)	(-9)	(1.0)	-2.3
HI	0.249	(2)	(-0.9)	26.1
HNO_3	0.123	(6)	(-1.6)	35.0
KBr	0.0508	8.6	-2.0	43.8
KCl	0.0408	6.9	-1.8	37.7
$KClO_3$	-0.076	(17)	(-3.3)	63.5
$KClO_4$	(-0.05)	(20)	(-3.8)	68.1
KF	0.0688	1.7	(-0.9)	22.2
KI	0.073	(12)	(-2.5)	50.6
KNO_3	-0.099	(16)	(-3.2)	59.5
LiBr	0.167	-1.51	-0.32	22.7
LiCl	0.149	-1.71	-0.25	16.6
$LiClO_4$	0.217	(11)	(-2.4)	47
LiI	0.229	(4)	(-1.3)	29.5
$LiNO_3$	0.135	(7)	(-1.7)	38.4
NaBr	0.1003	7.22	-2.1	33.7
NaCl	0.0803	7.02	-2.0	27.6
$NaClO_3$	0.031	(13)	(-2.7)	53.4

TABLE 3.12 (ctd)

Salt i	$B^{\bullet}$	$dB^{\bullet}/dT$ x10^4	$d^2B^{\bullet}/dT^2$ x10^5	$\bar{s}_i^o$ caldeg^{-1}mol^{-1}
$NaClO_4$ *1	0.063	15	(-3.0)	58.0
NaF	0.016	3.8	(-2.5)	12.1
$NaHCO_3$	(-0.099)	(7)	(-1.7)	37.1
NaI	0.1298	8.2	-2.1	40.5
$NaNO_3$	0.001	(12)	(-2.5)	49.4
NaOH	0.0908	4.59	-1.9	11.9
NH_4Cl	0.041	(8)	(-1.9)	40.0
NH_4NO_3	-0.044	(17)	(-3.3)	62.0
RbBr	0.020	9.67	-1.9	48.0
RbCl	0.023	7.42	-1.7	41.9
RbF	(0.092)	1.48	-1.4	26.4
RbI	0.017	11.81	-2.1	54.8
$RbNO_3$	-0.109	(18)	(-3.2)	63.7
$TlClO_4$	-0.129	(22)	(-4.2)	74.0
$TlNO_3$	-0.189	(18)	(-3.5)	65.4

Data from Pitzer and Brewer (1961),

Alkali halides - Fortier *et al* (1974).
*1 Vanderzee and Swanson (1963)
*2 Vanderzee, Nutter *et al* (1974).

enable $\bar{s}_i-\bar{s}_i^o$ (or $\Delta\bar{s}_i$) to be calculated for any temperature t. More simply, from eq 3.8.5,

$$(\Delta\bar{s}_i)_t-(\Delta\bar{s}_i)_{25} = (A_t-A_{25})_{jk} - C_{jk}\Big[2(dB^{\bullet}/dT)_{25}(t-25) + \tfrac{1}{2}(d^2B^{\bullet}/dT^2)(t-25)(3t+521.3)\Big] \qquad (3.8.11)$$

Values of the various terms of eq 3.8.11, for t=0°C are listed in Table 3.14 for the major seawater salts and for seasalt (i.e. the summed values weighted according to the relative concentrations of the salts). The

TABLE 3.13 B$^{\bullet}$ and the temperature derivatives at I = 0.723 (mol/kgwater) (equivalent to S = 35‰) 25°C 1-atm

2:1 and 1:2 Charge types Bracketed values estimated from Figures 3.4, 3.6 and 3.7.

Salt $\underline{i}$	$B_i^{\bullet}$	$dB_i^{\bullet}/dT$ x10^4	$d^2B_i^{\bullet}/dT^2$ x10^5	$\bar{s}_i^o$ caldeg^{-1}mol^{-1}
$BaBr_2$	0.409	(13)		41.6
$BaCl_2$	0.351	(9)		29.4
$Ba(ClO_4)_2$*3	0.452	26.0		90.2
$CaBr_2$	0.497	(10)		25.4
$CaCl_2$	0.420	5.8	-1.4	13.2
$Ca(ClO_4)_2$*5	0.584	13.5		74.0
$Ca(NO_3)_2$	0.252	(28)		56.8
$Co(ClO_4)_2$*2	(0.674)	11.9		60.2
Cs_2SO_4	0.101	(35)		68.0
K_2SO_4	-0.038	(26)		53.4
$La(ClO_4)_3$*4	(0.69)	35.0		
Li_2SO_4	0.180	(3)		10.9
$MgBr_2$	0.557	(6)		10.4
$MgCl_2$	0.473	(3.4)	(-1.2)	-1.8
$Mg(ClO_4)_2$*1	0.683	10.0		59.0
MgI_2	0.643	(11)		24.1
$Mg(NO_3)_2$	0.452	(11)		41.8
$Mn(ClO_4)_2$*5	(0.625)	9.8		67.2
Na_2SO_4	0.030	(23)	(-4)	33.2
$Ni(ClO_4)_2$*2	(0.70)	11.2		56.2
Rb_2SO_4	0.075	(28)		61.8
$SrBr_2$	0.461	(13)		27.2
$SrCl_2$	0.397	(8)	(-1.9)	17.0
$Sr(ClO_4)_2$*1	0.553	17.4		77.8

*1 Jongenburger and Wood (1965), *2 Gier and Vanderzee (1974a), *3 Vanderzee and Swanson (1972), *4 Vanderzee and Nutter(1974), *5 Gier and Vanderzee (1974b).

specific ion interaction term (2nd term of eq 3.8.11) is more sensitive to temperature than the Debye-Hückel term for 1:1 electrolytes and for sea-salt. The converse is true for the higher charge systems.

TABLE 3.14 Components of the function

$$(\Delta\bar{s}_i)_t - (\Delta\bar{s}_i)_{25} = (A_t - A_{25})_{jk} - C_{jk}\left[2(dB^{\bullet}/dT)_{25}(t-25) + (\tfrac{1}{2}d^2B^{\bullet}/dT^2)(t-25)(3t+521.3)\right]$$

for t = 0°C, S = 35‰ (I = 0.723mol/kg water)

$\mathrm{cal\,deg^{-1}mol^{-1}}$

Salt	$(\Delta\bar{s}_i)_{0^\circ} - (\Delta\bar{s}_i)_{25}$	$A_0 - A_{25}$	C_{jk}	f(t)
NaCl	-0.990	-0.3597	6.6177	0.0952
KCl	-0.908	"	"	0.0828
NaF	-1.312	"	"	0.1439
KBr	-0.938	"	"	0.0873
$NaHCO_3$	-0.861	"	"	0.0758
$CaCl_2$	-1.354	-1.0792	4.4118	0.0622
$MgCl_2$	-1.349	"	"	0.0612
$SrCl_2$	-1.448	"	"	0.0838
Na_2SO_4	-1.722	"	"	0.1457
Seasalt Soln A *	-1.072	-0.4881	$\Sigma f_i C_{jk} f(t) =$	0.5837

* See Table 4.16

$\Delta\bar{s}_i = (\bar{s}_i - \bar{s}_i^{\circ})$

This equation is to be compared with

$$(\Delta\bar{s}_i)_t - (\Delta\bar{s}_i)_{25} = \bar{J} \times 2.303\log((t+273.15)/298.15)$$

from eq (3.7.7)*.

The $B^{\bullet}$ equations should be useful when thermal data are lacking. The accuracy of prediction of $B^{\bullet}$ and its temperature derivatives should be sufficient to allow useful estimates of the entropy; such accuracy is not as critical as for $\bar{j}$.

As pointed out by Pitzer and Brewer (1961), it might be expected that the entropy of interaction of a pair of ions should be simply related to the free energy of interaction. Consequently, the heat of interaction could be expected to be simply related to the free energy of interaction. They illustrated this by the linear correlation of $B_i^{\bullet}$ with $dB_i^{\bullet}/dT$ for 36 electrolytes, *i*, of the 1:1 charge type, for concentrations $m \leq 0.1$. The data for $d^2B^{\bullet}/dT^2$ is much more restricted, however a similar correlation with $B^{\bullet}$ was demonstrated for 12 1:1 electrolytes. A similar correlation exists between $B^{\bullet}$ and $dB^{\bullet}/dT$ for an ionic strength equivalent to S=35‰ (0.723 mol/kgwater). However, the scatter is greater and the 1:1 data considerably depleted (around 20 electrolytes). In the next section, a more useful correlation will be presented, that allows estimates of higher charge systems to be made. Experimental data for these latter systems generally only extend to an ionic strength of 0.3*m*. Although the existing data have been fitted to polynomials in $m^{1/2}$, direct extrapolation via such functions is considered unreliable. The correlation therefore allows an indirect form of extrapolation that should provide more accurate values of $dB^{\bullet}/dT$. Data for $B^{\bullet}$ at the higher concentrations are more abundant. However, a method of estimating values at the concentration of "average" seawater (S=35‰) will also be given. In this way, a large number of electrolytes have been covered, for which the partial molal entropies at the ionic strength equivalent to S=35‰ can be estimated.

3.9 Correlations with the standard partial molal entropies (spme).

The spme of ions has proved a useful quantity for "scaling" various ion-ion and ion-solvent interaction quantities for electrolyte solutions. Leyendekkers (1970,1971, 1973, 1974) showed that, for strong electrolytes, simple correlations exist between the ionic spme and various interaction terms characterising aqueous electrolyte systems.

For example, the B˙ coefficient, which covers specific short-range ion-ion interactions, was simply related to the spme for a number of fixed anion - varying cation series. The Harned coefficient, α, which covers the extra interactions for a two-electrolyte system, was also linearly correlated with the spme of the cation for a common anion series.

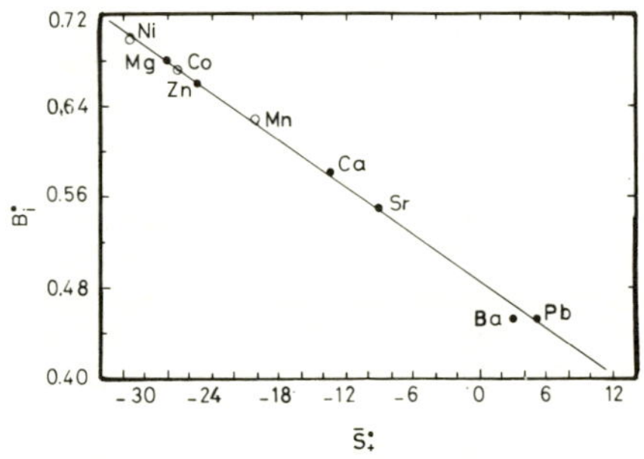

FIGURE 3.4

The specific ion interaction parameter B˙ for divalent chlorates versus the standard partial molal entropy of the cation.
Ionic strength 0.723 mol/kg water
(≡ S = 35‰)
t = 25°C, p = 1-atm.
Units: B˙ $kgmol^{-1}$, $\bar{s}^{o}_{+}$ $caldeg^{-1}mol^{-1}$

Figure 3.4 illustrates the correlation of $B^{\bullet}$ with $\bar{s}^{o}_{cation}$ for the chlorate series at I = 0.723mol/kgwater. Values of $B^{\bullet}$ for Co^{2+} and Ni^{2+} chlorates have been estimated on this basis, Table 3.13.

Values of $dB^{\bullet}/dT$ were also found to be simply related to the spme. The thermal data are quite extensive up to an ionic strength around 0.3m and the values of $dB^{\bullet}/dT$ at this ionic strength were plotted against the spme of the electrolytes (i.e. $\nu_{+}\bar{s}^{o}_{+} + \nu_{-}\bar{s}^{o}_{-}$). With a few exceptions, for 1:1 charge systems the fit was linear to within the experimental errors (ca. 1 to $2x10^{-4}$cal), Figure 3.5. The 2:1 and 1:2 charge systems could be normalised to fit on the same line by subtracting $12x10^{-4}$ units. The perchlorates were exceptional in that $6x10^{-4}$ units had to be added for normalisation.

A similar plot was drawn for I = 0.723m (S = 35‰ equivalent). The previous plot was used as a basis for estimates of $dB^{\bullet}/dT$ when no experimental data were available. The normalising increment was only $9x10^{-4}$ units in this case (Figure 3.6) (as indicated by the $CaCl_2$ data). The divalent perchlorates were 8 units below the line for this ionic strength. Values of $dB^{\bullet}/dT$, estimated on the basis of Figure 3.6 are listed in Tables 3.12 and 3.13 (bracketed values).

Data for $d^2B^{\bullet}/dT^2$ are not sufficient to warrant a prediction on the above analytical basis. However, a comparison with $dB^{\bullet}/dT$ values allowed some limited predictions (Figure 3.7 and Tables 3.12 and 3.13).

No experimental value of $\bar{L}_{Na_2SO_4}$ at the ionic strength of 35‰ seawater ($0.723molkg^{-1}$) is available and direct extrapolation from the lower concentrations is unreliable. The predicted value of $dB^{\bullet}/dT$ for this salt is probably accurate to $\pm$ $1x10^{-4}$ units (Table 3.13). The value of $d^2B^{\bullet}/dT^2$ was therefore estimated from eq 3.8.9 as follows: The

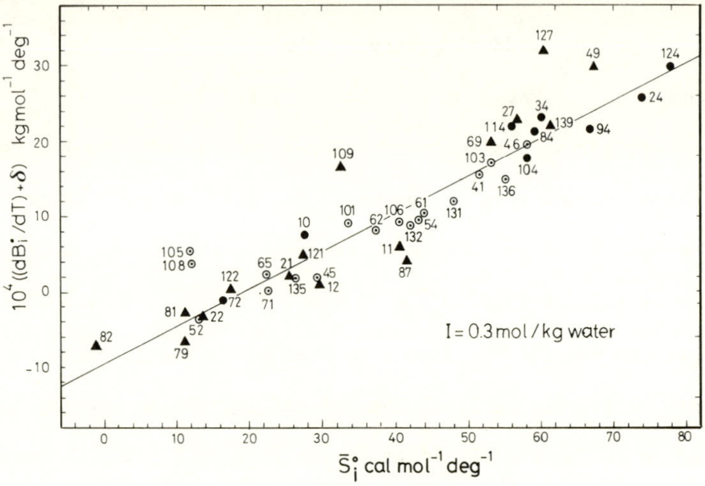

FIGURE 3.5

The temperature derivative of the specific ion interaction parameter $B_i^\bullet$ versus the standard partial molal entropy, $\bar{s}_i^o$, for an electrolyte i. 25°C, 1-atm pressure. Ionic strength 0.3mol/kg water

Number code:

1	Ba(2+) or Br(-)	7	Li(+) or NO_3(-)
2	Ca(2+) or Cl(-)	8	Mg(2+) or OH(-)
3	Co(2+) or ClO_3(-)	9	Mn(2+) or SO_4(2-)
4	Cs(+) or ClO_4(-)	10	Na+
5	H(+) or F(-)	11	Ni(2+)
6	K(+) or I(-)	12	Sr(2+)
		13	Rb(+)

Last number represents anion, e.g.
122 represents $SrCl_2$
δ is normalising factor as follows:

Charge type	δ
1:1	zero
divalent chlorates	+6
other 2:1, or 1:2	-12

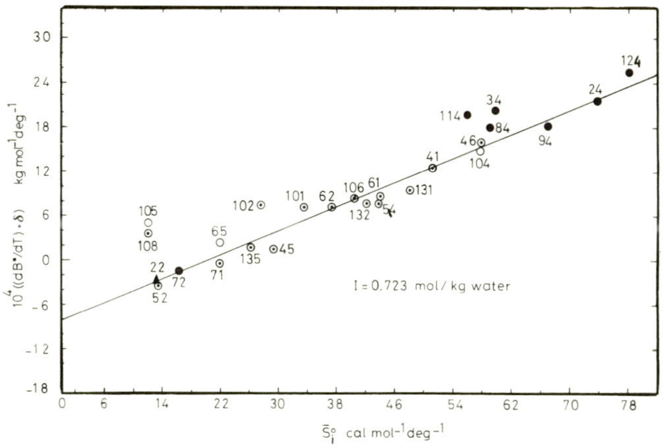

FIGURE 3.6

As for Figure 3.5, but ionic strength is 0.723 mol/kg water (corresponding to S = 35‰)

Charge type	δ
1:1	zero
divalent chlorates	+8
other 2:1, or 1:2	-9

---

value of $B^{\bullet}_{15}$ was estimated as 0.005 from the EMF data for $\gamma_{\pm Na_2SO_4}$ listed by Harned and Owen (Table 13-2-1), the value of this coefficient being taken as 0.346 at m = $0.241 molkg^{-1}$. The value of $B^{\bullet}_{25}$ and $(dB^{\bullet}/dT)_{25}$ were taken from Table 3.13. This calculation gave a value of $-4x10^{-5}$ $kgmol^{-1}deg^{-2}$ for $(d^2B^{\bullet}/dT^2)_{Na_2SO_4}$. On the other hand, with $\bar{j}_{Na_2SO_4}$ (at m = 0.241 $molkg^{-1}$ or I = 0.723) taken as $46.6caldeg^{-1}mol^{-1}$ (Harned and Owen 1958, p354 Table 8-4-3) and $dB^{\bullet}/dT$ from Table 3.13 a value of $-3.6x10^{-5}$ units was obtained for $d^2B^{\bullet}/dT^2$ (eq 3.8.6).

Since values of $B^{\bullet}$ and its temperature derivatives are now available for all the major components of seawater at I = $0.723molkg^{-1}$, the corresponding values of $\bar{s}_i$ for these components, i, can be estimated at temperatures 0-30°C. These values can then be weighted, in accordance with the seawater concentrations of the salts, and summed. The sum should give a close approximation of the solute contribution to the seawater entropy, as was the case for the volume (Chapters 4 and 6).

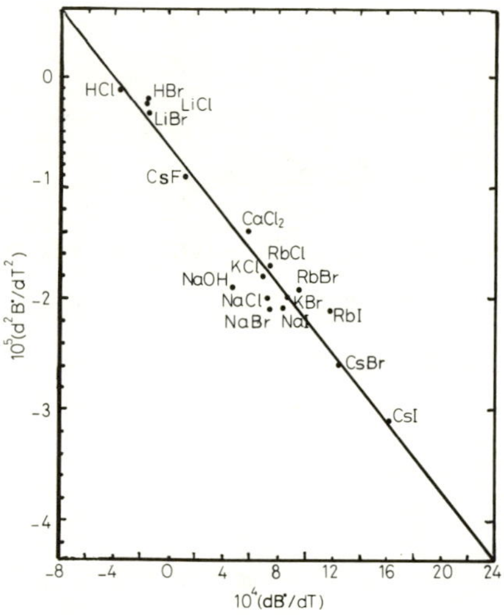

FIGURE 3.7

Relationship between the first and second temperature derivatives of the specific ion interaction parameter $B^{\bullet}$ (See Tables 12 and 13).

$dB^{\bullet}/dT$ in $kgmol^{-1}deg^{-1}$

$d^2B^{\bullet}/dT^2$ in $kgmol^{-1}deg^{-2}$

Ionic strength 0.723 mol/kg water (corresponds to S = 35‰).

t = 25°C, p = 1-atm.

3.10 The partial molal entropy (pme) of seawater components in binary solutions.

The method of calculating the pme's of components in binary solutions has been discussed in Section 3.8, where values for a number of electroytes at 25°C and an ionic strength 0.7 molkg^{-1} were given (Table 3.9). In this section, the values for the major components of seawater (solution A - Table 4.16) over a range of temperatures (0-30°C) will be considered.

The partial molal entropies were calculated from

$$\bar{s}_i = \bar{s}_i^o + A_{jk} - C_{jk}(B_i^{\cdot} + TdB_i^{\cdot}/dT) \qquad (3.85)^*$$

as outlined in Section 3.8, using the data in Tables 3.12 and 3.13.

The pme's of the water in the binary solutions were calculated as follows.

From eq 3.7.4, using the subscript w to represent water, as usual,

$$\bar{s}_w - \bar{s}_w^o = \bar{L}_w/T - R\ln a_w \qquad (3.10.1)$$

The partial molal enthalpy of water in a solution with a salt i, $\bar{L}_w^i$, at an ionic strength I, is given by

$$n_w\bar{L}_w^i = -I^{3/2}A_H(\sigma(I^{\frac{1}{2}})/3) + \frac{9.212RT^2\nu_1\nu_2I^2}{(z_1z_2\nu)^2}(\frac{dB^{\cdot}}{dT}) \qquad (3.10.2)$$

where n_w is the number of moles of water ($10^3/M_w$) 1 and 2 represent the ions into which i dissociates, values of A_H are given in Table 3.10.

and $\sigma(I^{\frac{1}{2}}/3) = (1/I^{3/2})\left[1 + I^{\frac{1}{2}} - (1/(1+I^{\frac{1}{2}})) - 2\ln(1+I^{\frac{1}{2}})\right]$

(Pitzer and Brewer 1961)

Also,

$$-R\ln a_w = (2RI/n_wz_1z_2)(1-z_1z_2 2.303A_\gamma I^{\frac{1}{2}}(\sigma(I^{\frac{1}{2}})/3) + 4.606\nu_1\nu_2IB^{\cdot}/\nu^2z_1z_2) \qquad (3.10.3)$$

Combining these relationships,

$$n_w(\bar{s}_w^i - \bar{s}_w^o) = (2RI/z_1z_2) - \left[1+I^{\frac{1}{2}} - \frac{I}{1+I^{\frac{1}{2}}} - 2\ln(1+I^{\frac{1}{2}})\right]\left[(A_H/T)+4.606RA_\gamma\right] + \frac{9.212R\nu_1\nu_2I^2}{(z_1z_2\nu)^2}\left[B_i^{\cdot} + TdB_i^{\cdot}/dT\right] \qquad (3.10.4)$$

TABLE 3.15 The partial molal entropies of the components of binary solutions at an ionic strength 0.723 $mol kg^{-1}$, 1-atm

cal $deg^{-1} mol^{-1}$

Salt	t^oC	0	5	10	15	20	25	30
NaCl	$\bar{s}$	33.57	32.84	32.13	31.45	30.80	30.18	29.60
	$\bar{s}^o$	31.98	31.07	30.18	29.31	28.45	27.60	26.77
	$\Delta\bar{s}_w$	0.060	0.059	0.058	0.057	0.056	0.054	0.053
KCl	$\bar{s}$	44.04	43.29	42.57	41.87	41.21	40.57	39.96
	$\bar{s}^o$	42.08	41.17	40.28	39.41	38.55	37.70	36.87
	$\Delta\bar{s}_w$	0.057	0.056	0.055	0.055	0.054	0.053	0.051
NaF	$\bar{s}$	18.81	18.14	17.49	16.87	16.30	15.74	15.23
	$\bar{s}^o$	16.48	15.57	14.68	13.81	12.95	12.10	11.27
	$\Delta\bar{s}_w$	0.054	0.053	0.052	0.051	0.049	0.047	0.046
KBr	$\bar{s}$	49.71	48.96	48.25	47.55	46.90	46.27	45.67
	$\bar{s}$	48.18	47.27	46.38	45.51	44.65	43.80	42.97
	$\Delta\bar{s}_w$	0.060	0.059	0.058	0.057	0.056	0.055	0.054
$NaHCO_3$	$\bar{s}$	44.39	43.63	42.90	42.19	41.52	40.87	40.26
	$\bar{s}^o$	41.48	40.57	39.68	38.81	37.95	37.10	36.27
	$\Delta\bar{s}_w$	0.051	0.050	0.049	0.049	0.048	0.047	0.046
$CaCl_2$	$\bar{s}$	28.97	28.30	27.67	27.04	26.49	25.94	25.45
	$\bar{s}^o$	17.58	16.67	15.78	14.91	14.05	13.20	12.37
	$\Delta\bar{s}_w$	0.023	0.023	0.023	0.022	0.022	0.022	0.021
$MgCl_2$	$\bar{s}$	14.06	13.39	12.75	12.13	11.57	11.03	10.53
	$\bar{s}^o$	2.58	1.67	0.78	-0.09	-0.95	-1.80	-2.63
	$\Delta\bar{s}_w$	0.023	0.023	0.023	0.022	0.022	0.021	0.021
$SrCl_2$	$\bar{s}$	32.49	31.84	31.22	30.61	30.08	29.56	29.09
	$\bar{s}^o$	21.38	20.47	19.58	18.71	17.85	17.00	16.17
	$\Delta\bar{s}_w$	0.024	0.024	0.023	0.023	0.022	0.022	0.022
Na_2SO_4	$\bar{s}$	48.06	47.45	46.88	46.33	45.86	45.40	45.01
	$\bar{s}^o$	37.58	36.67	35.78	34.91	34.05	33.20	32.37
	$\Delta\bar{s}_w$	0.025	0.025	0.024	0.024	0.023	0.023	0.022
water	$\bar{s}^o_w$	15.17	15.49	15.82	16.12	16.44	16.75	17.05

$\bar{s}^o$ Appendix 3 Table 3A.1 $\bar{s}$ eq 3.8.5 Tables 3.12 and 3.13

$\Delta\bar{s}_w = \bar{s}^i_w - \bar{s}^o_w$, from eq 3.10.4.

Values of A_H and A_γ for the temperature range 0-30°C have been given in Table 3.10. R is the gas constant (1.9872 caldeg^{-1}mol^{-1}). Values of $B_i^\bullet$ and $dB_i^\bullet/dT$ at 25°C are given in Tables 3.12, 3.13. For the values at other temperatures, eq 3.8.10 can be used. Values of $\bar{s}_w^o$ were taken from Pistorius and Sharp (1960).

The results are summarised in Table 3.15.

3.11 The entropies of seawater, water in seawater and the solute in seawater. S = 35‰, 1-atm

The entropy, $\$_{sw}$, of a solution of seawater containing 1 kg of water and m_sM_t g of solute is given by

$$\$_{sw} = n_w\bar{s}_w^{sw} + m_s \Sigma_i f_i\bar{s}_i^{sw} \qquad (3.11.1)$$

in the usual notation, $n_w = 10^3/M_w = 55.51\text{mol}$

$f_i = m_i/m_s$, $m_s = \Sigma_i m_i$.

From the results for other thermodynamic properties (see, e.g., chapters 4 and 6) it might be expected that the summed values of the binary-solution entropies, $\bar{s}_i$, (weighted according to the relative concentrations in seawater, i.e. by f_i) would give a good estimate of the seasalt entropy, $\bar{S}_s$.

The corresponding partial molal entropy of water could be obtained from the sum of $\bar{s}_w^i$ values, each weighted by I_i/I, the ionic strength fraction of the component in seawater (as for the partial molal volume, Chapter 4). In symbolic terms, the approximations are:

$$m_s\bar{S}_s \simeq m_s\Sigma_i f_i\bar{s}_i \qquad (3.11.2)$$

and $$\bar{s}_w^{sw} \simeq \frac{-I^2}{55.51W} \Sigma_i f_i(\partial\phi_{\$_i}/\partial I) + \bar{s}_w^o \qquad (3.11.3)$$

(cf the corresponding relationship for $\bar{v}_w$, chapter 4), where $\phi_{\$_i}$ is the apparent molal entropy of $\underline{i}$ and $W = \frac{1}{2}\Sigma_i f_i(z_+z_-\nu)_i = 1.3576$ for solution A (Table 4.16), N.B. $w_i = \frac{1}{2}(z_+z_-)_i\nu_i$, so that $W = \Sigma f_i w_i$.

Since $\phi_{\$_i}$ is rarely used, and as

$$w_i\bar{s}_w^i = \frac{-I^2}{55.51}(\partial\phi_{\$}/\partial I) + w_i\bar{s}_w^o \qquad (3.11.4)$$

eq 3.11.3 becomes

$$\bar{s}_w^{sw} = (1/W)\Sigma_i f_i w_i \bar{s}_w^i \qquad (3.11.5)$$

Values of $\bar{s}_i$ and $\bar{s}_w^i$ can be obtained from the equations given above. For example, values for I = 0.723molkg^{-1}(water), for t = 0-30°C are listed in Table 3.15.

Estimates of $\bar{S}_s$ and $\bar{s}_w^{sw}$ as well as the total entropy of the solution, $\$_{sw}$, were calculated from these relationships (Table 3.16).

A value of $\bar{s}_w^{sw}$ at 25°C was calculated independently in the following way.

Since

$$-R\ln a_w^{sw} = (R\phi/55.51)m_s\Sigma_i(f_i\nu_i) \qquad (3.11.6)$$

where ϕ is the osmotic coefficient of the seawater solution containing m_s moles of the salts, i, and 1 kg (i.e. 55.51moles) of water, (Harned and Owen 1958, p13). For solution A, m_s = 0.53274mol/kgwater (at S=35‰) and $\Sigma_i f_i\nu_i$ = 2.178796. The value of ϕ is taken as 0.907 (Whitfield 1974). Substituting these values into eq 3.11.6 gives

$$-R\ln a_w^{sw} = 0.03769$$

For this salinity, Bromley (1968) gives a value for $\bar{L}_w^{sw}$ of 0.032 cal/gwater or 32/55.51 cal/mol. This value was calculated on the assumption that seasalt was a single solute with the characteristics of a 1:1 electrolyte. Millero, Hansen and Hoff (1973) obtained a value of 0.029 cal/g water, using similar assumptions to analyse their experimental data. The value estimated from the summed binary $\bar{L}_w^i$ data is 0.054 cal/gwater. Substituting these values in turn into eq 3.10.1 gives (with T = 298.15)

$$\bar{s}_w^{sw} - \bar{s}_w^o = 0.039$$

and 0.041, with the latter $\bar{L}_w^{sw}$ estimate. The latter value is in excellent agreement with the value

TABLE 3.16 Components* of the entropy of seawater at 1 bar, pressure, S = 35‰ (I = 0.723mol/kgwater)

t^oC	$\bar{s}_w^{sw}$	$55.51\bar{s}_w^{sw}$	$m_s\Sigma_i f_i \bar{s}_i$	$\$_{sw}$	$\underline{\$}_{sw}$
0	15.215	844.603	17.305	861.91	0.832
3	15.405	855.130	17.069	872.20	0.842
5	15.535	862.329	16.922	879.25	0.848
7	15.644	868.421	16.766	885.19	0.854
10	15.861	880.441	16.554	896.99	0.866
15	16.163	897.218	16.196	913.41	0.881
20	16.485	915.100	15.866	930.97	0.898
25	16.791	932.092	15.545	947.64	0.914
30	17.090	948.691	15.245	963.94	0.930

* the effects of mixing the aqueous salt solutions to make seawater have been neglected.

$\bar{s}_w^{sw}$ partial molal entropy of water in seawater estimated from eq 3.11.5 cal/deg mol water

$55.51\bar{s}_w^{sw}$ ditto in cal/deg kg of water

$m_s\Sigma_i f_i \bar{s}_i$ the partial molal entropy of the solute in seawater cal/deg kg water see Tables 3.15, 3.16a

$\$_{sw}$ entropy of seawater cal/deg kg of water eq 3.11.1

$\underline{\$}_{sw}$ specific entropy of seawater cal/deg g of seawater

calculated from eq 3.11.5. This equation should therefore give reasonable values for the partial molal entropy of water in seawater at 1-atm.

Values of the various entropy terms contributing to the total entropy of seawater are listed in Tables 3.16 and 3.16a.

The values of $m_s\bar{S}_s$, the entropy contribution of the solute in seawater, are compared with the corresponding values for NaCl and $CaCl_2$ in their binary solutions at the same ionic strength (Figure 3.8)

Once the values at 1-atm have been established, it is only necessary to predict the changes with pressure, a straightforward procedure, as indicated by eq 3.5.2.

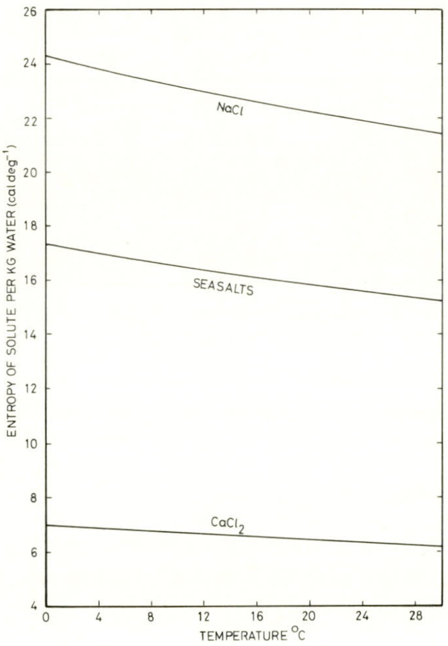

FIGURE 3.8

The entropies of various solutes ($m_i\bar{s}_i$ or $\Sigma m_i\bar{s}_i$, with i representing salt) in 1 kg water at an ionic strength of 0.723 mol/kg (corresponds to S = 35‰)
p = 1-atm
Number of moles of salt:

$CaCl_2$	0.2410
NaCl	0.7230
Seasalt	0.5327

(Solution A, Table 4.16)

TABLE 3.16a The entropy contributions of the components of seawater* to the standard solute entropy, and to the solute entropy at a salinity of 35‰, over a range of temperatures, pressure 1-atm.

t°C		0		5		10	
Component	f_i	$f_i\bar{s}_i^o$	$f_i\Delta\bar{s}_i$	$f_i\bar{s}_i^o$	$f_i\Delta\bar{s}_i$	$f_i\bar{s}_i^o$	$f_i\Delta\bar{s}_i$
NaCl	0.796440	25.470	1.2685	24.747	1.4053	24.038	1.5518
KCl	0.017666	0.743	0.0346	0.727	0.0374	0.712	0.0404
NaF	0.000139	0.002	0.0003	0.002	0.0004	0.002	0.0004
KBr	0.001602	0.077	0.0024	0.076	0.0027	0.074	0.0030
$NaHCO_3$	0.004539	0.188	0.0132	0.184	0.0139	0.180	0.0146
$CaCl_2$	0.020098	0.353	0.2289	0.335	0.2338	0.317	0.2389
$MgCl_2$	0.103630	0.267	1.1892	0.173	1.2144	0.081	1.2406
$SrCl_2$	0.000173	0.004	0.0019	0.004	0.0020	0.003	0.0020
Na_2SO_4	0.054895	2.063	0.5753	2.013	0.5917	1.964	0.6092
	SUMS	29.167	3.3143	28.261	3.5016	27.371	3.7009

t°C	15		20		25		30	
NaCl	23.341	1.7035	22.655	1.8781	21.982	2.0569	21.319	2.2536
KCl	0.696	0.0435	0.681	0.0470	0.666	0.0507	0.651	0.0547
NaF	0.002	0.0004	0.002	0.0005	0.002	0.0005	0.002	0.0006
KBr	0.073	0.0033	0.072	0.0036	0.070	0.0040	0.069	0.0043
$NaHCO_3$	0.176	0.0154	0.172	0.0162	0.168	0.0171	0.165	0.0181
$CaCl_2$	0.300	0.2439	0.282	0.2510	0.265	0.2561	0.249	0.2630
$MgCl_2$	-0.010	1.2663	-0.099	1.2981	-0.187	1.3290	-0.273	1.3642
$SrCl_2$	0.003	0.0021	0.003	0.0021	0.003	0.0022	0.003	0.0022
Na_2SO_4	1.916	0.6268	1.869	0.6483	1.823	0.6698	1.777	0.6940
SUMS	26.497	3.9052	25.637	4.1449	24.792	4.3863	23.962	4.6547

* Solution A (Table 4.16)

n.b. contribution of boric acid (f_i = 0.000818) has been neglected.

$\Delta\bar{s}_i = \bar{s}_i - \bar{s}_i^o$

3.12 The effect of pressure on the entropy of pure water.

The equation for calculating $(\alpha v)_w^{(p)}$, the temperature derivative of the specific volume, has been given in Section 5.71 viz

$$(\alpha v)_w^{(p)} = (\alpha v)_w^{(1)} - 0.315(\alpha v)_w^{(1)} \log\frac{(B+p)}{(B+1)}$$

$$- \frac{0.315 v_w^{(1)} \partial B/\partial t}{2.303(B+1)} \left[1/(B+p) - p/(B+p)\right] \qquad (5.71.6)^*$$

The specific entropy of water, is therefore given by: (in cm^3bar/deg g., the pressure being in bars, the superscript zero for pure water is omitted for simplicity)

$$-(\not{s}_w^{(p)} - \not{s}_w^{(1)}) = \int_1^p (\alpha v)_w dp$$

$$= (\alpha v)_w^{(1)}(1+0.315\log(B+1))(p-1)$$

$$- 0.315(\alpha v)_w^{(1)}\left[\log[(B+p)^{(B+p)}/(B+1)^{(B+1)}] - (p-1)/2.303\right]$$

$$- \frac{0.315 w_w^{(1)} \partial B/\partial t}{(B+1)} \left[(B+1)\log\frac{(B+p)}{(B+1)} - (p-1)/2.303\right] \qquad (3.12.1)$$

since $\int_1^p (p/(B+p))dp = (p-1) - B\ln\frac{(B+p)}{(B+1)}$

$$\int_1^p \ln(B+p)dp = (B+p)\ln(B+p) - (B+1)\ln(B+1) - (p-1)$$

and $\int_1^p (1/(B+p))dp = \ln\frac{(B+p)}{(B+1)}$

The specific volume at 1 bar and its temperature derivative are given by (from Chapter 5)

$$v_w^{(1)} = (1+bt)/P_5(t) \qquad (5.71.1)^*$$

and $$(\alpha v)_w^{(1)} = (1/P_5(t))\left[b - (1+bt)P_4(t)/P_5(t)\right] \qquad (5.71.2)^*$$

with the coefficients for the polynomials $P_4(t)$ and $P_5(t)$ given in Table 5.16.

TABLE 3.17 The molal entropy of water as a function of temperature and pressure

cal deg^{-1} mol^{-1}

$-10^3(S_w^{(p)} - S_w^{(1)})$

t p bars	0	5	10	15	20	25	30
50	-1.2	0.5	2.0 (1)	3.3	4.4 (4)	5.5	6.5 (7)
100	-2.1	1.3	4.2 (3)	6.8	9.1 (9)	11.2	13.1 (14)
150	-2.6	2.4	6.7 (6)	10.5	13.9 (13)	17.0	19.9 (21)
200	-2.8	3.8	9.5 (8)	14.4	18.9 (18)	22.9	26.7 (28)
250	-2.6	5.5	12.4 (11)	18.5	24.0 (23)	29.0	33.6 (34)
300	-2.0	7.4	15.6 (14)	22.8	29.2 (28)	35.1	40.5 (41)
350	-1.2	9.6	19.0 (18)	27.2	34.6 (33)	41.3	47.5 (48)
400	0.0	12.1	22.6 (21)	31.8	40.1 (38)	47.6	54.6 (55)
450	1.5	14.8	26.4 (25)	36.5	45.7 (43)	54.0	61.7 (62)
500	3.3	17.8	30.3 (29)	41.4	51.4 (49)	60.5	68.9 (69)
550	5.4	21.0	34.5	46.4	57.2	67.1	76.2
600	7.8	24.4	38.8 (37)	51.6	63.1 (60)	73.7	83.5 (83)
650	10.4	28.0	43.3	56.9	69.2	80.4	90.8
700	13.3	31.9	48.0 (46)	62.4	75.3 (71)	87.2	98.3 (98)
750	16.5	35.9	52.9	67.9	81.6	94.1	105.7
800	19.9	40.2	57.9 (56)	73.6	87.9 (83)	101.0	113.2 (113)
850	23.5	44.6	63.1	79.5	94.3	108.0	120.8
900	27.4	49.3	68.4 (66)	85.4	100.9 (95)	115.1	128.4 (128)
950	31.5	54.1	73.8	91.5	107.5	122.2	136.0
1000	35.8	59.1	79.5 (77)	97.6	114.2 (108)	129.4	143.7 (143)
1050	40.3	64.2	85.2	103.9	121.0	136.7	151.5
$10^3 S_w^{(1)}$	15170	15490	15817	16120	16443	16750	17050 ($\pm$30)

Bracketed values and $S_w^{(1)}$ values from Pistorius and Sharp (1960)

$-\Delta S_w^{(1,p)}$ from eq 3.12.1, with B values from eq 6.75.2, Chapter 6, Section 6.75.

The Tait parameter B, as a function of temperature, is given by

$$B = 2671.8 + 19.454t - 0.27028t^2 + 9.798x10^{-4}t^3 \qquad (6.75.2)^*$$

(see Section 6.7).

Values of the entropy changes with pressure are listed in Table 3.17 together with values for 1 bar. The corresponding values calculated by Pistorius and Sharp (1960), using the thermodynamic data of Hilsenrath _et al_ (1955) on steam at low pressures and specific volume data of Kennedy (1950) and Kennedy _et al_ (1958), are also listed.

The agreement of the two sets of results at 30°C is excellent. At this temperature, the "structural" perturbations characteristic of the experimental technique are minimal (Section 6.7, 3.144). The agreement at 10°C is also good, but at 20°C the earlier data give values up to 0.006caldeg^{-1}mol^{-1} higher (at 1000 bars). Such a large deviation is, however, less than the uncertainty in the 1-atm entropy values.

3.13 Change of entropy with pressure.

3.131 SEAWATER

If the pressure on seawater changes from 1 bar to p bars the corresponding change in entropy is given by

$$\emptyset_{sw}^{(p)} - \emptyset_{sw}^{(1)} = -\int_1^p (\alpha v)\, dp \qquad (3.5.2)^*$$

(from Section 3.5). If the usual values of αv (mldeg^{-1}g^{-1}) are substituted in, with pressure in bars, the entropy unit is mlbardeg^{-1}g^{-1}. This is converted to caldeg^{-1}g^{-1} by multiplying by 0.024218/1.01325. The units used here are the latter, in addition caldeg^{-1}mol^{-1} (the usual binary solution unit) will be used, the mol referring to the component water or one of the salts. The mol may also refer to the ions or to "seasalt" and this will be indicated. The temperature coefficient of the specific volume, αv, or $\frac{(\partial v)}{(\partial t)}_p$ has been given as a function of pressure in Chapter 5 (Section 5.72). The Tait-Gibson equation for αv (with the data of Ekman) was shown to give excellent agreement with the polynomial equation based on the data of Bradshaw and Schleicher (1970). Therefore, these two equations can be used to calculate the change of entropy with pressure, the values obtained being cross-checking. That is, computational errors can be assessed more readily.

Tait-Gibson Equation.

From Section 5.72

$$(\alpha v)_{sw}^{(p)} = (1-c)(\partial \Psi_w/\partial t)^{(p)} + c\partial \Psi_t/\partial t \qquad (5.72.10)^*$$

and (5.72.11)*

with $c=10^{-3}S(‰)$,

$$(\partial \Psi_w/\partial t)^{(p)} = \alpha_w^{(1)} \Psi_w^{(p)} + v_w^{(1)} 0.1368 D_1 \qquad (5.72.14)^*$$

$$D_1 = ((\partial B/\partial t)/(B+1)) - ((\partial B^*/\partial t)/(B^*+p))$$

B and B* being the Tait and Tait-Gibson parameters, respectively,

$$B^* = 2670.8 + 6.8966S + 19.39t - 0.0703St$$
$$-0.223t^2 \qquad (4.61.7)^*$$

$$B = B^*(S=0)$$

$$\Psi_w^{(p)} = v_w^{(1)}(1-0.315\log(B^*+p)/(B+1)) \qquad (5.72.12)^*$$

$$\partial \Psi_t/\partial t = -2.1896\times 10^{-3}$$

Using these relationships the entropy equation becomes

$$-(\phi_{sw}^{(p)} - \phi_{sw}^{(1)}) = \int_1^P (\alpha v)\,dP$$

$$= A_1\int_1^P \Psi_w\,dp + A_2\int_1^P D_1\,dp + A_3\int_1^P dp \qquad (3.131.1)$$

with $A_1 = (1-c)\alpha_w^{(1)}$, $A_2 = 0.1368(1-c)v_w^{(1)}$

$$A_3 = -2.1896c\times 10^{-3}$$

and

TABLE 3.18 The effect of pressure on the entropy of a volume of seawater containing 1 kg of water*

S = 35‰ (I= 0.723mol/kg water)

$-(S_{sw}^{(p)} - S_{sw}^{(1)})$ cal/deg kgwater

p bars	t°C	0	5	10	15	20	25	30
50	TG	0.0675	0.1397	0.2024	0.2584	0.3094	0.3568	0.4014
	B	0.0713	0.1399	0.2014	0.2570	0.3077	0.3545	0.3985
100	TG	0.1522	0.2950	0.4191	0.5299	0.6309	0.7247	0.8129
	B	0.1600	0.2954	0.4171	0.5270	0.6274	0.7203	0.8078
150	TG	0.2523	0.4629	0.6458	0.8091	0.9581	1.0963	1.2263
	B	0.2646	0.4636	0.6426	0.8047	0.9529	1.0902	1.2197
200	TG	0.3674	0.6429	0.8822	1.0958	1.2907	1.4715	1.6416
	B	0.3842	0.6439	0.8778	1.0898	1.2839	1.4640	1.6342
250	TG	0.4970	0.8347	1.1280	1.3898	1.6286	1.8502	2.0586
	B	0.5184	0.8360	1.1222	1.3821	1.6203	1.8420	2.0513
300	TG	0.6405	1.0379	1.3829	1.6908	1.9717	2.2323	2.4773
	B	0.6670	1.0394	1.3757	1.6813	1.9619	2.2230	2.4708
350	TG	0.7977	1.2521	1.6466	1.9986	2.3197	2.6176	2.8977
	B	0.8290	1.2539	1.6379	1.9874	2.3087	2.6080	2.8928
400	TG	0.9681	1.4771	1.9188	2.3130	2.6726	3.0061	3.3196
	B	1.0040	1.4789	1.9085	2.3000	2.6605	2.9970	3.3172
500	TG	1.3468	1.9580	2.4881	2.9610	3.3923	3.7923	4.1681
	B	1.3917	1.9590	2.4740	2.9442	3.3785	3.7850	4.1730
600	TG	1.7737	2.4782	3.0888	3.6333	4.1298	4.5902	5.0224
	B	1.8265	2.4788	3.0698	3.6123	4.1148	4.5870	5.0378
700	TG	2.2462	3.0355	3.7191	4.3285	4.8840	5.3989	5.8820
	B	2.3050	3.0316	3.6940	4.3029	4.8680	5.4010	5.9110
800	TG	2.7618	3.6280	4.3776	5.0455	5.6541	6.2180	6.7467
	B	2.8240	3.6182	4.3450	5.0144	5.6380	6.2270	6.7921
900	TG	3.3182	4.2538	5.0627	5.7831	6.4392	7.0468	7.6161
	B	3.3790	4.2350	5.0201	5.7456	6.4230	7.0641	7.6802
1000	TG	3.9133	4.9113	5.7732	6.5402	7.2386	7.8849	8.4899
	B	3.9694	4.8804	5.7183	6.4950	7.2220	7.9114	8.5748

* $\left[10^3 + 1.0049\frac{S}{(1-c)}\right]$ g of seawater TG eq 3.131.1

B eq 3.131.2 and (5.63.1)*

$$\int_1^p \Psi_w dp = v_w^{(1)} \left[1+0.315\log(B+1)\right](p-1)$$

$$-0.315v_w^{(1)} \left[(B^*+p)\log(B^*+p) -(B^*+1)\log(B^*+1) +(1-p)/2.303\right]$$

$$\int_1^p D_1 dp = (p-1)(\partial B/\partial t)/(B+1) -2.303 \ (\partial B^*/\partial t) \ \log(B^*+p)/(B^*+1)$$

$$\int_1^p dp = (p-1)$$

Values of $(\cancel{s}_{sw}^{(p)} - \cancel{s}_{sw}^{(1)})$ for a range of temperatures and pressures are shown in Tables 3.18, 3.19.

The values for $\alpha_w^{(1)}$ and $v_w^{(1)}$ were taken from (5.71.1)* and (5.71.2)* Section 3.12.

<u>Bradshaw & Schleicher's data - Bryden's polynomial</u> (from Chapter 5)

This equation is a polynomial of the form

$$\alpha v = a_o(p,S) + a_1(p,S)t + a_2(p,S)t^2 + a_3(p,S)t^3 \qquad (5.63.1)^*$$

with
$$10^6 a_n = b_o + b_1 p + b_2 p^2 + b_3 p^3$$
$$+ (d_o + d_1 p + d_2 p^2)(S-35)$$
$$+ e(S-35)^2 \qquad (n = 0,1,2,3)$$

(p represents pressure in bars, S, salinity in ‰).

The coefficients for this equation are listed in Table 5.11, integration with respect to pressure gives

$$10^6 \int_1^p a_n = \Big[b_o p + \tfrac{1}{2} b_1 p^2 +(1/3)b_2 p^3 + (1/4)b_3 p^4$$
$$+ (d_o p + \tfrac{1}{2} d_1 p^2 + (1/3)d_2 p^3)(S-35)$$
$$+ ep(S-35)^2\Big]_1^p \qquad (3.131.2)$$

Integration of eq (5.63.1)* in this manner gives $-(\cancel{s}_{sw}^{(p)}-\cancel{s}_{sw}^{(1)})$; the values so calculated are compared with those from the TG equation in Tables 3.18, 3.19.

TABLE 3.19 The effect of pressure on the entropy of 1 g of seawater for salinities 30,35 and 40‰

$-(s^{(p)}_{sw} - s^{(1)}_{sw})$ cal/deg(g of seawater) $\times 10^6$

p bars	Salinity	t^oC 0	5	10	15	20	25	30
50	30‰	49	122	185	242	293	340	385
	35‰	65	135	195	249	299	344	387
		(69)	(135)	(194)	(248)	(297)	(342)	(385)
	40‰	80	147	204	256	304	348	390
100	30‰	115	260	385	497	598	691	779
	35‰	147	285	404	511	609	699	784
		(155)	(285)	(402)	(509)	(605)	(695)	(779)
	40‰	176	308	422	525	619	706	789
200	30‰	294	573	815	1029	1225	1406	1575
	35‰	355	620	851	1057	1245	1420	1584
		(371)	(621)	(847)	(1052)	(1239)	(1413)	(1577)
	40‰	411	665	885	1083	1264	1432	1591
300	30‰	532	934	1282	1592	1874	2134	2378
	35‰	618	1001	1334	1631	1902	2154	2390
		(643)	(1003)	(1327)	(1622)	(1893)	(2145)	(2384)
	40‰	699	1064	1382	1667	1928	2170	2399
400	30‰	824	1339	1786	2183	2544	2877	3189
	35‰	934	1425	1851	2232	2579	2900	3203
		(969)	(1427)	(1841)	(2219)	(2567)	(2892)	(3200)
	40‰	1037	1505	1912	2276	2610	2920	3213

TABLE 3.19 (ctd)

t^oC		0	5	10	15	20	25	30
p bars	Salinity							
500	30‰	1168	1787	2323	2800	3233	3633	4007
	35‰	1299 (1343)	1889 (1890)	2400 (2387)	2857 (2841)	3273 (3260)	3659 (3652)	4022 (4026)
	40‰	1422	1984	2472	2909	3308	3681	4032
600	30‰	1561	2275	2892	3441	3940	4401	4832
	35‰	1711 (1762)	2391 (2391)	2980 (2962)	3506 (3485)	3985 (3970)	4429 (4426)	4846 (4861)
	40‰	1852	2499	3061	3563	4023	4451	4855
700	30‰	2000	2800	3492	4106	4665	5180	5662
	35‰	2167 (2224)	2929 (2925)	3588 (3564)	4176 (4152)	4712 (4697)	5209 (5211)	5675 (5703)
	40‰	2323	3048	3677	4239	4753	5232	5683
800	30‰	2482	3361	4119	4794	5406	5971	6498
	35‰	2665 (2724)	3500 (3491)	4224 (4192)	4868 (4838)	5455 (5440)	5999 (6008)	6509 (6553)
	40‰	2835	3629	4318	4934	5497	6021	6514
900	30‰	3005	3955	4774	5502	6162	6772	7340
	35‰	3202 (3261)	4104 (4086)	4885 (4844)	5580 (5544)	6213 (6197)	6799 (6816)	7348 (7410)
	40‰	3384	4241	4984	5648	6255	6819	7350
1000	30‰	3568	4581	5455	6230	6934	7582	8186
	35‰	3776 (3830)	4739 (4709)	5570 (5517)	6310 (6267)	6984 (6968)	7608 (7633)	8191 (8273)
	40‰	3969	4883	5674	6380	7025	7625	8189

Bracketed values from Bryden's polynomial, eq 3.131.2
other values from Tait-Gibson equation, 3.131.1

The entropy changes calculated from the two equations (3.131.1 and 3.131.2) are in good agreement, well within the errors of the 1-atm entropy values. This might be expected, in view of the excellent agreement of the expansibility values (Section 5.7).

The agreement at 5°C is particularly good, being generally better than 1×10^{-5} cal/deg for 1 g of seawater.

The changes of $\Delta\underline{\mathit{s}}_{sw}^{(1,p)}$ with salinity are only small. (Table 3.19). For p=1000 bars, t = 30°C, the change from S=30‰ to S=40‰ is less than 1×10^{-5} cal/deg g.sw, at 0°C it is 40×10^{-5} cal/deg.g.sw. The change in going from 0°C to 30°C at this pressure, for a salinity of 35‰ is, however, over 400×10^{-5} cal/deg g.sw (Table 3.19).

3.132 THE SOLUTE IN SEAWATER

As detailed in Chapter 4, the effect of pressure on the partial molal free energy (or chemical potential) of a component, $\underline{i}$, is given by the partial molal volume, $\bar{v}_i$, viz., at constant composition

$$(\partial\bar{G}_i/\partial p)_T = \bar{v}_i \qquad (4.1.1)^*$$

and since

$$(\partial\bar{G}_i/\partial T)_p = -\bar{s}_i \qquad (1.6.1)^*$$

$$\partial(\partial\bar{G}_i/\partial p)/\partial T = \partial(\partial\bar{G}_i/\partial T)/\partial p = \partial\bar{v}_i/\partial T$$

so that

$$-\partial\bar{s}_i/\partial p = \partial\bar{v}_i/\partial T = \bar{E}_i \qquad (3.132.1)$$

where $\bar{E}_i$ is the partial molal expansibility of $\underline{i}$. Values of $\bar{E}_i$ for a number of salts have been given in Chapter 5. The corresponding quantity for seasalt has also been given. In general, for seasalt,

$$\bar{E}_s = (\alpha v)_{sw}M_t + A_s(\partial(\alpha v)_{sw}/\partial S)_{T,p} \qquad (5.6.9)^*$$

where $A_s = (S(1-c)M_t + 10^3 M_t'(1-c)^2)$, with S representing the salinity, $c=10^{-3}S$, M_t is the molecular weight of the solute (i.e. $\Sigma m_i M_i/m_s$, $\underline{i}$ representing a salt as usual) M_t' is the corrected value of M_t (to allow for the difference in total salt mass and "salinity") - see Chapter 4, Table 4.16.

Thus,

$$\int_1^p \bar{E}_s dp = -(\bar{S}_s^{(p)} - \bar{S}_s^{(1)}) = -\Delta\bar{S}_s^{(1,p)}$$

$$= M_t \int_1^p (\alpha v)_{sw} dp + A_s \int_1^p (\partial(\alpha v)_{sw}/\partial S)\, dp \qquad (3.132.2)$$

$(\alpha v)_{sw}$ and its salinity differential, as functions of pressure, are given in Chapter 5, (e.g. eqs 5.63.1 and 5.63.2). The Tait-Gibson form of $\bar{E}_s$ (eq 5.6.2 - Ekman's data) and the polynomial of Bryden based on the data of Bradshaw and Schleicher (eqs 5.63.1 and 5.63.2 and Table 5.11) will both be used, as for seawater. This provides a check on the derivations, rounding errors and deficiencies in the salinity coefficients.

Tait-Gibson. From Chapter 5, eq (5.6.2)*

$$\bar{E}_s = M_t d\psi/dt - K(1-c)^2 v_w^{(1)} [\partial(\partial B^*/\partial S)/\partial t + \alpha_w^{(1)} \partial B^*/\partial S - (\partial B^*/\partial S)(\partial B^*/\partial t)/(p+B^*)] / (p+B^*)$$

with $K = 434.3 \times 0.315 M_t'$, $M_t'^A = 68.0811$ g (solution A, Table 4.16)

$M_t^A = 68.4148$ g

$$B^* = 2670.8 + 6.8966S + 19.39t - 0.0703St - 0.223t^2 \qquad (4.61.7)^*$$

$$d\psi/dt = -0.0021896$$

so that

$$\int_1^p \bar{E}_s dp = -\Delta\bar{S}_s^{(1,p)}$$

$$= -2.1896 \times 10^{-3} M_t (p-1)$$

$$- K(1-c)^2 v_w^{(1)} \Big[(\alpha_w^{(1)} A_1 - A_2)\, 2.3031 \log\frac{(B^*+p)}{(B^*+1)}$$

$$- A_1 A_3 (p-1)/[(B^*+p)(B^*+1)]\Big] \qquad (3.132.3)$$

with $A_1 = 6.8966 - 0.0703t$

$A_2 = 0.0703$

$A_3 = 19.39 - 0.0703S - 0.446t$

Bryden's polynomial. This equation has the form for αv as given above (eq 5.63.1*) so that differentiating w.r.t. salinity

$$\partial(\alpha v)/\partial S = g_o(p,S) + g_1(p)t + g_2(p)t^2 \qquad (5.63.2)*$$

with $10^6 g_n = 10^6(\partial a_n/\partial S)$

$$= d_o + d_1 p + d_2 p^2 + 2e(S-35)$$

(see equation for a_n in previous section). The coefficients a,b,d,e are listed in Table 5.11). The integrals of eq (5.63.1)*, Section 3.131, are therefore

$$\int_1^p \alpha v dp = \int_1^p a_o dp + t\int_1^p a_1 dp + t^2 \int_1^p a_2 dp + t^3 \int_1^p a_3 dp \qquad (3.132.4)$$

with $10^6 \int_1^p a_n dp$ as given in previous section, and

$$10^6 \int_1^p g_n dp = d_o(p-1) + \tfrac{1}{2}d_1(p^2-1) + (1/3)d_2(p^3-1)$$
$$+ 2e(S-35)(p-1) \qquad (3.132.5)$$

Thus

$$-\Delta\bar{S}_s^{(1,p)} = M_t \sum_{n=o}^{3} t^n \int_1^p a_n dp + A_s \sum_{n=o}^{2} t^n \int_1^p g_n dp \qquad (3.132.6)$$

Selected values of $\Delta\bar{S}_s^{(1,p)}$ from eq 3.132.6 are compared with the TG values in Table 3.20.

The discrepancies noted for the corresponding $\bar{E}_s$ values (Section 5.6) naturally carry over and are somewhat emphasised. In particular, the uncertainties in the salinity differentials of the polynomial are amplified in this case. Since the polynomial (Bryden 1970, Bradshaw and Schleicher 1970) is based on only three salinity values (30,35 and 40‰), the salinity differential cannot be expected to be very accurate. The discrepancies between $\bar{E}_s$ calculated from this polynomial and the "acoustic" equation of state (Wang and Millero 1973), illustrate this (Section 5.6 Table 5.14). At 0°C, where the Sxtxp coefficients drop out, the agreement is good

TABLE 3.20 The effect of pressure on the entropy of the solute in a volume of seawater containing 1 kg of water (S = 35‰)

$10^3 \, m_s(\bar{S}_s^{(p)} - \bar{S}_s^{(1)})$ cal/deg for 1036.447 g of seawater

p bars	t°C	0	5	10	15	20	25	30
50	TG	110	90	73	59	48	38	31
	B	(115)	(97)	(80)	(66)	(54)	(44)	(37)
100		218	177	144	116	94	75	60
200		420	341	276	223	179	142	113
		(440)						
300		606	491	396	318	254	201	158
400		776	627	505	404	321	252	196
		(811)						
500		932	752	603	481	379	296	227
600		1074	865	692	549	430	332	252
		(1120)						
700		1205	967	771	608	474	362	272
800		1324	1059	841	660	510	386	285
		(1373)						
900		1432	1143	903	705	541	405	294
1000	TG	1530	1217	958	743	565	417	297
	B	(1579)	(1350)	(1144)	(961)	(803)	(668)	(559)
1000	B-TG	49	133	186	218	238	251	262
	B-BS	-35	0	-4	-24	-39	-26	-36

TG eq 3.132.3 with data of Ekman (as given by Li 1967) and data of Knudsen et al (Dorsey 1940, Table 108)

B Bryden's (1973) polynomial using the data of Bradshaw and Schleicher (1970), eq 3.132.6 Table 5.11.

BS polynomial of Bradshaw and Schleicher (1970)

(Table 3.20). The results from the polynomial of Bradshaw and Schleicher (1970) give the same order of deviation. At increased temperatures, and the higher pressures, the deviations are in accord with those for $\bar{E}_s$, relative to the "acoustic" results (Table 5.14). It should be noted that the unit cm^3bar/degmol (from above eqs 3.132.3, 3.132.6) is converted to cal/degmol by multiplying by 0.023901, see Chapter 2, Section 2.5.

3.133 The mean apparent molal entropy of seasalt, $\Phi_{\not{s}}$

As for the other mean apparent molal quantities (Sections 4.62 and 6.9) the value of $m_s\Phi_{\not{s}}$ is obtained from the solution and pure water data, viz., using the specific entropies of seawater, $\not{s}_{sw}$ and water, $\not{s}_w$.

$$m_s\Phi_{\not{s}} = w_s\not{s}_{sw} - 10^3\not{s}_w \qquad (3.133.1)$$

taking a solution of seawater containing 1 kg of water as usual. The quantity m_s represents the moles of salt, and w_s represents the mass of the solution $(10^3+1.0049S/(1-c))$ grams. The change in $\Phi_{\not{s}}$ for a pressure change of 1 to p bars, is given by

$$m_s\Delta\Phi_{\not{s}}^{(1,p)} = w_s\Delta\not{s}_{sw}^{(1,p)} - 10^3\Delta\not{s}_w^{(1,p)} \qquad (3.133.2)$$

The same equation can be derived from

$$m_s\Delta\Phi_{\not{s}}^{(1,p)} = -m_s\int_1^p \Phi_E dp$$

via eq 3.142.1

or, from Section 5.61, using

$$\bar{E}_s = \Phi_E + m_s(\partial\Phi_E/\partial m_s) \qquad (3.6.3)^*$$

and

$$m_s\int_1^p(\partial\Phi_E/\partial m_s)dp = \frac{10^3}{M_w m_s}\left(\Delta\bar{s}_w^{sw\,(1,p)} - \Delta\not{s}_w^{(1,p)}\right)$$

with the entropies given per mole of water, and

since

$$w_s \Delta \underline{\$}_{sw}^{(1,p)} = \frac{10^3 \Delta \bar{s}_w^{sw(1,p)}}{M_w} + m_s \Delta \bar{S}_s^{(1,p)} \qquad (3.6.1)^*$$

TG Equation. In terms of this equation,

$$m_s \Delta \Phi_{\$} = 315(\alpha v)_w^{(1)} \log Z_1 + 315 v_w^{(1)} Z_2 + 2.1896(p-1)c/(1-c) \qquad (3.133.3)$$

where $Z_1 = (x_1^{x_1} y^{y})/(x^{x} y_1^{y_1})$

with $x_1 = (B+1)$, $x = (B+p)$, $y_1 = (B^*+1)$, and $y = (B^*+p)$

and $Z_2 = \frac{dB^*}{dt}\log(y/y_1) - \frac{dB}{dt}\log(x/x_1)$

which can be derived from eqs 3.133.2, 3.131.1 and 3.12.1.

3.14 WATER IN SEAWATER

3.141 Calculation of values

The entropy, $\$_{sw}$, of a quantity of seawater containing m_s moles of salts and $10^3/M_w$ moles of water, is given by

$$\$_{sw} = (10^3/M_w)\bar{S}_w^{sw} + m_s \bar{S}_s \qquad (3.6.1)^*$$

where $\bar{S}_w^{sw}$, $\bar{S}_s$ are the partial molal entropies of the water and solute respectively. The entropy of the water is obtained therefore, most simply from

$$\bar{S}_w^{sw} = 10^{-3} M_w (\$_{sw} - m_s \bar{S}_s) \qquad (3.141.1)$$

using the values of $\bar{S}_s$ and $\$_{sw}$ as obtained above. It should be noted that $\underline{\$}_{sw}$ is for 1 g of seawater, so that for a solution containing 1 kg water,

$$\$_{sw} = (10^3 + m_s M_t)\underline{\$}_{sw}$$
$$= [10^3 + 1.0049S/(1-c)]\underline{\$}_{sw}$$

The general equation for the partial molal entropy of water in seawater ($-(\bar{S}_w^{sw(p)} - \bar{S}_w^{sw(1)})$) is given by

$$-(10^3/M_w)\Delta \bar{S}_w^{sw(1,p)} = 10^3 \int_1^p (\alpha v)_{sw} dp - m_s A_s \int_1^p (\partial(\alpha v)_{sw}/\partial S) dp \qquad (3.141.2)$$

TABLE 3.21 Water in seawater

The effect of pressure on the partial molal entropy of water in seawater

(S = 35‰, I = 0.723mol/kg water)

$-10^4(\bar{s}_w^{sw(p)} - \bar{s}_w^{sw(1)})$ cal/deg (mol of water)

p bars		t°C 0	5	10	15	20	25	30
50	TG	-8	9	23	36	47	57	67
	B	-8	8	22	34	46 (45)	56	65
100	TG	-12	21	50	74	97	117	136
	B	-12	19	47	71	94	114	132
200	TG	-10	54	109	157	200	240	276
	B	-10	50	103	151	194	233	268
300	TG	6	99	178	247	309	366	418
	B	6	91	168	237	299	355	407
400	TG	35	153	255	344	424	496	563
	B	35	143	241	329	409	482	548
500	TG	75	217	340	447	543	630	710
	B	76	205	322	427	523	611	692
600	TG	126	291	432	556	667	767	859
	B	127	276	410	532	642	744	838
700	TG	188	373	531	670	795	907	1011
	B	190	355	505	641	765	880	986
800	TG	259	463	637	790	927	1051	1164
	B	261	442	606	755	892	1019	1137
900	TG	340	561	749	915	1063	1197	1319
	B	342	536	712	874	1023	1161	1290
1000	TG	429	666	867	1044	1202	1345	1476
	B	431	636	824	997	1156	1305	1444
	BS	(425)	(636)	(823)	(992)	(1149)	(1300)	(1450)

TG eq 3.141.1 data of Ekman (1908), as given by Li (1967) via eqs 3.132.3 and 3.131.1 and data of Knudsen et al (Dorsey 1940, Table 108)

B Bryden's (1973) polynomial based on the data of Bradshaw and Schleicher (1970) (see Table 5.11)

BS Bradshaw and Schleicher's polynomial based on their data (1970) eq 3.141.1 via eqs 3.5.2, 3.131.2, 3.132.6.

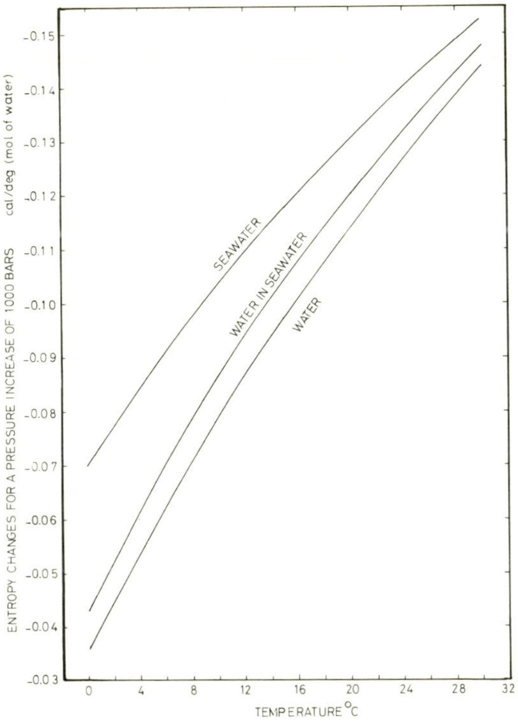

FIGURE 3.9

Entropy changes of water and seawater (S = 35‰) when undergoing a pressure increase from 1 to 1000 bars at various temperatures. The corresponding changes for the partial molal entropy of water-in-seawater are also shown. See Tables 3.18, 3.21 and 3.17.

which follows from eqs 3.141.1, 3.132.2, 3.5.2.

The salinity function $m_s A_s$ reduces to $(0.0049S^2 + 10^3S)$ or (ca. 10^3S).

For Bryden's polynomial, the terms of eq 3.141.2 are given by eqs. 3.132.4 and 3.132.5.

Values of $\Delta\bar{s}_w^{sw(1,p)}$ have been calculated from the two equations of state via eq 3.141.1 and are compared in Table 3.21.

The agreement is good, as for the seawater. The high mole fraction of water minimises the errors from the salinity derivatives of the polynomial.

The temperature changes of the entropies of seawater, water-in-seawater, and pure water are compared in Figure 3.9.

3.142 Relationship with the mean apparent molal expansibility

It may be of interest to note the relationship of $\Phi_E^{(p)}$, the mean apparent expansibility of seasalt, at a pressure p bars, to the pressure effects on pure water and on seawater. Since, (see Section 5.61)

$$w_s(\alpha v)_{sw}^{(p)} = m_s\Phi_E^{(p)} + 10^3(\alpha v)_w^{(p)} \qquad (3.142.1)$$

from eq 5.6.8, with $w_s = (10^3 + m_s M_t)$,

$$w_s\int_1^p (\alpha v)_{sw}dp = m_s\int_1^p \Phi_E dp + 10^3\int_1^p (\alpha v)_w dp$$

so that

$$-(w_s\Delta\underline{s}_{sw}^{(1,p)} - 10^3\Delta\underline{s}_w^{(1,p)}) = k'm_s\int_1^p \Phi_E dp \qquad (3.142.2)$$

the constant k' (= 0.024218/1.01325) converts from cm^3bar to cal, so that the entropies will be in cal/deg.g. The volume of seawater is that containing 1 kg of water and m_s moles of solute.

Thus, the Φ_E integral gives the difference in the pressure increments of water, and of seawater containing the same number of moles of water (i.e. $10^3/M_w$).

3.143 Relationship with the entropy of pure water

Another interesting comparison is that of water-in-seawater, and pure water.

In the usual way (Section 4.2) the partial molal quantity (per mole) for water in the solution is related to the pure water value, $\not{s}_w$, by

$$\bar{s}_w^{sw} = \not{s}_w - 10^{-3} M_w m_s^2 (\partial\Phi_{\not{s}}/\partial m_s) \qquad (3.143.1)$$

with $m_s = S/(1-c)M_t'$, as usual (Table 4.16) and M_w is the molecular mass of water. Note that the superscript zero for pure water is omitted here for simplicity, as in Sections 3.12, 3.133 and 3.142.

The effect of pressure on the gradient is given by

$$\left(10^{-3} M_w m_s^2\right) \frac{\partial\Delta\Phi_{\not{s}}^{(1,p)}}{\partial m_s} = \Delta\not{s}_w^{(1,p)} - \Delta\bar{s}_w^{sw(1,p)} \qquad (3.143.2)$$

(in cal/deg.mol of water)

Values for the gradient can be obtained from the corresponding values for the entropies as given in a previous section (Tables 3.17 and 3.21). The gradient can be calculated directly from

$$m_s^2 \frac{\partial\Delta\Phi_{\not{s}}^{(1,p)}}{\partial m_s} = 10^3\left[-0.315(\alpha v)_w^{(1)} \log Z_1 + 0.315 v_w^{(1)} \frac{dB}{dT} \log Z_3\right] - 10^3\left[S(1-c)\frac{\partial p_e}{\partial S}\frac{\partial\Delta\Psi_w^{(1,p)}}{\partial t} + c\frac{\partial p_e}{\partial t}\Delta\Psi_w^{(1,p)}\right] \qquad (3.143.3)$$

where $Z_1 = (x_1^{x_1} y^{y})/(x^{x} y_1^{y_1})$

with $x_1 = (B+1)$, $x = (B+p)$, $y_1 = (B^*+1)$ and $y = (B^*+p)$

and $Z_3 = xy_1/x_1y$

$$\Delta\Psi_w = -0.315 v_w^{(1)} \log\frac{(B^*+p)}{(B^*+1)} \qquad (5.72.12)^*$$

and $p_e = B^* - B = 6.8966S - 0.0703St$ bars (3.143.4)

from the data of Ekman (1908) as given by Li (1967) with salinity, S, in p.p.t. and t in °C.

3.144 Relationships with the apparent specific entropy of water and the entropy of pure water under an additional pressure due to the solute

It might be expected, from Tammann's principle (Chapter 6) that water in seawater would respond to pressure in the same way as water subjected to an additional pressure of P_e bars. This pressure P_e is due to the compressional effect of the solute on the water, and has the values given by eq 3.143.4.

The apparent specific volume of water Ψ_w, has been shown to be equivalent to the specific volume of pure water subject to the solute compression of p_e bars, i.e.

$$v_w^{(p+p_e)} = \Psi_w^{(p)} \qquad \text{(Sections 4.7, 6.53)}$$

The quantity corresponding to $(\alpha v)_w$ to be integrated is therefore

$$(\alpha v)_w^{(p+p_e)} = (\alpha v)_w^{(1)}\left[1-0.315\log\frac{(B^*+p)}{(B+1)}\right] + \frac{0.315 v_w^{(1)}}{2.303}\left[\frac{dB/dt}{(B+1)} - \frac{dB^*/dt}{(B^*+p)}\right] \qquad (3.144.1)$$

which is the same as $(\partial\Psi_w/\partial t)$ (Section 3.131)

The pressure change of the apparent specific entropy of water-in-seawater $(\Psi_s)_w$ is therefore given by

$$-\Delta(\Psi_s)_w^{(1,p)} = \int_1^p (\partial\Psi_w/\partial t)\,dp = (\alpha v)_w^{(1)}(1+0.315\log(B+1))(p-1)$$
$$- 0.315(\alpha v)_w^{(1)}\left[\log\frac{(B^*+p)}{(B^*+1)}^{(B^*+p)}{}^{(B^*+1)} + \frac{(1-p)}{2.303}\right]$$
$$+ \frac{0.315}{2.303}v_w^{(1)}\left[\frac{dB/dt}{(B+1)}(p-1) - 2.303(\partial B^*/\partial t)\log\frac{(B^*+p)}{(B^*+1)}\right] \qquad (3.144.2)$$

with $c = 10^{-3}S(‰)$ and the units being $(cm^3 bar/deg\ g)$.

The value of B*, B and the necessary derivatives are obtained from eq (5.6.2a)* Section 3.132. Values of $(\alpha_w)^{(1)}$ and $v_w^{(1)}$ are obtained from eqs 1, 2 or 3 and 4 of Section 5.71 together with Table 5.18.

It should be noted that $(\Delta\Psi_{\phi})_w^{(1,p)}$ is not the same as $(\phi_w^{(p_e+1,\ p_e+p)} - \phi_w^{(p_e+1)})$ or $\Delta\phi_w^{(p_e, p_e+p)}$.

This is because of the difference in the temperature coefficients of the parameters B (Tait) and B* (Tait-Gibson), i.e. $\partial p_e/\partial t$ (from eq 3.143.4).

From eqs 3.12.1 and 3.144.2 with ϕ representing the molal entropy,

$$\Delta(\Psi_{\phi})_w^{(1,p)} \quad - \quad \frac{\Delta\phi_w^{(p_e,p_e+p)}}{M_w}$$

$$= \quad (\partial p_e/\partial t)\left[0.315 v_w^{(1)} \log\frac{(B^*+p)}{(B^*+1)}\right]$$

$$= \; - \quad (\partial p_e/\partial t)(\Delta\Psi_w)^{1,p} \qquad (3.144.3)$$

since $(\Psi_w^{(p)} - \Psi_w^{(1)}) \; = \; -v_w^{(1)} 0.315 \log\frac{(B^*+p)}{(B^*+1)}$

(see Section 6.53)

The pressure change of the partial molal entropy of water-in-seawater, $\Delta\bar{s}_w^{sw(1,p)}$, has much the same sensitivity to temperature as pure water (see Figure 3.9).

This is in contrast to the corresponding apparent specific entropy as reflected in the trends of the $\Delta\phi_{sw}^{(1,p)}$ curve.

When the pressure increment of the partial molal entropy of water-in-seawater is compared with the value of $\Delta\phi_w^{(p_e,p_e+p)}$ it is found that the difference in the two quantities is related to the structural properties of water, and is zero at 30°C, (Table 3.22).

This is illustrated in Figure 3.10 where $(\Delta\bar{s}_w^{sw(1,p)} - \Delta\phi_w^{(p_e,p_e+p)})$ is plotted against the A parameter of the equation $\phi_v^o \; = \; Ar^3 - B'z^2/r$ (Section 4.5). The temperature changes of A are related to a water-structure term (Curthoys and Mathieson 1970). See also, Figure 6.18a.

The value of the "structural" entropy term can be formulated from the above set of equations, viz., with S representing salinity (‰), $c=10^{-3}S$, as usual.

TABLE 3.22 Comparison of the effect of pressure on water under a solute pressure of p_e bars and of water in seawater (S = 35‰)

Changes of entropy with pressure cal/deg mol of water

p bars	$-\Delta \not{s}_w$ (1)	$-\Delta \bar{s}_w^{sw}$ (2)	(1)-(2)	$-\Delta \not{s}_w$ (3)	$-\Delta \bar{s}_w^{sw}$ (4)	(3)-(4)
	t = 0°C, p_e = 241 bars			t = 20°C, p_e = 192 bars		
100	0.00131	-0.00118	0.0025	0.01029	0.00968	0.0006
200	0.00389	-0.00095	0.0048	0.02108	0.02003	0.0011
300	0.00767	0.00630	0.0070	0.03236	0.03094	0.0014
400	0.12570	0.00346	0.0091	0.04408	0.04237	0.0017
500	0.18528	0.00748	0.0111	0.05624	0.05428	0.0020
600	0.02548	0.01260	0.0129	0.06879	0.06665	0.0021
700	0.03338	0.01876	0.0146	0.08172	0.07945	0.0023
800	0.04216	0.02591	0.0163	0.09501	0.09268	0.0024
900	0.05179	0.03398	0.0178	0.10864	0.10626	0.0024
1000	0.06221	0.04293	0.0193	0.12259	0.12023	0.0024
	t = 10°C, p_e = 217 bars			t = 30°C, p_e = 168 bars		
100	0.00628	0.00496	0.0013	0.01374	0.01357	0.0002
200	0.01338	0.01092	0.0025	0.02775	0.02755	0.0002
300	0.02127	0.01777	0.0035	0.04200	0.04179	0.0002
400	0.02989	0.02547	0.0044	0.05649	0.05628	0.0002
500	0.03921	0.03395	0.0053	0.07119	0.07100	0.0002
600	0.04918	0.04318	0.0060	0.08610	0.08593	0.0002
700	0.05976	0.05311	0.0067	0.10120	0.10107	0.0001
800	0.07093	0.06371	0.0072	0.11649	0.11640	0.0001
900	0.08266	0.07493	0.0077	0.13194	0.13192	0.0000
1000	0.09491	0.08674	0.0082	0.14756	0.14760	0.0000

$$\Delta \not{s}_w = \not{s}_w^{(p_e+p)} - \not{s}_w^{(p_e)}, \quad \Delta \bar{s}_w^{sw} = \bar{s}_w^{sw(p)} - \bar{s}_w^{sw(1)}$$

$\not{s}_w^{(p)}$ from eq 3.12.1,

B parameter from eq 6.75.2

$\bar{s}_w^{sw(p)}$ from eq 3.141.1 (see Table 3.21)

p_e from eq 3.143.4.

$$(1/M_w)(\Delta\bar{s}_w^{sw}(1,p) \;-\; \Delta\phi_w(p_e,p_e+p))$$

$$= 0.315v_w^{(1)}\log\frac{(B^*+p)}{(B^*+1)}\left[\partial p_e/\partial t - S(1-c)(\alpha_w^{(1)}\frac{\partial B^*}{\partial S} + \frac{\partial(\partial B^*/\partial S)}{\partial t})\right]$$

$$+ \frac{0.315}{2.303}S(1-c)v_w^{(1)}\left[\frac{\partial B^*}{S}\frac{\partial B^*}{t}(p-1)/(B^*+p)(B^*+1)\right]$$

(3.144.4)

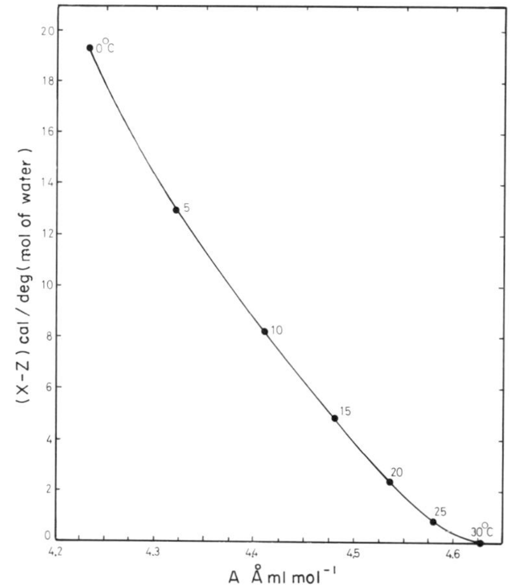

FIGURE 3.10

The "structural" parameter A, versus the differences in the partial molal entropy of water-in-seawater and the entropy of water under an initial pressure of p_e bars (eq 3.143.4).

$X = \bar{s}_w^{sw}(p) - \bar{s}_w^{sw}(1)$ (eq 3.141.1)

and $Z = \phi_w(p_e+p) - \phi_w(p_e)$ (eq 3.12.1)

See Table 3.22.

A is from the equation,

$\phi_v^o = Ar^3 - B'z^2/r$ (Section 4.5)

The molecular weight of water, M_w, converts the molal quantities to the specific entropies.

Since $$\Delta\Psi_w^{(1,p)} = -0.315 v_w^{(1)} \log \frac{(B^*+p)}{(B^*+1)} \qquad (5.72.12)^*$$

and $$\frac{\partial \Delta\Psi_w^{(1,p)}}{\partial t} = -0.315(\alpha v)_w^{(1)} \log \frac{(B^*+p)}{(B^*+1)} + \frac{0.315}{2.303} v_w^{(1)} \left[\frac{\partial B^*}{\partial t} \frac{(p-1)}{(B^*+p)(B^*+1)} \right] \qquad (5.72.14)^*$$

and $\frac{\partial B^*}{\partial S} = \frac{\partial p_e}{\partial S}$, $\frac{\partial(\partial B^*/\partial S)}{\partial t} = \frac{1}{S}\frac{\partial p_e}{\partial t}$ from eq 3.143.4

thus, eq 3.144.4 becomes

$$(1/M_w)\left[\Delta\bar{s}_w^{sw(1,p)} - \Delta \not{s}_w^{(p_e,p_e+p)}\right] = S(1-c)(\partial p_e/\partial S)\frac{\partial \Delta\Psi_w^{(1,p)}}{\partial t} - c(\partial p_e/\partial t)\Delta\Psi_w^{(1,p)} \qquad (3.144.5)$$

And, from eq 3.144.3

$$(1/M_w)\ \Delta\bar{s}_w^{sw(1,p)} = \Delta(\Psi_{\not{s}})_w^{(1,p)} + S(1-c)(\partial p_e/\partial S)\left[\frac{\partial \Delta\Psi_w^{(1,p)}}{\partial t}\right] + (1-c)(\partial p_e/\partial t)\Delta\Psi_w^{(1,p)} \qquad (3.144.6)$$

Alternatively, since $\partial p_e/\partial S = 6.8966 + (t/S)(\partial p_e/\partial t)$ with $\partial p_e/\partial t = -0.0703S$,

$$\left[\Delta\bar{s}_w^{sw(1,p)} - M_w \Delta(\Psi_{\not{s}})_w^{(1,p)}\right] = M_w(1-c)\left[(6.8966S\left(\frac{\partial \Delta\Psi_w^{(1,p)}}{\partial t}\right) - 0.0703S\left[\left(\frac{\partial \Delta\Psi_w^{(1,p)}}{\partial t}\right)t + \Delta\Psi_w^{(1,p)}\right]\right]$$

(in cm^3bar/deg molwater) (3.144.7)

Thus, at constant salinity, the difference between the pressure increments of the partial specific and apparent specific entropies of water is a function of $\Delta\Psi_w^{(1,p)}$ and its temperature derivative. The temperature changes of this quantity are related to a temperature sensitive structural factor of water. In fact, this quantity is proportional to the structural factor A as shown in Figure 3.11. This "structural" factor, therefore, can be conveniently represented by eq 3.144.7. This function should be more useful than the partial entropy function of water, $\Delta\bar{s}_w^{sw(1,p)}$. This latter quantity merely gives the change, with pressure, of the total entropy contribution of the water to the seawater solution. For those marine studies where the structure of water is important, it would be interesting to see correlations with the values from eq 3.144.7 and important controlling factors in the system being studied.

As noted above,

$$\Delta\Psi_w^{(1,p)} = \Delta v_w^{(p_e,p_e+p)}$$

so that the change in the specific volume of pure water can be used instead of Ψ_w with the appropriate pressure range of p_e to (p_e+p). The value of p_e is now known very accurately (probably to better than 0.5 bars - see Section 6.84) and $v_w^{(p)}$ is accurate to within 20ppm or better, so that $\Delta v_w^{(p_e,p_e+p)}$ is reliable to 20-30 ppm (1-1000 bars, 0-30°C).

From eq 3.144.3

$$M_w\Delta(\Psi_{\not{s}})_w^{(1,p)} = \Delta\not{s}_w^{(p_e,p_e+p)} - M_w \frac{\partial p_e}{\partial t} \Delta v_w^{(p_e,p_e+p)} \qquad (3.144.8)$$

in cm^3bar/deg molwater

so that the "structural" term, given by eq 3.144.7 is probably reliable to 0.001 cal/deg molwater.

TABLE 3.23 Differences in the partial molal and apparent molal entropy of water-in-seawater. The values are relative changes to those at 1 bar pressure*

S = 35‰

$10^4[\Delta\bar{s}_w^{sw\,(1,p)} - M_w\Delta(\Psi_\phi)_w^{(1,p)}]$ cal/deg mol of water

p (bars) / t°C	500	1000
0	339	619
5	307	562
10	280	516
15	258	477
20	240	445
25	225	418
30	212	397

* $\Delta Y^{(1,p)} = Y^{(p)} - Y^{(1)}$ with $Y = \bar{s}_w^{sw}$, or $(\Psi_\phi)_w$

Calculated from eq 3.144.7

From eq 3.144.8 it can be seen that it is the term $M_w(\partial p_e/\partial t)\Delta v_w^{(p_e,p_e+p)}$ that linearises the curve shown in Figure 3.10 (as illustrated by Figure 3.11).

Note that the values have been converted from ccbar to calories in Figure 3.10 and Table 3.23 which lists values of the "structural" parameter from eq 3.144.7.

There is one final point that should be mentioned. The value of p_e given by eq 3.143.4 is based on Ekman's data (as given by Li 1967). If other data are used, say the acoustic data, then p_e should have the same value. However, the B and B* values should be from the same or compatible data sets, otherwise p_e will not be as accurate (Equation 3.143.4 is recommended).

In Table 3.22 the values of $\not{s}_w$ were calculated from the Tait equation. The B parameter of this equation is based on the "smoothed" acoustic data (more correctly, the way of deriving B smooths the data - see Sections 6.73 and 6.75). However, agreement with B of Ekman is good at the lower temperatures (Table 6.15). The values of $\bar{s}_w^{sw}$ on the other hand are based on the B* of Ekman. This is a slight mismatch of data. However, since the Tait equation smooths out the small discrepancies in the acoustic data, the values of $(\Delta\bar{s}_w^{sw(1,p)} - \Delta\not{s}_w^{(p_e,p_e+p)})$ calculated from eq 3.144.5 should not differ very much from those in Table 3.22. Calculations have been made, using eq 3.144.5 for pressures of 500 bars and 1000 bars. The deviations from the values in Table 3.22 are negligible (less than 0.001 cal/degmolwater) except at 1000 bars, 10°C and 20°C where they are 0.002 cal/degmol water lower. In other words, in the oceanographic P-T range the differences are within the expected accuracy of 0.001 cal/degmol water for the "structural" function.

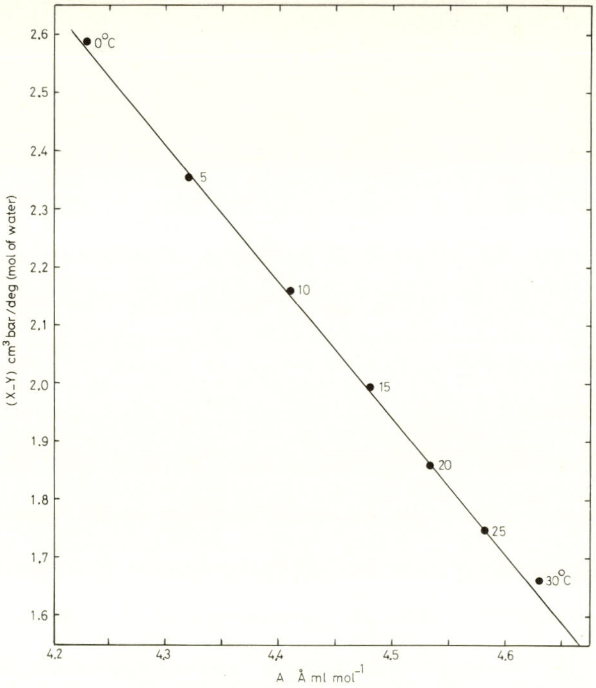

FIGURE 3.11

The difference in the pressure increments of the partial molal entropy and apparent molal entropy of water-in-seawater (S=35‰), versus the "structural" parameter of water, A. A is from the equation $\phi_V^o = Ar^3 - Bz^2/r$ (Section 4.5).

$$X = \bar{s}_w^{sw(p)} - \bar{s}_w^{sw(1)}$$

$$Y = M_w \left[(\Psi_\phi)_w^{(p)} - (\Psi_\phi)_w^{(1)} \right]$$

$\bar{s}_w^{sw}$ is the partial molal entropy, and $(\Psi_\phi)_w$ is the apparent specific entropy of water-in-seawater.

M_w is the number of grams in a mole of water.

P = 1000 bars

The ordinate values are multipled by 0.023901 to convert cm^3 bars to cals, see Table 3.23.

REFERENCES Chapter 3

Bradshaw A. and K.E.Schleicher 1970.

Direct measurement of thermal expansion of sea-water under pressure.
Deep-Sea Res. 17: 691-706.

Bromley L.A. 1968.

Relative enthalpies of sea salt solutions at 25°C.
J.chem.Engng.Data. 13(3): 399-401.

Bryden H.L. 1973.

Polynomials for thermal expansion, adiabatic temperature gradient and potential temperature of sea water.
Deep-Sea Res. 20: 401-408.

Cobble J.W. 1953.

Empirical considerations of entropy.
I The entropies of the oxy-anions and related species.
J.Chem.Phys. 21: 1443-1446.

Cobble J.W. 1953.

II the entropies of inorganic complex ions.
J.Chem.Phys. 21: 1446-1450.

Cobble J.W. 1953.

III A structural approach to the entropies of aqueous organic solutes and complex ions.
J.Chem.Phys. 21: 1451-1456.

Couture A.M. and K.J. Laidler 1957.

Entropies of ions in aqueous solutions.
II Empirical equation for oxy-anions.
Can.J.Chem., 35: 202-206.

Curthoys G. and J.G. Mathieson 1970.

Partial molal volumes of ions.
Trans.Faraday Soc. 66(1): 43-50

Dorsey N.E. 1940.

"Properties of ordinary water-substance",
Reinhold, New York. 673 pp.

Ekman V.W. 1908.

Die Zusammendrueckbarkeit des Meerwassers.
Publ.Circonst.perm.Explor.Mer, 43: 1-47.

Eley D.D. and M.G. Evans 1938.

Heats and entropy changes accompanying the solution of ions in water.
Trans.Faraday Soc. 34: 1093-1112.

Erlander Stig R. 1968a.

Determination of the structure and effective dielectric constant of hydrated ions.
J.Macromol.Sci.-Chem.A2(4): 833-858.

Erlander Stig R. 1968b.

Interaction of aqueous ions: a correlation between the hydration and solubility properties of inorganic, organic, and polymeric salts having mono- or polyatomic ions.
J.Macromol.Sci.-Chem. A2(3): 623-643.

Erlander Stig R. 1968c.

The structure of water and its relationship to hydrocarbon-water interactions.
J.Macromol.Sci.-Chem. A2(3): 595-621.

Fine R.A. and F.J. Millero 1973.

The compressibility of water as a function of temperature and pressure.
J.Chem.Phys. 59(10): 5529-36.

Chapter 3.

Fortier, J.-L., P.-A. Leduc, and J.E. Desnoyers 1974.

Thermodynamic properties of alkali halides. II. Enthalpies of dilution and heat capacities in water at 25°C.
J. Solution Chem. 3(4): 323-349.

Gier L.J. and C.E. Vanderzee 1974a.

Enthalpies of dilution and relative apparent molar enthalpies of aqueous cobalt and nickel perchlorates.
J.Chem.Engng. Data 19(4): 315-319.

Gier L.J. and C.E. Vanderzee 1974b.

Enthalpies of dilution and relative apparent molar enthalpies of aqueous calcium and manganous perchlorates.
J.Chem.Engng.Data. 19(4): 323-325.

Harned H.S. and B.B. Owen 1958

"The Physical chemistry of electrolytic solutions", 3rd ed. ACS monograph, Reinhold, New York, 803pp.

Hasted J.B., D.M. Ritson, and C.H. Collie, 1948.

Dielectric properties of aqueous ionic solutions. Part I - II.
J.Chem.Phys. 16: 1-21.

Hilsenrath, J., C.W. Beckett, W.S. Benedict, L.Fano, H.J.Hoge, J.F. Masi, R.L. Nuttall, Y.S. Touloukian and H.W. Woolley 1955.

Tables of thermal properties of gases.
U.S. Natl. Bur. Standards. Circ. 564.

Jongenburger H.S. and R.H. Wood 1965.

The heats and entropies of dilution of the perchlorates of magnesium and strontium.
J.Phys.Chem. 69(12): 4231-4238.

Chapter 3.

Kearns E.R. 1966.

Compressibilities of some dilute aqueous solutions. Ph.D. Thesis, Yale Univ., University Microfilms Inc., Ann Arbor, Michigan, Ord. No. 66-4902.

Kelley K.K. 1950.

Contributions to the data on theoretical metallurgy. XI. Entropies of inorganic substances, Revision (1948) of data and methods of calculation: U.S. Bur.Mines Bull. 477.

Kelley K.K. and E.G. King 1961.

Critical summary of entropy data and references to original sources. U.S. Bur.Mines Bull. 592. (as given by Pitzer and Brewer 1961, p137).

Kennedy, G.C. 1950.

Pressure-volume-temperature relations in water at elevated temperatures and pressures. Am.J.Sci. 248: 540-564.

Kennedy, G.C., Knight, W.L. and Holser, W.T. 1958.

Properties of water. Part III. Specific volume of liquid water to 100°C and 1400 bars. Am.J.Sci. 256: 590-595.

King E.L. 1959.

Some comments on the values of Δs^o for ionic reactions and the values of the entropy of solution of ions. J.Phys.Chem. 63: 1070-1072.

Krishnan C.V. and H.L. Friedman 1974.

Model calculations for Setchenow coefficients. J.Sol.Chem. 3(9): 727-744.

Laidler K.J. and C. Pegis 1957.

The influence of dielectric saturation on the thermodynamic properties of ions. Proc.Roy.Soc. (London). A241: 80-92.

Larson J.W. 1970.

Entropies of sodium and hydroxide ions and the thermodynamics of ionization of water. J.Phys.Chem. 74(3): 685-686.

Chapter 3.

Leyendekkers J.V. 1970.

Activity coefficients in mixed solutions.
Prediction of Harned Coefficients
from ionic entropies.
J.Phys.Chem. 74(10): 2225-2229.

Leyendekkers J.V. 1971.

Thermodynamics of mixed electrolyte solutions.
Ionic entropy correlations and volume fraction
statistics.
J.Phys.Chem. 75(7): 946-956.

Leyendekkers J.V. 1973.

The ionic activity function of water and the
activity coefficient of the hydrogen ion
in seawater.
Limnol. Oceanogr. 18(5): 784-787.

Leyendekkers J.V. 1974.

Prediction of the partial molal volumes
of electrolytes in seawater.
Mar. Chem. 2(2): 89-110.

Li Y. -H. 1967.

Equation of state of water and seawater.
J.Geophys.Res. 72(10): 2665-2678.

Masterton W.L. 1975.

Salting coefficients for gases in
seawater from scaled-particle theory.
J.Sol.Chem. 4(6): 523-534.

Millero F.J., E.V. Hoff and L. Kahn 1972.

The effect of pressure on the
ionization of water at various
temperatures from molal-volume data.
J.Sol.Chem. 1: 309-327.

Millero F.J., Lee D. Hansen and Edward V.Hoff 1973.

The enthalpy of seawater from
0 to 30°C and from 0 to 40‰ salinity.
J.Mar.Res. 31(1): 21-39.

Chapter 3.

Nancollas G.H. 1966.

"Interactions in electrolyte solutions - Metal complex and ion-pair formation in solution", Elsevier, Amsterdam, 214pp.

Parsons R. 1959.

"Handbook of electrochemical constants", Butterworths, London 113pp.

Pauling Linus 1960.

The Nature of the Chemical Bond 3rd ed. p514, Cornell University Press, Ithaca, N.Y.

Pistorius Carl W.F.T. and W.E. Sharp 1960.

Properties of water. Part VI. Entropy and Gibbs Free Energy of water in the range 10-1000°C and 1-250,000 bars. Am.Jour.Sci. 258: 757-768.

Pitzer K.S. and L. Brewer 1961.

revised edition of "Thermodynamics" by G.N.Lewis and M.Randall. McGraw-Hill, New York, N.Y. 723pp.

Pitzer K.S. and G. Mayorga 1973.

Thermodynamics of electrolytes. II. Activity and osmotic coefficients for strong electrolytes with one or both ions univalent. J.Phys.Chem. 77(19): 2300-2308.

Powell R.E. and W.M. Latimer 1951.

The entropy of aqueous solutes. J.Chem.Phys. 19: 1139-1141.

Chapter 3.

Vanderzee C.E. and J.D. Nutter 1974.

Relative apparent molar enthalpies of aqueous lanthanum perchlorate.
J.Chem.Engng. Data. 19(3): 268-270.

Vanderzee, C.E., J.D. Nutter, W.W. Rodenburg, M.L.N. Rodenburg, and D.L. King 1974.

Enthalpies of dilution and related thermodynamic properties of aqueous hydrobromic acid.
J.Chem.Engng.Data. 19(4): 320-23.

Vanderzee C.E. and J.A. Swanson 1963.

Heats of dilution and relative apparent molal enthalpies of aqueous sodium perchlorate and perchloric acid.
J.Phys.Chem. 67: 285-291.

Vanderzee C.E. and J.A. Swanson 1972.

Enthalpies of dilution and relative apparent molal enthalpies of aqueous barium perchlorate.
J.Chem.Engng. Data. 17(4): 488-490.

Wang Dong-Ping and F.J. Millero 1973.

Precise representation of the P-V-T properties of water and seawater determined from sound speeds.
J.Geophys.Res. 78(3): 7122-7128.

Whitfield Michael 1972.

Self-ionization of water in dilute sodium chloride solutions from 5-35°C and 1-2000 bars.
J.Chem.Engng.Data 17: 124-8.

Whitfield Michael 1974.

A comprehensive specific interaction model for seawater. Calculation of the osmotic coefficient.
Deep-sea Research. 21: 57-67.

APPENDIX 3. ENTROPY

The partial molal entropy, $\bar{s}_i$, of an electrolyte, i, may be calculated from the equations:

$$\bar{s}_i - \bar{s}_i^o = \bar{L}_i/T - \nu R \ln \gamma_{\pm_i} m_i - R \ln(\nu_1^{\nu_1} \nu_2^{\nu_2}) \qquad (3.8.1)*$$

$$(\bar{s}_i - \bar{s}^o)_t = (\bar{s}_i - \bar{s}_i^o)_{25} + 2.303\bar{j}\log(T/298.15) \qquad (3.7.7)*$$

Values of $\bar{L}_i$, $\bar{j}_i$, $\gamma_\pm$ are listed herein, by Parsons(1959), and in Part II.

OR from

$$\bar{s}_i - \bar{s}_i^o = A_{jk} - C_{jk}(B_i^{\cdot} + TdB_i^{\cdot}/dT) \qquad (3.8.5)*$$

with

$$A_{jk} = \nu z_1 z_2 (I^{\frac{1}{2}}/(1+I^{\frac{1}{2}}))(A_H/2T + 2.303RA_\gamma) - 2.303R(\nu\log 2I + \log(\nu_1^{\nu}1\nu_2^{\nu}2)/(\nu z_1 z_2)^{\nu})$$

$$C_{jk} = (9.212R\nu_1\nu_2/\nu z_1 z_2)I,$$ (j,k represent the charges (z_1, z_2) on the cation and anion, respectively).

In terms of salinity, the ionic strength, I, is given by

$$I = 0.01994S/(1-10^{-3}S) \quad \text{(I in mol/kg water S in ‰)}$$

Values of A_{jk} and C_{jk} are given in Table 3.11.

Values of A_H and A_γ and R, are given in Table 3A.1a.

Values of $B_i^{\cdot}$, $dB_i^{\cdot}/dT$, $d^2B_i^{\cdot}/dT^2$ for 25°C are given in Tables 3A.2a, b and c.

Values of $B^{\cdot}$ and $dB^{\cdot}/dT$ at other temperatures may be obtained from the following equations, on the assumption that changes in $d^2B^{\cdot}/dT^2$ with temperature are negligible.

$$B_t^{\cdot} = B_{25}^{\cdot} + \left(\frac{dB^{\cdot}}{dT}\right)_{25}(t-25) + \tfrac{1}{2}\left(\frac{d^2B}{dT^2}\right)(t-25)^2 \qquad (3.8.9)*$$

$$\left(\frac{dB^{\cdot}}{dT}\right)_t = \left(\frac{dB^{\cdot}}{dT}\right)_{25} + \left(\frac{d^2B^{\cdot}}{dT^2}\right)(t-25) \qquad (3.8.8)*$$

Also,

$$\left(B^{\cdot} + T\frac{dB^{\cdot}}{dT}\right)_t = B_{25}^{\cdot} + \left(\frac{dB^{\cdot}}{dT}\right)_{25}(2t+248.15) + \tfrac{1}{2}\left(\frac{d^2B^{\cdot}}{dT^2}\right)(t-25)(3t+521.3) \qquad (3.8.10)*$$

APPENDIX 3 (ctd)

or

$$(\bar{s}_i-\bar{s}_i^o)_t = (\bar{s}_i-\bar{s}_i^o)_{25} + (A_t-A_{25})_{jk}$$

$$-C_{jk}\left[2\left(\frac{dB^{\cdot}}{dT}\right)_{25}(t-25) + \tfrac{1}{2}\left(\frac{d^2B^{\cdot}}{dT^2}\right)(t-25)(3t+521.3)\right] \qquad (9.8.11)^*$$

Values of $\bar{s}_i^o$ for 25°C are given in Table 3A.1.

The relative values at 0°C are given in Table 3.6.

When $d^2B^{\cdot}/dt^2$ data for the range I = 0.1 to 1 m are unavailable, the values for I = 0.723m in Table 3.12 or 3.13 are useful approximations, since changes with ionic strength are not very great (see, for example, Table 3A.2a).

TABLE 3A.1 The standard partial molal entropies of ions at 25°C 1-atm. $\text{cal deg}^{-1}\text{mol}^{-1}$ based on $\bar{s}^o_{H+} = 0$

Cation	$\bar{s}^o$	Cation	$\bar{s}^o$	Anion	$\bar{s}^o$
Ag(+)	17.67	Ni(2+)	(-31)*	CrO_4(2-)	(10.6)
$Ag(NH_3)_2$(+)	57.8	Pb(2+)	5.1	F(-)	-2.3
Al(3+)	-74.9	Pu(3+)	(-39)	$HAsO_4$(2-)	0.9
Ba(2+)	3	PU(4+)	(-87)	H_2AsO_4(-)	28
Be(2+)	(-27)*	Rb(+)	28.7 (29.7)	H_2BO_3(-)	7.3
Ca(2+)	-13.2	Sn(2+)	-5.9	HCOO(-)	21.9
Cd(2+)	-14.6	Sn(4+)	(-95)*	HCO_3(-)	22.7 (22.2)
Co(2+)	(-27)*	Sr(2+)	-9.4	HF_2(-)	(26)
Cr(3+)	(-73.5)	Tl(+)	30.4	HPO_4(2-)	-8.6
Cs(+)	31.8	U(3+)	(-36)	H_2PO_4(-)	21.3 (21.6)
Cu(2+)	-23.6	U(4+)	(-78)	HS(-)	15.0
Fe(2+)	-27.1	UO_2(2+)	-17	$HSeO_4$(-)	22.0
Fe(3+)	-70.1	Zn(2+)	-25.45	HSO_3(-)	(32.6)
Fe(OH)(2+)	-23.2			HSO_4(-)	30.52
FeNO(2+)	-10.6	**Anion**		I(-)	26.14
Ga(3+)	(-83)	$Ag(CN)_2$(-)	49.0	I_3(-)	57.1
Gd(3+)	-43	BF_4(-)	40.0	IO_3(-)	(28.0)
H(+)	0	Br(-)	19.29 (19.25)	MnO_4(-)	45.4 (46.7)
Hg(2+)	-5.4	BrO_3(-)	(38.5)	NO_2(-)	29.9
Hg_2(2+)	(17.7)	CH_3COO(-)	20.8	NO_3(-)	35.0
In(3+)	(-62)	Cl(-)	13.2	OH(-)	-2.52
K(+)	24.5	ClO(-)	(10.0)	PO_4(3-)	-52
Li(+)	3.4	ClO_2(-)	24.1	$PtCl_4$(2-)	42
Mg(2+)	-28.2	ClO_3(-)	39 (39.4)	$PtCl_6$(2-)	52.6
Mn(2+)	-20	ClO_4(-)	43.2 (43.6)	S(2-)	-4 (-6.4)
Na(+)	14.4	CN(-)	28.2 (25)	SeO_3(2-)	3.9
NH_4(+)	26.97 (26.4)	CNO(-)	31.1	SeO_4(2-)	5.7
N_2H_6(2+)	19	CO_3(2-1)	-12.7 (-13.0)	SH(-)	(14.9)
N_2H_5(+)	31	C_2O_4(2-)	10.6	SO_3(2-)	(3)
NH_2OH_2(+)	37	C_5H_5COO	(27.9)	SO_4(2-)	4.1

From the compilation given by Pitzer and Brewer (1961, p400)
Bracketed values from Parsons (1959)
( )* approximate, calculated value

TABLE 3A.1a

Debye-Huckel Limiting
Constants for water at 1-atm
on molal scale (mol/kg water)

$t^{o}C$	A_{γ}	A_{H} $calmol^{-1}$	
0	0.492	464	
10	0.499	545	1 cal
20	0.507	637	= 4.1840joules
25	0.511	688	
30	0.517	743	

Gas constant $\underline{R}$ = (8.3147±0.00005) joule$deg^{-1}mol^{-1}$
= (1.9872±0.00001) cal$deg^{-1}mol^{-1}$
= (0.082057±0.0000005) l.atm.$deg^{-1}mol^{-1}$

TABLE 3A.2a The 1st and 2nd temperature derivatives of $B_i^{\cdot}$, the ion interaction parameter.

25°C 1-atm

Alkali halides

I mol/kg		0.1	0.3	0.5	0.723	1.0
S‰		5	14.9	24.6	35.00	48.0
NaCl	a	8.34	7.74	7.36	7.02	6.56
	b	-2.40	-2.22	-2.11	-2.03	-1.95
LiBr	a	1.85	0.70	0.18	-0.13	-0.37
	b	-0.72	-0.51	-0.39	-0.32	-0.24
LiCl	a	-0.22	-0.59	-0.90	-1.13	-1.29
	b	-0.51	-0.34	-0.24	-0.17	-0.11
NaF	a	3.37	3.83	3.87	3.81	3.66
	b	-2.72	-2.61	-2.53	-2.47	-2.41
NaI	a	10.52	9.38	8.78	8.23	7.48
	b	-2.53	-2.31	-2.19	-2.10	-1.99
RbCl	a	11.06	8.99	8.01	7.41	6.97
	b	-2.21	-1.93	-1.78	-1.69	-1.60
RbBr	a	15.48	12.22	10.76	9.67	9.57
	b	-2.61	-2.25	-2.07	-1.94	-1.88
RbI	a	-18.34	14.92	13.30	11.80	9.66
	b	-2.86	-2.49	-2.30	-2.14	-1.95
CsBr	a	20.11	15.53	13.21	12.36	13.30
	b	-3.38	-2.92	-2.70	-2.59	-2.60
CsI	a	25.10	19.69	17.26	16.07	15.84
	b	-3.97	-3.47	-3.22	-3.09	-3.02
CsF	a	2.10	1.76	1.46	1.25	1.10
	b	-1.2	-1.04	-0.93	-0.87	-0.80
RbF	a	2.01	1.95	1.72	1.47	1.20
	b	-1.72	-1.58	-1.48	-1.41	-1.34

a $dB^{\cdot}/dT \times 10^4$ $kg\,mol^{-1}deg^{-1}$

b $d^2B^{\cdot}/dT^2 \times 10^5$ $kg\,mol^{-1}deg^{-2}$

from the data of Fortier, Leduc and Desnoyers (1974) - Chapter 3

TABLE 3A.2b The 1st temperature derivative of $B_i^{\cdot}$ the ion interaction parameter of electrolyte i.

25°C 1-atm

Perchlorate salts

$dB^{\cdot}/dT \times 10^4$ kgmol^{-1}deg^{-1}

I mol/kg	0.1	0.2	0.3	0.4	0.5	0.6	0.7	0.8	0.9	1.0
$Ba(ClO_4)_2$	47.25	39.24	34.67	31.60	29.36	27.64	26.27	25.14	24.18	23.36
$Ca(ClO_4)_2$	28.12	22.62	19.75	17.71	16.10	14.77	13.69	12.84	12.21	11.81
$Co(ClO_4)_2$	26.51	20.95	17.75	15.68	14.21	13.09	12.14	11.27	10.40	9.48
$Mg(ClO_4)_2$	23.11	17.91	15.28 (15.57)	13.50	12.16	11.09 (11.06)	10.20	9.45	8.80 (8.83)	8.23
$Mn(ClO_4)_2$	24.59	18.53	15.73	13.85	12.34	11.06	9.99	9.11	8.45	8.02
$Ni(ClO_4)_2$	29.49	18.54	16.12	14.98	13.86	12.65	11.46	10.47	9.85	9.71
$Sr(ClO_4)_2$	33.93	27.80	24.21 (23.69)	21.81	20.05	18.70 (18.77)	17.61	16.71	15.95 (15.91)	15.30
$HClO_4$	12.94	10.52	9.56	8.92	8.37	7.85	7.37	6.94	6.57	6.29
$NaClO_4$	20.26	18.64	17.59	16.79	16.12	15.53	15.00	14.52	14.08	13.66
$La(ClO_4)_3$	81.48	63.70	53.42	46.65	41.86	38.29	35.51	33.23	31.27	29.51

Data from Jongenburger and Wood (1965), Gier and Vanderzee (1974a) Vanderzee and Swanson (1972), Vanderzee and Nutter (1974) and Gier and Vanderzee (1974b) - see Table 3.13.

A "least-squares" fit was made of the ϕ_L (see Table 3A.3b) bracketed values: calculated from the fitted values of Jongenburger and Wood (1965, Tables III and IV, $\bar{L}_2$ values); their ϕ_L equation covered a much larger ionic strength range (up to 9 m).

TABLE 3A.2c The 1st temperature derivative of $B_i^{\cdot}$ the ion interaction parameter of electrolyte i. 25°C 1-atm

Miscellaneous divalent salts

$dB^{\cdot}/dT \times 10^4$ $kg mol^{-1} deg^{-1}$

I mol/kg	0.068	0.120	0.150*	0.188	0.270
$BaBr_2$	29.7	24.4	22.9	20.8	18.2
$BaCl_2$	15.5	14.8	14.9	14.1	13.2
$CaBr_2$	23.4	18.6	16.7	15.9	14.2
$Ca(NO_3)_2$	49.3	43.2	39.5	38.3	35.1
Cs_2SO_4	69.3	58.2	53.6	49.1	53.6
Li_2SO_4	0.5	3.5	4.5	4.5	5.9
$MgBr_2$	14.7	12.0	11.8	10.3	9.7
$Mg(NO_3)_2$	23.7	21.2	20.4	18.4	16.4
Rb_2SO_4	52.6	45.8	43.8	39.2	34.5
$SrBr_2$	27.8	22.7	22.3	19.7	17.3
$Sr(NO_3)_2$	65.7	56.8	53.6	49.9	45.0

Imol/kg	0.160	0.250	0.360
$ZnSO_4$	-62.3	-28.8	-12.0

Data from Parsons (1959)

* Data from Pitzer and Brewer (1961)

TABLE 3A.3a The relative apparent molal enthalpy of alkali halides

Coefficients of the equations

$$\phi_{L_i} = A_L(d_o m)^{1/2} + B_L m + C_L m^{3/2} + D_L m^2 + E_L m^{5/2} + F_L m^3$$

and

$$\bar{L}_i = 1.5A_L(d_o m)^{1/2} + 2B_L m + 2.5C_L m^{3/2} + 3D_L m^2 + 3.5E_L m^{5/2} + 4F_L m^3$$

25°C, 1-atm J-mole^{-1} (1 cal = 4.1840 J)

i	B_L	C_L	D_L	E_L	F_L	initial conc. mol/kg
LiCl	-1669 (25)	1383 (58)	-400 (34)			0.06-0.96
LiBr	-2463 (76)	3011 (338)	-1925 (521)	509 (265)		0.02-0.79
NaF	-1983 (28)	480 (27)				0.008-0.72
NaCl	-3331 (28)	2252 (109)	-1508 (147)	515 (67)		0.01-1.0
NaI	-4000 (65)	3368 (288)	-2631 (447)	974 (231)		0.01-0.72
RbF	-1911 (43)	948 (92)	-159 (53)			0.01-0.90
RbCl	-4335 (80)	3916 (335)	-2297 (499)	537 (250)		0.02-0.80
RbBr	-7322 (315)	15902 (1760)	-25956 (3903)	22099 (3881)	-7342 (1433)	0.04-0.96
RbI	-6483 (15)	7615 (72)	-6776 (117)	2699 (64)		0.04-0.72
CsF	-2057 (48)	1365 (94)	-398 (49)			0.02-1.12
CsBr	-5973 (42)	2709 (193)	2675 (299)	-2543 (150)		0.04-0.94
CsI	-7789 (26)	6895 (87)	-2858 (67)			0.03-0.72

( ) standard errors

A_L = 1975 (the limiting slope of the Debye Huckel theory for 1:1 electrolytes)

Data and least squares analysis from Fortier, Leduc and Desnoyers (1974)

TABLE 3A.3b The relative apparent molal enthalpy of perchlorate salts and hydrobromic acid

Coefficients of the equations

$$\phi_{L_i} = a_o + a_1 I^{1/2} + a_2 I + a_3 I^{3/2}$$

cal/mol

$$\bar{L}_i = a_o + 1.5a_1 I^{1/2} + 2a_2 I + 2.5a_3 I^{3/2}$$

$t = 25°C$, p=1-atm $0.1 \leq I \leq 1.3$

Salt i	a_o	a_1	a_2	a_3	Std. Dev. ±	No. exp points
$HClO_4$	13.4063	319.928	-579.936	197.092	0.9	27
$NaClO_4$	13.5453	346.190	-995.793	276.544	0.4	18
$Ba(ClO_4)_2$	98.0496	622.215	-859.955	181.307	0.8	29
$Ca(ClO_4)_2$	67.8843	860.476	-758.358	222.760	1.4	26
$Co(ClO_4)_2$	43.4579	953.033	-835.042	282.109	2.1	17
$Mg(ClO_4)_2$	84.7185	789.748	-529.664	150.373	0.2	8
$Mn(ClO_4)_2$	71.4685	856.055	-647.341	214.821	1.3	16
$Ni(ClO_4)_2$	61.4796	839.067	-573.236	129.741	3.4	15
$Sr(ClO_4)_2$	102.481	678.208	-627.462	133.891	0.6	8
$La(ClO_4)_3$	147.710	1516.75	-1203.29	341.783	3.0	24
HBr	-0.2	635.464	-444.527	312.399	0.9	10

$I = (0.01994S)/(1-10^{-3}S)$

Data source, see Table 3A.2b

TABLE 3A.3c **Divalent electrolytes**

Coefficients for the equations

$$dB^{\bullet}/dT = A + Bm^{1/2} + Cm$$

$$\bar{L}_2 = a + bm^{1/2} + cm + dm^{3/2} \qquad 0.06 \leq I \leq 0.3$$

Ionic strength = 3m 25°C 1-atm

Coeff. \ Salt	$BaBr_2$	$BaCl_2$	$CaBr_2$	$Ca(NO_3)_2$	Cs_2SO_4
10^4A	51.006	28.703	48.170	90.111	119.048
10^2B	-1.8801	-1.1736	-2.2759	-3.6109	-4.2147
10^2C	2.6758	2.2766	3.8686	6.0215	5.6164
$\pm 10^4$st.dev.	1	1	0.8	1.2	2
a	15.4594	-12.9893	32.4	43.569	26.0892
$10^{-3}b$	3.0521	3.7372	2.583	2.201	2.6418
$10^{-3}c$	-8.1359	-9.3426	-3.442	-5.712	-13.7710
$10^{-3}d$	8.9249	9.0945	-	-	18.8558
$\pm$ st.dev.	2.0	1.0	2.0	3.0	3.0

	Li_2SO_4	$MgBr_2$	$Mg(NO_3)_2$	Rb_2SO_4	$SrBr_2$
10^4A	-20.392	22.45	32.573	55.910	49.093
10^2B	1.7727	-0.4735	-0.5603	0.0749	-1.8838
10^2C	-3.0010	-	-	-2.7376	2.8215
$\pm 10^4$st.dev.	1	1	1	1	1
a	18.145	29.3	-3.12	28.786	6.747
$10^{-3}b$	3.3863	2.790	3.582	3.0338	3.251
$10^{-3}c$	-4.6165	-3.365	-10.522	-13.8503	-9.010
$10^{-3}d$	-	-	12.658	18.9084	10.432
$\pm$ st.dev.	3.5	2.5	1.5	4.0	2.5

Least square analysis of data compiled by Parsons (1959)
$dB^{\bullet}/dT$ $kg\,mol^{-1}deg^{-1}$, $\bar{L}_2$ $cal\,mol^{-1}$

TABLE 3A.4 Partial molal entropies of perchlorate salts $cal deg^{-1} mol^{-1}$

at I = 0.723 mol/kg water

(Salinity equivalent is 35‰)

$25^{o}C$ 1-atm

Salt	$\bar{s} - \bar{s}^{o}$ *1	$\bar{s}^{o}$	$\bar{s}$	$\bar{L}$
$Ba(ClO_4)_2$	9.95	89.4	99.4	-70.7
$Ca(ClO_4)_2$	11.01	73.2	84.2	420.2
$Co(ClO^4)_2$	10.82	58.4	69.2	480.3
$HClO_4$	28.22	43.2	71.4	-111.0
$La(ClO_4)_3$	23.80	-	-	869.1
$Mg(ClO_4)_2$	11.03	58.2	69.2	555.5
$Mn(ClO_4)_2$	11.32	66.4	77.7	565.5
$NaClO_4$	27.85	57.6	85.5	-559.8
$Ni(ClO_4)_2$	10.80	55.4	66.2	508.9
$Sr(ClO_4)_2$	10.63	77.0	87.6	266.6

*1 calculated from eq (3.8.1)

Data source - see Table 3A.2b

$\bar{s}^{o}$ values from Pitzer and Brewer (1961) - see Table 3A.1

The value of $\bar{s}^{o}_{ClO_4}$ - given by Parsons (1959) is 0.4 cal/deg mol higher

(see Table 3.13)

TABLE 3A.5a The ion interaction parameter B' for 1:1 electrolytes 25°C 1-atm

$kg\,mol^{-1}$

Ionic strength mol/kg water	0.1	0.5	1.0
$AgNO_3$	-0.123	-0.119	-0.113
CsBr	-0.003	-0.017	-0.014
CsCl	+0.002	-0.014	-0.010
CsI	-0.013	-0.025	-0.019
$CsNO_3$	-0.128	-0.132	-0.120
CsOAc	0.237	0.187	0.1579
CsOH	0.307	0.177	0.150
HBr	0.287	0.217	0.196
HCl	0.235	0.181	0.1634
$HClO_4$	0.257	0.193	0.170
HI	0.337	0.268	0.237
HNO_3	0.207	0.137	0.115
KBr	0.097	0.058	0.045
$KBrO_3$	-0.053	-0.093	-
KCl	0.087	0.047	0.0359
$KClO_3$	-0.033	-0.069	-0.082
KCNS	0.08	0.043	0.032
KF	0.117	0.075	0.064
KH_2AsO_4	-0.023	-0.073	-0.079
KH_2PO_4	-0.138	-0.131	-0.121
KI	0.132	0.082	0.065
KNO_3	-0.093	-0.103	-0.099
KOAc	0.227	0.174	0.149
KOH	0.117	0.127	0.121
LiBr	0.227	0.177	0.160
LiCl	0.197	0.157	0.144
$LiClO_4$	0.307	0.237	0.203
LiI	0.327	0.251	0.213
$LiNO_3$	0.187	0.144	0.126
LiOAc	0.157	0.111	0.092
LiOH	-0.183	-0.043	-0.024
NaBr	0.152	0.109	0.092
$NaBrO_3$	0.007	-0.016	-0.021
NaCl	0.136	0.090	0.0724
$NaClO_3$	0.087	0.040	0.024
$NaClO_4$	0.117	0.072	0.054
NaCNS	0.177	0.127	0.107
NaF	0.057	0.024	0.013
NaHCOO	0.132	0.094	0.075
NaH_2AsO_4	0.067	-0.006	-0.028
NaH_2PO_4	-0.053	-0.083	-0.076
NaI	0.187	0.141	0.122
$NaNO_3$	0.037	0.003	-0.006
NaOAc	0.237	0.167	0.136
NaOH	0.057	0.097	0.086
NH_4Cl	0.087	0.047	0.035
NH_4NO_3	-0.083	-0.043	-0.044
RbBr	0.047	0.024	0.017
RbCl	0.057	0.027	0.011
RbI	0.037	0.020	0.015
$RbNO_3$	-0.123	-0.122	-0.112
RbOAc	0.227	0.177	0.154
$TlClO_4$	-0.143	-0.133	
$TlNO_3$	-0.313	-0.233	
TlOAc	-0.023	-0.037	-0.033

Data from the compilation of Pitzer and Brewer (1961)

B' values at I = 0.723 mol/kg water (≘S=35‰) are listed in Table 3.12

TABLE 3A.5b The ion interaction parameter B' for 2:1 and 1:2 electrolytes

25°C 1-atm

$kg mol^{-1}$

Ionic strength Mol/kg water	0.3	0.723	1.5		0.3	0.723	1.5
$BaBr_2$	0.563	0.409	0.307	$MgCl_2$	0.63	0.473	0.366
$BaCl_2$	0.50	0.351	0.250	$Mg(ClO_4)_2$	0.87	0.683	0.54
$Ba(ClO_4)_2$	0.61	0.452	0.343	MgI_2	0.83	0.643	0.508
BaI_2	0.68		0.414	$Mg(NO_3)_2$	0.60	0.452	0.351
$Ba(NO_3)_2$	-0.03		-0.092	$Mg(OAc)_2$	0.173		0.133
$Ba(OAc)_2$	0.195		0.171	$MnCl_2$	0.57		0.312
$CaBr_2$	0.66	0.497	0.379	Na_2CO_3	0.225		0.088
$CaCl_2$	0.57	0.420	0.3209	Na_2CrO_4	0.315		0.096
$Ca(ClO_4)_2$	0.75	0.584	0.464	Na_2HAsO_4	0.375		0.081
CaI_2	0.73		0.448	Na_2HPO_4	0.23		-0.012
$Ca(NO_3)_2$	0.375	0.252	0.186	Na_2SO_4	0.13	0.030	-0.009
$CdBr_2$	-2.69		-0.82	$Na_2S_2O_3$	0.225		0.058
$CdCl_2$	-2.11		-0.654	$(NH_4)_2SO_4$	-0.09		-0.062
CdI_2	-4.40		-1.292	$NiCl_2$	0.60		0.333
$Cd(NO_3)_2$	0.555		0.291	$Ni(ClO_4)_2$		(0.70)	
$CoBr_2$	0.705		0.426	$Pb(ClO_4)_2$	0.615		0.345
$CoCl_2$	0.60		0.340	$Pb(NO_3)_2$	-0.23		-0.174
CoI_2	0.825		0.508	Rb_2SO_4	0.18	0.075	0.028
$Co(NO_3)_2$	0.59		0.321	$SrBr_2$	0.62	0.461	0.348
$CuCl_2$	0.52		0.268	$SrCl_2$	0.548	0.397	0.299
$Cu(NO_3)_2$	0.53		0.289	$Sr(ClO_4)_2$	0.66	0.553	0.403
Cs_2SO_4	0.21	0.101	0.051	SrI_2	0.73		0.426
$FeCl_2$	0.583		0.326	$Sr(NO_3)_2$	0.308		0.119
K_2CrO_4	0.225		0.058	UO_2Cl_2	0.70		0.408
K_2HAsO_4	0.46		0.156	$UO_2(ClO_4)_2$	1.07		0.666
K_2HPO_4	0.25		0.036	$UO_2(NO_3)_2$	0.72		0.433
K_2SO_4	0.008	-0.038	-0.031	$ZnBr_2$	0.75		0.406
Li_2SO_4	0.308	0.180	0.114	$ZnCl$	0.57		0.238
$MgBr_2$	0.72	0.557	0.441	$Zn(ClO_4)_2$	0.87		0.527
				ZnI_2	0.89		0.516
				$Zn(NO_3)_2$	0.645		0.353

Data from the compilation of Pitzer and Brewer (1961) See also Table 3.13.

TABLE 3A.6 Entropy of hydration of complex ions $cal\,deg^{-1}\,mol^{-1}$

$25^{\circ}C$ 1-atm

	$\bar{s}^{o}$(solution)	$\bar{s}^{o}$(gas)	$-\Delta\bar{s}^{o}_{hyd}$
BaOH+	9.3	58.4	49.1
CaOH+	- 4.8	53.8	58.6
CrOH(2+)	-16.4	53.8	70.2
FeOH(2+)	-23.2	53.9	77.1
SrOH+	- 4.3	56.0	60.3
UOH(3+)	-28.5	58.5	86.0
$AgCl^{o}$	36.5	58.9	22.4
$CaSO_4^{o}$	7.0	69.5	62.5
$CoSO_4^{o}$	- 0.8	70.2	71
$MgSO_4^{o}$	2.0	68.2	66.2
$NiSO_4^{o}$	3	70.2	67
$ZnSO_4^{o}$	3.0	70.3	67.3

From the compilation by Nancollas (1966, p 181)

See Section 3.73.

TABLE 3A.7 Entropy changes for ion association

25°C 1-atm

Complex	K $\ell.mol^{-1}$	$\Delta\bar{G}^o$ $kcalmol^{-1}$	$\Delta\bar{H}^o$ $kcalmol^{-1}$	$\Delta\bar{S}^o$ $caldeg^{-1}mol^{-1}$
$AgCl^o$	$2.0x10^3$	-4.5	-2.7	6.4
CdCl+	95	-2.68	1.10	12.7
	91	-2.67	0.98	12.2
PbCl+	41	-2.20	1.25	11.6
$TlCl^o$	4.76	-0.93	-1.43	1.7
$AgIO_3^o$	6.8	-1.13	5.14	20.3
BaOH+	4.4	-0.87	1.75	8.8
CaOH+	25	-1.91	1.19	10.4
	23	-1.86	1.25	11.3
FeOH(2+)	$6.4x10^{11}$	-16.1	-3.0	44
SrOH+	6.7	-1.12	1.15	7.6
$TlOH^o$	6.67	-1.12	0.37	5.1
UOH(3+)	$2.5x10^{13}$	-18.2	-2.5	52
$PbNO_3+$	15.1	-1.62	-0.57	3.5
$CaSO_4^o$	200	-3.15	1.65	16.1
$CoSO_4^o$	230	-3.21	1.74	16.6
$MgSO_4^o$	234	-3.22	4.55	26.1
$MnSO_4^o$	181	-3.07	3.37	22.6
$NiSO_4^o$	211	-3.16	3.31	21.7
$ZnSO_4^o$	240	-3.25	4.01	24.4
$LaCo(CN)_6^o$	$5.8x10^3$	-5.13	1.33	21.7

From the compilation of Nancollas (1966, p 179)

See Section 3.75

REFERENCES Appendix 3

Fortier, J.-L., P.-A. Leduc, and J.E. Desnoyers 1974.

Thermodynamic properties of alkali halides.
II. Enthalpies of dilution and heat capacities in water at 25°C.
J. Solution Chem. 3(4): 323-349.

Gier L.J. and C.E. Vanderzee 1974a.

Enthalpies of dilution and relative apparent molar enthalpies of aqueous cobalt and nickel perchlorates.
J.Chem.Engng. Data 19(4): 315-319.

Gier L.J. and C.E. Vanderzee 1974b.

Enthalpies of dilution and relative apparent molar enthalpies of aqueous calcium and manganous perchlorates.
J.Chem.Engng. Data. 19(4): 323-325.

Jongenburger H.S. and R.H. Wood 1965.

The heats and entropies of dilution of the perchlorates of magnesium and strontium.
J.Phys.Chem. 69(12): 4231-4238.

Nancollas G.H. 1966.

"Interactions in electrolyte solutions - Metal complex and ion-pair formation in solution", Elsevier, Amsterdam, 214pp.

Parsons R. 1959.

"Handbook of electrochemical constants", Butterworths, London, 113pp.

Pitzer K.S. and L. Brewer 1961.

revised edition of "Thermodynamics" by G. N. Lewis and M. Randall. McGraw-Hill, New York, N.Y., 723pp.

Appendix 3

Vanderzee C.E. and J.D. Nutter 1974.

Relative apparent molar enthalpies of aqueous lanthanum perchlorate.
J.Chem.Engng.Data. 19(3): 268-270.

Vanderzee C.E. and J.A. Swanson 1972.

Enthalpies of dilution and relative apparent molal enthalpies of aqueous barium perchlorate.
J.Chem.Engng.Data. 17(4): 488-490.

CHAPTER 4

THE PARTIAL MOLAL VOLUME

The Pressure Derivative of the Chemical Potential

4.1 BASIC EQUATIONS

The partial molal free energy is the prime indicator of the solution state since it governs the direction of transfer of chemical substances. In oceanography, where the systems studied are commonly at pressures greater than atmospheric, the pressure derivative of the chemical potential of component *i*, $\bar{G}_i$, is of special interest.

At constant composition, temperature, and charge

$$(\partial\bar{G}_i/\partial p)_T = \frac{\partial(\partial G/\partial n_i)_{P,T}}{\partial p} = \frac{\partial(\partial G/\partial p)_T}{\partial n_i}$$

$$= (\partial V/\partial n_i)_{P,T} = \bar{v}_i \qquad (4.1.1)$$

so that,

$$(\partial \ln a_i/\partial p)_T = (\bar{v}_i - \bar{v}_i^o)/RT \qquad (4.1.2)$$

where $a_i = a_\pm^\nu = (\gamma_\pm m_\pm)^\nu$, Section 1.9

for the electrolyte components. The symbol $\bar{v}_i$ represents the partial molal volume of the component *i*, the superscript zero indicates the reference state, as usual.

(For convenience, the subscripts indicating composition variables held constant have been omitted).

The total volume, V , is given by

$$(\partial G/\partial P)_T = V = \Sigma_i n_i \bar{v}_i \qquad (4.1.3)$$

so that the equation covering all the components, including the solvent, pure water, is:

$$RT\left[\frac{n_1\partial\ln a_1 + n_2\,\partial\ln a_2 + \ldots}{\partial P}\right]_T = n_1(\bar{v}_1-\bar{v}_1^o)$$

$$+ n_2(\bar{v}_2-\bar{v}_2^o) + \ldots \qquad (4.1.4)$$

Equation 1.1 illustrates mathematically what the name of $\bar{v}_i$ implies, i.e.

the rate of change of the total volume of the solution with the amount of component _i_. In physical terms $\bar{v}_i$ is the change in volume of a very large amount of solution when one mole of substance _i_ is added. The volume may increase or decrease depending on the component added, e.g. if one mole of $CaCO_3$ or of NaOH is added to a large amount of seawater, at 1 bar pressure, the volume will decrease, whereas at 1000 bars it will increase (see Sections 4.2, 6.11).

This is because the volume change is the resultant of two changes. The total volume is increased by the volume occupied by the added solute molecules, but is diminished by the decrease in the volume of the water, arising from the solute's effect on it. The sign of the volume change therefore depends on which effect has the larger magnitude.

In this chapter, unless otherwise indicated, the systems considered will be at constant temperature (usually 25°C) and pressure (1-bar). Thus, $(\ )_{T,P}$ or $(\ )_T$, as applicable, is tacitly implied for all the derivatives. The effects of temperature and pressure on the pmv will be considered in Chapters 5 and 6 respectively.

4.2 Apparent molal volume (amv)

Binary solutions. For the purpose of determining $\bar{v}_i$, it is convenient to define a quantity related to $\bar{v}$ as is a ratio of finite increments to the corresponding differential coefficient. This quantity is the apparent molal volume of _i_, ϕ_{v_i}. For a binary solution such as aqueous sodium chloride, with water as component 1 and the salt as component 2, the amv is given by

$$\phi_v = (V - n_1 V_1^o)/n_2 \qquad (4.2.1)$$

where n is the number of moles of the component

V the volume of the solution

$n_1 V_1^o$ the volume of the pure solvent .

If the solvent is water and the mass is fixed at 1000 g, then $n_1 = 55.51$, $n_2 = m$ and the water volume equals $10^3/d_w$, where d_w is the density of the water. Since water is the only solvent dealt with in this text, the subscript w is used to represent water, rather than the usual subscript 1. However, the latter subscript is retained in a number of the basic definitions to conform with the usual notation, which emphasises the generality of such equations.

Thus $$V = m\phi_v + 55.51V_1^o \quad (4.2.2)$$

and $$\bar{v}_2 = \partial V/\partial m = \phi_v + m\partial\phi_v/\partial m \quad (4.2.3)$$

or $$\bar{v}_2 = \phi_v + I\partial\phi_v/\partial I \quad (4.2.4)$$

in terms of the ionic strength I (= $\frac{1}{2}z_+z_-\nu m$ = wm).

This gives the pmv of the salt. The pmv of the water in the solution can be obtained by combining equations 4.1.3, 4.2.2 and 4.2.3 to give

$$\bar{V}_1 = V_1^o - (m^2/55.51)\,\partial\phi_v/\partial m \quad (4.2.5)$$

or $$\bar{V}_1 = V_1^o - (I^2/55.51w)\,\partial\phi_v/\partial I \quad (4.2.6)$$

(where $w = \frac{1}{2}z_+z_-\nu$)

More generally, $$\bar{V}_1 = V_1^o - \frac{n_2}{n_1}\frac{I\partial\phi_v}{\partial I} \quad (4.2.6a)$$

Thus, from eq 4.2.5 it can be seen that $m\partial\phi_v/\partial m$ or the difference between $\bar{v}_2$ and ϕ_v is simply the difference per mole of salt between the volume contribution of the water with and without the salt, i.e.

$$\bar{v}_2 - \phi_v = 55.51(V_1^o - \bar{V}_1)/m \quad (4.2.7)$$

Figure 4.1 shows the volume changes as sodium sulfate is added to a kilogram of water up to an ionic strength of 1 molal (i.e. 1/3 moles of Na_2SO_4). The volume contribution of the salt $(1/3)\bar{v}_2$ is 7.215ml, yet the change in volume is only 6.138ml. This is due to the decrease in the volume contribution of the water of 1.077 ml, as indicated in Fig. 4.1. This reduction in volume is assumed to be due to the pressure effect of the salt which compresses the water. The usefulness of this assumption will be illustrated later in this chapter. If points along the lower curve are divided by m (or I/3), then ϕ_V is obtained (eqs 4.2.1, 4.2.2), whilst the gradient of the curve gives the pmv (since $\partial V/\partial I = (\frac{1}{3}\bar{v}_2)$ the moles of water being constant. Fig. 4.1 thus illustrates all the equations (4.2.1) - (4.2.7).

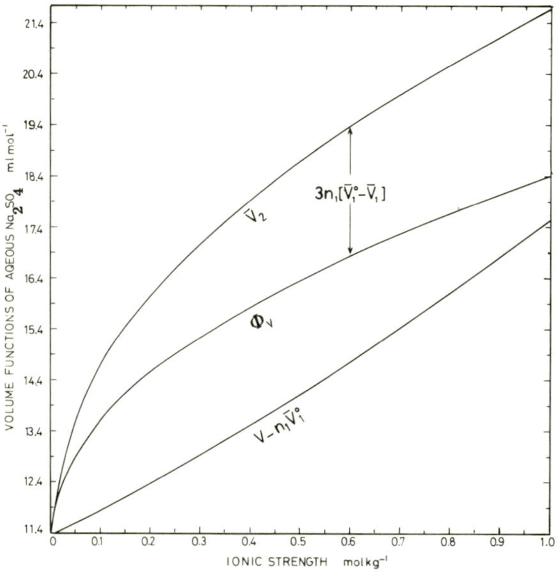

The partial molal volume, $\bar{v}_2$, and the apparent molal volume, ϕ_V, of Na_2SO_4 as a function of concentration.

FIGURE 4.1

The amv may also be expressed in terms of the density of the solution, d, and of the pure water, d_w, since, density = mass/ volume. The mass of the water is 10^3 g, so that the volume of pure water is $10^3/d_w$, whilst the mass of salt is mM_2 where M represents the molecular weight. Thus,

$$V = (10^3 + mM_2)/d$$

Substituting these quantities into eq 4.2.1, gives

$$\phi_v = 10^3(d_w - d)/mdd_w + M_2/d \qquad (4.2.8)$$

or, in terms of the specific volume v = 1/d,

$$\phi_v = 10^3(v - v_w)/m + vM_2 \qquad (4.2.9)$$

These forms of eq 4.2.1 will be found very useful for the discussion of multicomponent solutions.

Much of the ϕ_v data have been obtained from density measurements. At high dilution ϕ_v is very sensitive to experimental uncertainties in the density, e.g. an error of 1×10^{-5} in the density would correspond to an uncertainty of about 1 ml in ϕ_v when m = 0.01; a corresponding uncertainty at 1 molal would be associated with an error in d of 10^{-3}. Nevertheless, very accurate density measurements have been made in dilute solutions (Lee 1966, Harned and Owen 1958), so that reliable values of ϕ_v^o, the value of ϕ_v at infinite dilution, are available for many salts.

The tabulated data for ϕ_v are generally in the form of the Masson equation (molar scale - moles per litre),

$$\phi_v = \phi_v^o + s_v \underline{c}^{\frac{1}{2}} \qquad (4.2.10)$$

and values of ϕ_v^o may vary, according to whether the data covers dilute or concentrated (ca. m ≥ 0.2) solutions. Generally, the ϕ_v^o used here will correspond to the latter (an ionic strength range of 0.2-1 molal corresponds to S = 11-48‰.

Equation 4.2.10, while useful for summarising the experimental data, is not very acceptable theoretically. The theoretical limiting equations for apparent molal properties are discussed by Harned and Owen (p390 Section 8) and by Millero (1971). The slope s_v^L (limiting) is given by

$$s_v^L = k\underline{w}^{3/2} \qquad (4.2.11)$$

where $k = N^2e^3(8\Pi/1000D^3RT)^{\frac{1}{2}}(\partial(\ln D)/\partial p - \beta/3)$ and

where β is the compressibility of the solvent and D the dielectric constant, the other symbols have their usual meaning (see Chapter 2)

$$w = 0.5\Sigma\nu_k z_k^2 = 0.5z_+z_-\nu_i \qquad (4.2.12)$$

with ν_k representing the number of ions of species k and valency z_k formed by one molecule of the electrolyte, i.

With the same value of ϕ_v^o, the experimental slope s_v is now equivalent to $(s_v^L + b_v c^{\frac{1}{2}})$ where b_v is the "deviation constant". If ϕ_v^o has an error, $\delta\phi_v^o$, then the apparent b_v, b_v', is given by

$$b_v' = (b_v + \delta\phi_v^o/c) \quad (4.2.13)$$

For most simple 1:1 electrolytes b_v is only small (Table 4.1). Millero (1971) found that some 46 electrolytes (mostly alkali and alkaline earth salts, and alkyl ammonium salts with the anions: F^-, Cl^-, Br^-, I^-, OH^-, NO_3^-, SO_4^{2-}, ClO_3^-, $O_2CCH_3^-$) could be represented by

$$b_v(25°C) = -0.412 + 2.772\text{x}10^{-2}\phi_v^o - 2.697\text{x}10^{-4}(\phi_v^o)^2 \quad (4.2.14)$$

However, as will be shown later (Section 4.21, Table 4.4) the values calculated from this equation give only rough estimates. This is partly due to insufficiencies in the data and to the smallness of the b_v values. In addition, b_v may be a function of concentration, as recently shown for the alkaline earth halides (Perron <u>et al</u> (1974)). The ϕ_v values of these salts, for the ionic strength range 0.03-1.2mol/kg at 25°C, could be represented by

$$\phi_v = \phi_v^o + s_v^L(d_w m)^{\frac{1}{2}} + B_v m + C_v m^{3/2} \quad (4.2.15)$$

with d_w representing the density of water, so that

$$b_v \simeq (B_v + C_v m^{\frac{1}{2}}) \quad (4.2.16)$$

since the values of B_v and C_v on the <u>c</u> and <u>m</u> scales differ by only about 1-2%. These values, and the values of b_v at I = 0.723 mol/kg (m = 0.241 mol/kg) are shown in Table 4.1<u>a</u>.

Values of k for the limiting slope, as a function of temperature are given by

$$(s_v^L/w^{3/2}) = k = 1.444 + 1.6799\text{x}10^{-2}t - 8.4055\text{x}10^{-6}t^2 + 5.5153\text{x}10^{-7}t^3 \quad (4.2.17)$$

(Redlich and Meyer 1964).

TABLE 4.1 Deviation constant* b_v (Redlich ϕ_v equation) for electrolytes in water.

25°C.

Salt	b_v cm^3 l.mol^{-2}	Salt	b_v cm^3 l.mol^{-2}
$BaCl_2$	-3.21	LiCl	-0.36
$CaCl_2$	-1.25	LiF	1.1
CsBr	0.09	LiI	-0.76
CsCl	0.12	$MgCl_2$	-6.3
CsF	0.25	NaBr	-0.22
CsI	0.11	NaCl	0.02
HCl	-1.03	NaF	0.54
KBr	-0.06	NaI	-0.39
KCl	0.11	NaOH	0.95
$KClO_3$	0.42	Na_2SO_4	2.96
KF	0.52	NH_4Br	0.55
KI	-0.30	RbBr	-0.26
KNO_3	0.72	RbCl	0.17
LiBr	-0.55	RbF	0.55
		RbI	-0.05
		$NaB(OH)_4$	4.708**

* Millero (1969)
See Table 4.1a

** Ward and Millero (1974).

Values of s_v^L at 25°C for various charge types are listed in Table 4.2. These values have been experimentally confirmed (Millero 1971). The appropriate equation for ϕ_v is therefore

$$\phi_v = \phi_v^o + s_v^L c^{\frac{1}{2}} + b_v c \qquad (4.2.18)$$

which is usually referred to as the Redlich equation. This equation may be expressed in terms of m, a more convenient concentration unit than c, and may be extended to include higher terms, for example, eq 4.2.15 as derived by Perron *et al* (1974). Fortier, *et al* (1974) showed that the same ϕ_v^o is obtained (within the experimental error) whether the data are expressed in c or m. As can be seen from Table 4.1*a*, in the concentration range

TABLE 4.1a Coefficients of the equation

$$\phi_V = \phi_V^o + s_V^L(d_w m)^{1/2} + B_V m + C_V m^{3/2} \quad cm^3 mol^{-1} \qquad (4.2.15)$$

Salt	ϕ_V^o	B_V	C_V	b_V at I=0.723mol/kg = $(B_V + 0.5C_V)$
$BaCl_2$	22.98	-3.28(-3.24)		B_V
$CaCl_2$	17.65	-10.45(-10.55)	8.38(8.58)	-6.2
$MgCl_2$	14.02	- 8.32(- 8.38)	5.36(5.48)	-5.6
$SrCl_2$	18.26	-11.25(-11.37)	10.16(10.42)	-6.2

Data of Perron et. al (1974), ( ) values for molar scale, d_w taken as 0.997047 g cm^{-3}; s_V^L = 9.706 (See Table 4.2)

TABLE 4.2 Values of Debye-Huckel theoretical limiting law slope for ϕ_V (aqueous solutions) eq 4.2.11

	25°C	
Charge type	$w^{3/2}$	$s_V^L(cm^3 l^{1/2} mol^{-3/2})$
1:1	1.000	1.868*
2:1	5.196	9.706
2:2	8.000	14.944
3:1	14.697	27.454
4:1	31.623	59.072

* Redlich and Meyer (1964)
Other data from Millero (1971a)

of seawater, this is also true (within a few percent) for the other parameters.

Since s_V^L is known and b_V is small the Redlich form of equation is more useful for predicting values of ϕ_V than the Masson equation.

In addition to the correlation of b_V with ϕ_V^o (eq 4.2.14 and Table 4.4), we found that b_V values for alkali halides (and NaOH) were simply correlated with the corresponding values of the compressibility parameter

s_k (Table 4.2b). This quantity is approximately equal to the pressure derivative of s_v, viz

$$s_k = -((\partial s_v/\partial p)_T + 0.5\beta s_v) \qquad (4.2.19)$$

with β representing the compressibility of the solution. The compressibilities of electrolyte solutions and seawater will be discussed in detail in Chapter 6. The linear correlation of b_v with s_k appears to hold for higher charge systems (the coefficients are different, however). The data are not extensive enough at present to investigate this further. However, such a correlation might be expected in view of the compressional effect of the solute (Section 6.3). (See also Section 6.21).

The interpretation of ϕ_v^o ($\bar{v}^o$) will be considered in Section 4.5.

TABLE 4.2b Comparison of b_v (experimental) and b_v calculated from $b_v = -1.2 + 10^3 s_k$

Salt	b_v exp	b_v calc	s_kx10^3	b_v(calc - exp)
CsBr	0.09	-0.27	0.93	-0.36
CsCl	0.12	-0.05	1.15	-0.17
CsI	0.11	-0.45	0.75	-0.56
HCl	-1.03	-0.90	0.30	0.13
KBr	-0.06	-0.31	0.89	-0.25
KCl	0.11	0.04	1.24	-0.07
KF	0.52	0.34	1.54	-0.18
KI	-0.30	-0.55	0.65	-0.25
LiBr	-0.55	-0.49	0.71	0.06
LiCl	-0.36	-0.29	0.91	0.07
LiI	-0.76	-0.69	0.51	0.07
NaBr	-0.22	-0.40	0.8	-0.18
NaCl	0.02	-0.06	1.14	-0.08
NaF	0.54	0.42	1.62	-0.12
NaI	-0.39	-0.71	0.49	-0.32
NaOH	0.95	0.89	2.09	-0.06
RbBr	-0.26	-0.30	0.90	-0.04
RbCl	0.17	-0.08	1.12	-0.25
RbI	-0.05	-0.55	0.65	-0.50

exp (Table 4.1)

s_k from Parsons (1959)

units: b_v based on ϕ_v in $cm^3 mol^{-1}$, concentration moles/litre

s_k based on ϕ_k in $cm^3 bar^{-1} mol^{-1}$ ditto

The practical interest in ϕ_v lies in its use for predicting the values of the partial molal volume, $\bar{v}$, for electrolytes in seawater, and for comparing the binary solute effects and seasalt effects. The values at ionic strengths 0.1 to 0.9 $molkg^{-1}$ will therefore be of especial interest. Since a range of temperatures and pressures are to be covered, the molal scale is more convenient than the molar (or volume) scale. A very useful relationship for converting eq 4.2.10 or 4.2.18 to the molal scale is

$$m = \underline{c} / \left[d_w (1 - 10^{-3} \underline{c} \phi_v) \right] \tag{4.2.20}$$

or

$$\underline{c} = m d_w / (1 + 10^{-3} m \phi_v d_w)$$

Values of ϕ_v calculated via eq 4.2.10 have been fitted to the corresponding molal ionic strength to give equations of the form

$$\phi_v = \phi_v^o + sI^{\frac{1}{2}} + tI + \delta' \tag{4.2.21}$$

where δ' represents higher terms in I which were necessary for a few electrolytes. Lee (1966) has fitted his very accurate data to equations of this form. He studied NaCl, $MgCl_2$, $CaCl_2$, Na_2SO_4, NaBr and $MgSO_4$, so that all the major electrolytes of seawater (on the basis of Kester et al's (1967) grouping of the ions) are covered with the exception of KCl, $SrCl_2$, NaF and H_3BO_3. The values of the parameters for eq 4.2.21 are listed in appendix 4 (Table 4A.3).

Values for $\bar{v}_2$ (2 representing the solute) can be calculated from

$$\bar{v}_2 = \phi_v^o + (3/2)sI^{\frac{1}{2}} + 2tI + \delta' + I\partial\delta'/\partial I \tag{4.2.22}$$

obtained from eqs 4.2.4 and 4.2.21

Most of the data are for 25°C and 1-atm pressure, but this gives a good reference state when effects of temperature and pressure are to be considered, as in Chapters 5 and 6.

4.21 Additivity of ionic volumes

When no data are available for a particular electrolyte, can an estimate be made from the existing data? It has been shown (Harned and Owen 1958, p398) that for the alkali halide solutions, the individual ionic volumes are additive over a wide concentration range.

Thus, $\phi_v(LiCl) + \phi_v(NaBr) - \phi_v(LiBr) = \phi_v(NaCl)$ (4.21.1)

and so on.

The principle should hold for all electrolytes at high dilution where the ions might be expected to be the only contributors to ϕ_v. At concentrations where the dissociation of the electrolyte is incomplete, combinations such as eq 4.21.1 cannot be expected to give an accurate estimate. For example, consider the estimate of ϕ_v for magnesium sulfate, from data on Na_2SO_4, $MgCl_2$ and NaCl: (the subscript v is omitted for convenience).

$$\phi_{Na_2SO_4} = (1-\alpha_1)\phi_{Na-SO_4^-} + \alpha_1(2\phi_{Na^+} + \phi_{SO_4^{2-}}) \qquad (4.21.2)$$

and

$$\begin{aligned}(\phi_{MgSO_4})_A &= \phi_{MgCl_2} + \phi_{Na_2SO_4} - 2\phi_{NaCl} \\ &= \phi_{Mg^{2+}} + \alpha_1\phi_{SO_4^{2-}} + 2(\alpha_1-1)\phi_{Na^+} \\ &\quad + (1-\alpha_1)\phi_{Na-SO_4^-}\end{aligned} \qquad (4.21.3)$$

where α_1 is the degree of dissociation of Na_2SO_4, and it is assumed that for NaCl and $MgCl_2$ this is complete, i.e. $\alpha = 1$, the subscript A indicates that ϕ_v has been estimated via the principle of additivity.

TABLE 4.3 Prediction of the apparent molal volumes of electrolytes from the principle of additivity

25°C

Salt	ϕ_v^o exp (ml mol^{-1})	ϕ_v^o calc (ml mol^{-1})	ϕ_v (I=0.72 mol kg^{-1}) exp (ml mol^{-1})	ϕ_v (I=0.72 mol kg^{-1}) calc (ml mol^{-1})	Components
$MgSO_4$	-7.157	-7.157	0.543	-1.768	$MgCl_2$+Na_2SO_4-2NaCl
		-7.023		0.321	$MgCl_2$+K_2SO_4-2KCl
		-7.071		-1.253	$CaSO_4$+$MgCl_2$-$CaCl_2$
		-7.19[a]			
Na_2SO_4	11.420	11.420	17.357	19.668	$MgSO_4$+2NaCl-$MgCl_2$
		12.04		19.446	K_2SO_4+2NaCl-2KCl
		11.56[a]			
$CaSO_4$	-3.956	-3.956	2.77	1.749	$CaCl_2$+Na_2SO_4-2NaCl
		-2.51		3.838	$CaCl_2$+K_2SO_4-2KCl
		-3.956		4.06	$MgSO_4$+$CaCl_2$-$MgCl_2$
		-3.87[a]			
$MgCl_2$	14.657	14.657	17.321	19.632	$MgSO_4$+NaCl-Na_2SO_4
	14.02[b]	14.49[a]			
KNO_3	38.18	38.10	40.11	39.67	NH_4NO_3-NH_4Cl+KCl
		38.02[a]			
NaOH	-6.7	-6.4	-3.22	-3.52	LiOH+NaCl-LiCl
		-5.25[a]			
KI	45.36	44.96	46.66	46.39	KCl+RbI-RbCl
		45.24[a]			

Data for $MgCl_2$, Na_2SO_4, NaCl, $MgSO_4$ and $CaCl_2$ from Lee (1966) other from Parsons (1959).

a Sum of the ionic values (Appendix Table 4A.1)

b Perron *et al* (1974)

The actual value for magnesium sulfate, however, is

$$\phi_{MgSO_4} = (1-\alpha_2)\phi_{Mg-SO_4} + \alpha_2\phi_{Mg^{2+}} + \alpha_2\phi_{SO_4^{2-}} \qquad (4.21.4)$$

α_2 being the degree of dissociation of $MgSO_4$.

Thus, the error in the calculated value will be

$$(\phi_{MgSO_4})_A - \phi_{MgSO_4} = \phi_{Mg^{2+}}(1-\alpha_2) + \phi_{SO_4^{2-}}(\alpha_1-\alpha_2) - 2(\alpha_1-1)\phi_{Na^+} + (1-\alpha_1)\phi_{Na-SO_4^-} - (1-\alpha_2)\phi_{Mg-SO_4} \qquad (4.21.5)$$

Table 4.3 shows the results of such estimates for a number of salts at an ionic strength equivalent to S = 35‰, and for an infinitely dilute solution.

Lee (1966) estimated α_2 as 0.59 over a range of 0.1-0.8 moles of $MgSO_4$ per kg of water, and ϕ_{Mg-SO_4} as 4.2-11.64 $ml\,mol^{-1}$ over the same range (ϕ_{Mg-SO_4} was given by $8.6I^{1/2}$). Equations of the form of 4.21.2 can be used in a reverse sense, i.e. to study ion-pairing effects. Millero (1971a) has illustrated a general method of calculating association constants in this way from $\bar{v}_2$ data. More recently, Millero and Masterton (1974) have used eq 4.21.4 to estimate the apparent molal volume of the ion pair $Mg\text{-}SO_4^{\,o}$ for the temperatures 0, 25, 50°C. They also calculated the volume change for the ion pair formation. They give a fairly detailed account of this approach in the analysis of ion pairing. They also cite most of the relevant literature.

As can be seen from Table 4.3 good estimates of ϕ_v^o values can be obtained from the summed ionic values (Appendix 4, Table 4A.1). Such data cover a wide range of ions so that, if the corresponding values of s_v can be predicted, ϕ_v can be estimated for a large number of salts. The use of s_v^L as a first approximation will be found especially useful as illustrated in Table 4.4. In many cases the value of b_v is small enough for it to be neglected (the error of prediction is often larger than its actual value

TABLE 4.4 Prediction of the apparent molal volumes of electrolytes ($ml mol^{-1}$) on the basis of the Redlich equation. (eq 4.2.18)

25°C

Salt	ϕ_v^o	b_v exp	b_v calc eq 4.2.14	b_v calc linear*	$\phi_v^o + s_v^L c^{1/2}$	ϕ_v (I=0.72 $mol kg^{-1}$) eq 4.2.18**	ϕ_v (I=0.72 $mol kg^{-1}$) exp
$CaSO_4$	-3.956		-0.53	1.36	2.38	2.65	2.77
KNO_3	38.18	0.72	0.25	0.19	39.77	39.91	40.11
KOH	2.9		-0.33	1.4	4.49	5.50	6.59
$MgCl_2$	14.657	-6.3	-0.06	1.0	19.41	19.65	17.32
$MgSO_4$	-7.157		-0.62	1.75	-0.82	-0.47	0.543
NaCl	16.617	0.02	-0.03	0.93	18.20	18.87	18.22
NaOH	-6.7	0.95	-0.61	1.27	-5.11	-4.20	-3.22
Na_2SO_4	11.420	2.96	-0.13	1.11	16.18	16.46	17.357
NH_4NO_3	47.56		0.30	-0.14	49.15	49.05	48.38

* $b_v = 1.5 - 3.44\times10^{-2}\phi_v^o$ (Millero 1971a)

** linear b_v used,

charge type	1:1	2:1	2:2
$s_v^L c^{1/2}$	1.585	4.755	6.34 (with s_v^L from Table 4.2)

- Table 4.4). As more accurate data become available, and the theoretical significance of b_v becomes more fully understood, it should be possible to predict accurately the values of ϕ_v via this method.

4.3 Multicomponent solutions.

For solutions of three or more components it is convenient to define the mean apparent molal volume, which can easily and directly be determined from the experimental data. The mean apparent molal volume for a solution containing n_1 moles of water, n_2 moles of electrolyte 2, n_3 moles of electrolyte 3, and so on, is given by

$$\begin{aligned}\Phi_v &= (V-n_1V_1^o)/(n_2+n_3+n_4+\ldots) \\ &= (V - 55.51V_1^o)/(m_2+m_3+\ldots) \\ &= (V - 10^3/d_w)/m_s \qquad (4.3.1)\end{aligned}$$

where V is the volume of the solution and $m_s = \Sigma_i m_i$, d_w is the density of pure water, $n_1V_1^o$ the volume. The ionic strength of the solution, I, is simply the sum of the ionic strengths of the electrolytes.

$$I = \Sigma_i I_i = \tfrac{1}{2}\Sigma_i (z_+z_-\nu)_i m_i = \Sigma_i w_i m_i \qquad (4.3.2)$$

or $$I = Wm_s \quad (\text{with } W = \Sigma_i m_i w_i/m_s = \Sigma_i f_i w_i;\ \ f_i = m_i/m_s) \qquad (4.3.3)$$

4.31 Ternary Solutions.

The ratio of the ionic strength to total molality in a binary solution is fixed but equations for multicomponent solutions must take such a change into account. e.g. For a solution of two electrolytes, in water, at molalities m_2 and m_3 respectively, suppose that m_2 is held constant and m_3 is changed. The ionic strength is increased by the amount $dI = w_3dm_3$ so that, if binary solution quantities are being used, then these must change accordingly. In the two binary solutions at the ionic strength, I, of the solution, $dm_3 = dI/w_3$ and $dm_2 = dI/w_2$, so that $dm_3 = w_2dm_2/w_3$.

On mixing the two binary solutions, suppose a volume change of ΔV occurs. Since the ionic volumes for the alkali halides have been shown to be additive, it might be expected that ΔV would be zero for mixtures of these salts. This has been shown to be the case for systems KCl-NaCl-H_2O and KBr-NaCl-H_2O (Young and Smith 1954). However, this will not be generally true, since many ionic volumes are not additive, as shown in Table 4.3, so that, omitting the subscript v from ϕ_{v_i}, for convenience.

$$(m_2+m_3)\Phi_v = m_2\phi_2 + m_3\phi_3 + (m_2+m_3)\,\Delta\Phi_v \qquad (4.3.4)$$

where ϕ_i represents the amv in the binary solution (of the electrolytes 2 or 3) at the same ionic strength I of the ternary solution. The quantity $\Delta\Phi_v$, or $(\Phi_v - \Sigma_i f_i\phi_i)$ may be called the excess apparent molal volume. It should be noted that, in general

$$m_s\Delta\Phi_v = \Sigma_i m_i\Delta\phi_i \qquad (\text{or } \Delta\Phi_v = \Sigma_i f_i\Delta\phi_i\ )$$

so that

$$\Delta\Phi_v + m_s(\partial\Delta\Phi_v/\partial m_i)_{m_j} = \Delta\Phi_v + m_i(\partial\Delta\phi_i/\partial m_i)_{m_j} + \Sigma_j\, m_j(\partial\Delta\phi_j/\partial m_i)_{m_j} \qquad (4.3.5)$$

where *i j* are solute components of the multicomponent solution.

From eqs 4.3.1 and 4.3.4

$$V = 10^3/d_w + m_2\phi_2 + m_3\phi_3 + (m_2+m_3)\Delta\Phi_v \qquad (4.3.6)$$

Differentiating with respect to m_3 at constant m_2, gives the pmv of the salt (3) in the mixed solution.

$$(\partial V/\partial m_3)_{m_2} = \bar{v}_3^{mix} = \phi_3 + m_3(\partial\phi_3/\partial m_3)_{T,p} + m_2(\partial\phi_2/\partial m_3)_{T,p} + \Delta\Phi_v + (m_2+m_3)(\partial\Delta\Phi_v/\partial m_3)_{T,p,m_2}$$

$$= \phi_3 + m_3(\partial\phi_3/\partial m_3)_{T,p} + (m_2w_3/w_2)(\partial\phi_2/\partial m_2)_{T,p} + \delta$$

where

$$\delta = \Delta\Phi_v + (m_2+m_3)(\partial\Delta\Phi_v/\partial m_3)_{T,p,m_2} \qquad (4.3.7)$$

since the $(\partial\phi_i/\partial m_3)$ terms refer to the binary solutions which are at the same ionic strength, i.e. $w_2m_2=w_3m_3$, and $\partial m_3 = \frac{w_2}{w_3}\partial m_2$

In the mixture,

$$I = w_2m_2 + w_3m_3$$

$$m_3 = I/w_3 - (w_2/w_3)m_2$$

and $m_3(\partial\phi_3/\partial m_3) = I(\partial\phi_3/\partial I) - (m_2w_2/w_3)(\partial\phi_3/\partial m_3)$

Thus,

$$\bar{v}_3^{mix} = \bar{v}_3 - (m_2w_2/w_3)(\partial\phi_3/\partial m_3)_{T,p} + (m_2w_3/w_2)(\partial\phi_2/\partial m_2)_{T,p} + \delta \quad (4.3.8)$$

since $\bar{v}_3 = \phi_3 + I\partial\phi_3/\partial I$ (eq 4.2.4).

The main problem in calculating $\bar{v}_3^{mix}$ from this equation will be in determining the last term. Even when Φ_v data are available, so that $\Delta\Phi_v$ can be determined from eq 4.3.4, the partial molal quantities for the components are unknown. Nevertheless, this equation should be more useful, in general, than the one derived by differentiating eq 4.3.1, viz.

$$\bar{v}_3^{mix} = \Phi_v + m_s(\partial\Phi_v/\partial m_3)_{T,p,m_2} \quad (4.3.9)$$

This is because the partial molal quantities for $\Delta\Phi_v$ should be much smaller than for Φ_v so that the errors of estimate will be less significant for eq 4.3.8. When experimental data are available for multicomponent solutions, both equations will be useful for studying the terms $\Delta\Phi_v$ and $(\partial\Delta\Phi_v/\partial m_i)_{m_j}$.

Lee (1966) has measured $\bar{v}_3^{mix}$ for sodium chloride solutions with component 3 as Na_2SO_4, $MgCl_2$, $CaCl_2$, $MgSO_4$ and NaBr, and with $m_{NaCl} = 0.16$ and 0.725 mol/kg. This data should provide a useful guide to our study of $(\partial\Delta\Phi_v/\partial m_i)m_j$ for seawater solutions.

4.32 The excess apparent molal volume.

Lee (1966) presents his results in the form of equations

$$y = a_o + a_1m_3 + a_2m_3^2 + a_3m_3^3 \quad (4.32.1)$$

TABLE 4.5 The mean and excess apparent molal and partial molal volumes of Na_2SO_4 (in $ml\,mol^{-1}$) in 0.725m NaCl

25°C, 1 bar

$m_{Na_2SO_4}$	I $mol\,kg^{-1}$	Φ_v	$\bar{v}^{mix}_{Na_2SO_4}$	$\bar{v}^{mix'}_{Na_2SO_4}$	$\Delta\Phi$	$(\partial\Delta\Phi/\partial m_{Na_2SO_4})_{NaCl}$
0.00	0.725	18.192	20.646	19.755	-0.036	1.279
0.001	0.728	18.196	20.979	19.775	-0.035	1.707
0.01	0.755	18.231	21.084	19.942	-0.020	1.582
0.05	0.875	18.391	21.570	20.650	+0.036	1.141
0.10	1.025	18.603	22.213	21.460	+0.082	0.814
0.20	1.325	19.068	23.624	22.894	+0.156	0.621
0.30	1.625	19.588	25.200	24.153	+0.220	0.807
0.40	1.925	20.164	26.940	25.285	+0.323	1.184

Data from Lee (1966)

where y is Φ_v^{mix}, or $\bar{v}^{mix}$. The limiting values for the pmv, $\bar{v}^{o\,mix}_{3,m_3 \to 0}$ were also determined. These were not equivalent to a_o.

Values of $\Delta\Phi_v$ were estimated from this data via eq 4.3.4, using binary solution data as listed in Table 4A.3. The partial molal quantities could then be obtained by differentiation, as m_2 (m_{NaCl}) was constant. With the $\Delta\Phi_v$ terms (δ) omitted from eq 4.3.8, we get $\bar{v}_3^{mix'}$, say, and if this is compared with the experimental value of $\bar{v}_3^{mix}$, an estimate of the partial molal value of $\Delta\Phi_v$ can be made.

i.e. $$\bar{v}_3^{mix} - \bar{v}_3^{mix'} = \Delta\Phi_v + m_s(\partial\Delta\Phi_v/\partial m_3)_{m_2} \tag{4.32.2}$$

and since $\Delta\Phi_v = \Phi_v - \Sigma m_i\phi_i/m_s$

$$m_s(\partial\Delta\Phi_v/\partial m_3)_{m_2} = (\bar{v}_3^{mix}) - (\bar{v}_3^{mix})' + \Sigma m_i\phi_i/m_s - \Phi_v \tag{4.32.3}$$

These are the general equations.

Table 4.5 lists the results for the NaCl-Na_2SO_4-H_2O system.

TABLE 4. 6 The mean and excess apparent molal and partial molal volumes of salts (in $mlmol^{-1}$) in 0.725m NaCl 25°C, 1 bar.

Component $\underline{i}$	m_i $molkg^{-1}$	Φ_v	$\bar{V}_i^{mix}$	$\bar{V}_i^{mix'}$	$\Delta\Phi$	$(\partial\Delta\Phi/\partial m_i)_{m_{NaCl}}$
$CaCl_2$	0.00	18.191	23.079	23.292	-0.037	-0.176
	0.01	18.256	22.978	23.391	-0.042	-0.500
Na_2SO_4	0.00	18.192	20.646	19.755	-0.036	1.279
	0.01	18.231	21.084	19.942	-0.020	1.582
$MgSO_4$	0.00	18.192	2.90	3.811	-0.036	-0.864
	0.01	17.986	3.126	3.869	-0.035	-0.963
$MgCl_2$	0.00	18.191	19.308	19.76	-0.037	-0.415
	0.01	18.207	19.395	19.864	-0.043	-0.426
NaBr	0.00	18.191	25.792	25.777	-0.037	+0.052
	0.01	18.287	25.266	25.829	-0.053	-0.507

Data from Lee (1966)

In Table 4.6 the values for the different salts at m_{NaCl} = 0.725 and m_3 = 0.01 and 0.0 are compared.

Since a sodium chloride solution of 0.725m represents a rough approximation to seawater, the values of $(\partial\Delta\Phi_v/\partial m_3)_{m_2}$ in Tables 4.5 and 4.6 should be close to those for the salts in seawater. An interesting feature of the results is the sign of the gradients. The relatively large positive value for sodium sulfate could indicate that ion association has increased on addition of this salt. This interpretation is based on the fact that the electrostriction of water molecules around the ions would cause a net contraction of the system on ionization. Such a picture is compatible with the seawater results in regard to ionpairing (about 2/3rds of the association of the sulfate ion is with sodium).

The pmv's of many salts in seawater have been measured, so that the corresponding values of $(\partial\Delta\Phi_v/\partial m_i)_{m_j}$ can be estimated via eq 4.3.8.

However, before going on to the complicated seawater solutions, it will be of value to consider further the mixing of two electrolytes. The equations that can be derived will serve as a basis for a more quantitative interpretation of the mixing effects.

4.33 The volume change on mixing two electrolytes.

Consider the mixing of electrolytes A and B with compositions as shown in diagram

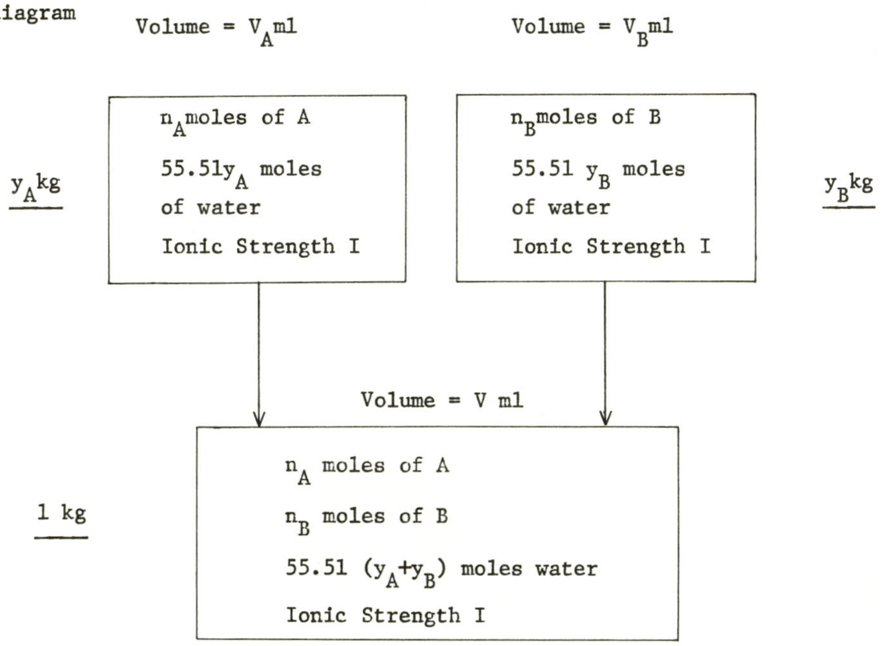

The values of y_A and y_B are chosen so that $(y_A+y_B) = 1$.

If ΔV is the volume change on mixing, $\bar{V}_w$ the pmv of the water and $\bar{v}_i^{mix}$ the pmv of electrolyte i in the mixture.

$$V = V_A + V_B + \Delta V \qquad (4.33.1)$$

and

$$V = 55.51\bar{V}_w + m_A\bar{v}_A^{mix} + m_B\bar{v}_B^{mix} \qquad (4.33.2)$$

since $n_A = m_A$ and $n_B = m_B$ (there being 1 kg of water in the final solution).

For the binary solutions

$$V_A = 55.51 y_A \bar{V}_{w_A} + m_A \bar{v}_A \qquad (4.33.3)$$

$$V_B = 55.51 y_B \bar{V}_{w_B} + m_B \bar{v}_B \qquad (4.33.4)$$

The corresponding molalities are m_A/y_A and m_B/y_B.

Also, $I = w_A m_A / y_A = w_B m_B / y_B = w_A m_A + w_B m_B = I_A + I_B$

where I_A and I_B are the ionic strengths of the electrolytes in the mixture and $w_i = \frac{1}{2}(\nu z_+ z_-)_i$, as usual.

From these relationships the volume change on mixing is given by

$$\Delta V = 55.51(\bar{V}_w - (1/I)(I_A \bar{V}_{w_A} + I_B \bar{V}_{W_B})) + m_A(\bar{v}_A^{mix} - \bar{v}_A) + m_B(\bar{v}_B^{mix} - \bar{v}_B) \qquad (4.33.5)$$

or

$$\Delta V = 55.51\, \Delta\bar{V}_w + \sum_i m_i \Delta\bar{v}_i \qquad (4.33.6)$$

Where $\Delta\bar{V}_w = \bar{V}_w - (y_A \bar{V}_{w_A} + y_B \bar{V}_{w_B})$, from eq (4.33.5) (4.33.7)

It might be remembered that the free-energy change on mixing is just the sum of the changes in the chemical potential for the various components. Thus, in an analogous manner the volume change on mixing is the sum of the changes in the partial molal volumes of the various components.

The change in the <u>pmv</u> of the water is obtained as follows:

From eq 4.2.6, for Solution A,

$$\bar{V}_{w_A} = V_w^o - \frac{n_A}{n_{w_A}} \left[I \frac{\partial \phi_A}{\partial I} \right] \qquad (4.2.6)^*$$

and, since $n_A = m_A$ and $n_{w_A} = 55.51 y_A$,

$$\bar{V}_{w_A} = V_w^o - \frac{m_A}{55.51 y_A} I \frac{\partial \phi_A}{\partial I} \qquad (4.33.8)$$

and similarly for B, so that, with $(y_A + y_B) = 1$,

$$(y_A \bar{V}_{w_A} + y_B V_{w_B}) = V_w^o - \frac{I}{55.51} \left[m_A \frac{\partial \phi_A}{\partial I} + m_B \frac{\partial \phi_B}{\partial I} \right]$$

For the mixture,

$$\bar{V}_w = V_w^o - \frac{m_s}{55.51} \left[I \frac{\partial \Phi_v}{\partial I} \right] \qquad (4.33.9)$$

Thus, eq 4.33.7 becomes

$$55.51\Delta\bar{V}_w = I\left[\Sigma_i m_i \frac{\partial \phi_i}{\partial I} - m_s \frac{\partial \Phi_v}{\partial I}\right] \quad (4.33.10)$$

or

$$55.51\Delta\bar{V}_w = \Sigma_i m_i(\bar{v}_i - \phi_i) - m_s(\bar{V}_s - \Phi_v) \quad (4.33.11)$$

where $m_s\bar{V}_s = \Sigma_i m_i \bar{v}_i^{mix}$, $m_s = m_A + m_B$.

The change in the pmv's of each salt is given by eq 4.3.8 viz.

$$m_A\Delta\bar{v}_A = -\frac{m_A m_B w_B}{w_A}\left[\frac{\partial \phi_A}{\partial m_A}\right] + \frac{m_A m_B w_A}{w_B}\left[\frac{\partial \phi_B}{\partial m_B}\right] + m_A\delta_A$$

and similarly for B, so that

$$m_A\Delta\bar{v}_A + m_B\Delta\bar{v}_B = m_A(\partial\Delta V/\partial m_A)_{m_B} + m_B(\partial\Delta V/\partial m_B)_{m_A} \quad (4.33.12)$$

From eq 4.3.6 $\Delta V = (m_A + m_B)\Delta\Phi_v = m_s\Delta\Phi_v$ (4.33.13a)

so that

$$(\partial\Delta V/\partial m_A)_{m_B} = m_s(\partial\Delta\Phi_v/\partial m_A)_{m_B} + \Delta\Phi_v \quad (4.33.13b)$$

and

$$(\partial\Delta V/\partial m_B)_{m_A} = m_s(\partial\Delta\Phi_v/\partial m_B)_{m_A} + \Delta\Phi_v \quad (4.33.13c)$$

Combining eqs 4.33.6, 4.33.12, and 4.33.13 (a, b and c) gives

$$-55.51\Delta\bar{V}_w/m_s = m_A(\partial\Delta\Phi_v/\partial m_A)_{m_B} + m_B(\partial\Delta\Phi_v/\partial m_B)_{m_A}$$

$$= \Sigma_i m_i(\partial\Delta\Phi_v/\partial m_i)_{m_j} \quad (4.33.14)$$

The changes may also be expressed in terms of the pressure derivatives of the chemical potential changes, e.g. Suppose that, for components A and B, such changes can be represented by the simple Harned's rule (Harned and Owen 1958, p 603), then

$$\log\gamma_{\pm A}^{mix} = \log\gamma_{\pm A} - \alpha_{AB} w_B m_B \qquad (4.33.15)$$

where α_{AB} is the Harned coefficient and is independent of m_A and m_B, i.e. it is a function of the ionic strength, I, only.

Differentiating eq 4.33.15 w.r.t. pressure at constant temperature and composition, and since

$$2.303\nu_A(\partial\log\gamma_{\pm A}/\partial p)_{T,m_A,m_B} = (\bar{v}_A - \bar{v}_A^o)/RT \qquad (4.33.16)$$

from eq 4.1.2, then

$$\bar{v}_A^{mix} = \bar{v}_A - 2.303\nu_A RTw_B m_B(\partial\alpha_{AB}/\partial p)_{T,I} \qquad (4.33.17)$$

and similarly for B, so that

$$m_A\Delta\bar{v}_A + m_B\Delta\bar{v}_B = -2.303RTm_A m_B(\nu_A w_B G_{AB} + \nu_B w_A G_{BA}) \qquad (4.33.18)$$

where $G_{AB} = (\partial\alpha_{AB}/\partial p)_{T,I}$ and with $\Delta\bar{v}_i = \bar{v}_i^{mix} - \bar{v}_i$.

Thus, at a given ionic strength, pressure and temperature

$$\Sigma m_i\Delta\bar{v}_i = Km_A m_B \qquad (4.33.19)$$

where K is a constant.

Eq 4.33.6 may now be expressed as

$$\Delta V = m_s\Delta\Phi_v = 55.51(\Delta\bar{V}_w) + Km_A m_B \qquad (4.33.20)$$

which, combined with eq 4.33.14 gives

$$\Delta\Phi_v = -\Sigma m_i(\partial\Delta\Phi_v/\partial m_i)_{m_j} + Km_A m_B/m_s \qquad (4.33.21)$$

For mixtures of strong 1:1 electrolytes (e.g. $HCl-NaCl-H_2O$), at constant ionic strength, the simple relationship

$$m_s\Delta\Phi_v = K'm_A m_B \qquad (4.33.22)$$

is found to hold (Harned and Owen 1958, p400) where K' is a constant. Differentiating eq (4.33.22) w.r.t. m_B (m_A constant) and vice versa gives

$$m_A(\partial\Delta\Phi_v/\partial m_A)_{m_B} = K'm_Am_B/m_s - \Delta\Phi_v m_A/m_s \qquad (4.33.23)$$

$$m_B(\partial\Delta\Phi_v/\partial m_B)_{m_A} = K'm_Am_B/m_s - \Delta\Phi_v m_B/m_s \qquad (4.33.24)$$

and, with (4.33.21), it follows

$$K' = \tfrac{1}{2}K, \text{ i.e.}$$

$$K' = (-2.303RT/2)(\nu_A w_B G_{AB} + \nu_B w_A G_{BA}) \qquad (4.33.25)$$

For ternary systems where eq 4.33.16 contains additional terms in the concentration, the form of eq 4.33.22 can also be expected to be more complicated, e.g. if

$$\nu_A \beta_{AB}(w_B m_B)^2$$

is added to eq 4.33.16

$$Km_Am_B \text{ becomes } Km_Am_B + K'm_Am_B^2$$

4.4 Seawater solutions.

Returning to eq 4.3.6 and increasing the components to any number desired, gives

$$V = 10^3/d_w + \Sigma_i m_i\phi_i + (\Sigma m_i)\,\Delta\Phi \qquad (4.4.1)$$

(with the subscript v of ϕ_{v_i} omitted for convenience, as in Section 4.3) and differentiating in the same way as before

$$(\partial V/\partial m_i)_{m_j} = \bar{v}_i^{mix} = \phi_i + m_i(\partial\phi_i/\partial m_i)_{T,p}$$

$$+ \sum_j m_j(\partial\phi_j/\partial m_i)_{T,p} + (\partial m_s\Delta\Phi_v/\partial m_i)_{T,p,m_j}$$

In the binary solutions at the same ionic strength (4.4.2)

$m_3w_3 = m_2w_2 = \ldots$ so that $\partial m_3 = (w_2/w_3)\partial m_2$ and so on.

$$\bar{v}_i^{mix} = \phi_i + m_i(\partial\phi_i/\partial m_i)_{T,p}$$

$$+\Sigma_j (w_i/w_j)m_j(\partial\phi_j/\partial m_j)_{T,p}$$

$$+ (\partial m_s\Delta\Phi_v/\partial m_i)_{m_j,T,p} \qquad (4.4.3)$$

Also, $m_i = I/w_i - (1/w_i)\Sigma_j w_j m_j$

and $$m_i(\partial\phi_i/\partial m_i)_{T,p} = I(\partial\phi_i/\partial I)_{T,p} - (\Sigma_j w_j m_j)(\partial\phi_i/\partial I)_{T,p}$$

$$\bar{v}_i^{mix} = \bar{v}_i - (\Sigma_j w_j m_j)(\partial\phi_i/\partial I)_{T,p} + w_i \Sigma_j m_j(\partial\phi_j/\partial I)_{T,p} + (\partial m_s \Delta\Phi_v/\partial m_i)_{m_j,T,p} \qquad (4.4.4)$$

When the ϕ_i data is in the form of eq 4.2.21

$$\partial\phi_i/\partial I = \tfrac{1}{2}sI^{-\frac{1}{2}} + t + \partial\delta'/\partial I$$

For seawater, from eq 4.3.4

$$\Phi_v = \Sigma_i f_i \phi_i + \Delta\Phi_v \qquad (4.4.5)$$

where $f_i = m_i/m_s$, and is assumed independent of I so that

$$(\partial\Phi_v/\partial I)_{T,p,f_i} = f_i(\partial\phi_i/\partial I)_{T,p} + \Sigma_j f_j(\partial\phi_j/\partial I)_{T,p} + (\partial\Delta\Phi_v/\partial I)_{T,p,f_i} \qquad (4.4.6)$$

with $W = w_i f_i + \Sigma_j w_j f_j$, and $Wm_s = I$, $W = \tfrac{1}{2}\Sigma_i f_i(z_+ z_- \nu)_i$

Equation 4.4.4 may be written, at constant temperature and pressure

$$\bar{v}_i^{sw} = \bar{v}_i - I\partial\phi_i/\partial I + (Iw_i/W)(\partial\Phi_v/\partial I - \partial\Delta\Phi_v/\partial I) + \Delta\Phi_v + (I/W)(\partial\Delta\Phi_v/\partial m_i)_{m_j} \qquad (4.4.7)$$

or $$\bar{v}_i^{sw} = \phi_i + (Iw_i/W)(\partial\Phi_v/\partial I - \partial\Delta\Phi_v/\partial I) + \Delta\Phi_v + (I/W)(\partial\Delta\Phi_v/\partial m_i)_{m_j} \qquad (4.4.8)$$

Both eqs 4.4.4 and 4.4.8 have been used to estimate the pmv's of salts in seawater (Leyendekkers 1974).

The value of Φ_v was calculated from seawater density data and $\Delta\Phi_v$ from Φ_v and the binary solution data. Details of these calculations are given in Section 4.62. See, also, Appendix 4, Table 4A.7 and Table 5.3 The $(\partial\Delta\Phi_v/\partial m_i)_{m_j}$ term was calculated by comparing the experimental values of $\bar{v}_i^{sw}$ with the corresponding values of $\bar{v}_i^{sw}$, from eqs 4.4.4 and 4.4.8, (see eq 4.32.3 for the ternary systems with NaCl). It was found, Leyendekkers (1974), that the values of this term could be correlated with the ionic entropy of the anion, $\bar{s}^o_{ion}$, (cubic equation in $\bar{s}^o_{ion}$). This is illustrated in Fig. 4.2 where $([m_s(\partial\Delta\Phi_v/\partial m_i)_{m_j}]/\nu_-\bar{s}^o_{ion})$ is plotted against $\bar{s}^o_{ion}$; ν_- is the number of moles of anion from one mole of the salt (fully dissociated). The values in NaCl solutions are shown for comparison. These results indicated:

1. that for most salts the value of $m_s(\partial\Delta\Phi_v/\partial m_i)_{m_j}$, ($I=0.72 \text{molkg}^{-1}$), would be within ±1 mlmol^{-1}, since most values of $\bar{s}^o_{ion}$ for anions lie between -4 and 36 $\text{caldeg}^{-1}\text{mol}^{-1}$ at 25°C (excepting some oxyanions, e.g. PO_4^{3-}, MnO_4^- and ClO_4^-, Appendix 3, Table 3A.1).

2. The anion is the critical determinant of the value of the partial molal <u>emamv</u> (excess <u>mean</u> apparent molal volume). For Na_2SO_4, the influence of $NaSO_4^-$ appeared dominant. Since neutral ion pairs (such as $MgSO_4^o$ and $CaSO_4^o$ and $CaCO_3^o$) have entropies very similar to the anion values, their contributions to $(\partial\Delta\Phi_v/\partial m_i)_{m_j}$ would not be clearly distinguishable from that of the anion.

The curve in Fig. 4.2 is given by

$$Q'/\bar{s}^o_{ion} = -0.268+0.026\bar{s}^o_{ion} - 5.31\text{x}10^{-4}(\bar{s}^o_{ion})^2 \qquad (4.4.9)$$

Values of $(\partial\Delta\Phi_v/\partial m_i)_{m_j}$ for a number of salts are listed in Table 4.7. The very large deviation of the carbonate ion is interesting and indicates strong interaction with the solvent, water, (structurally) and with the cations (ion-pairing). Both effects are related, since ion pairing is associated with the disturbance of the electrostatic ordering of the water around the cation. This disturbance gives rise to the observed volume changes. A consistency check on the $(\partial\Delta\Phi_v/\partial m_i)_{m_j}$ values can be made from the relationship

$$I\ \Sigma_i\ (f_i/W)(\partial\Delta\Phi_v/\partial m_i)_{m_j} = I(\partial\Delta\Phi_v/\partial I)$$ which follows from eq 4.4.8

since $\bar{V}_s = \Sigma_i f_i \bar{v}_i^{sw}$

$$= \Sigma f_i \phi_i + \Delta\Phi_v + I(\partial\Phi_v/\partial I)$$

with $(\Sigma f_i \phi_i + \Delta\Phi_v) = \Phi_v$, from eq 4.3.4

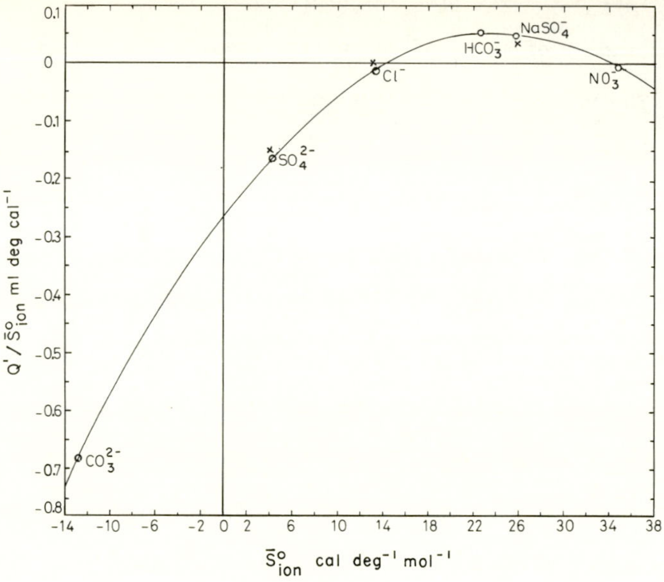

FIGURE 4.2

The partial molal emamv of seasalt as a function of the conventional ionic entropy. $Q' = m_s(\partial\Delta\Phi_v/\partial m_i)_{m_j}/\nu_-$ Crosses represent the corresponding quantity for a 0.725m sodium chloride solution containing 0.01m of each of the salts (separately) $CaCl_2$, $MgCl_2$, $MgSO_4$ and Na_2SO_4 (see Tables 4.6 and 4.7). Curve represents eq 4.4.9

For 25°C, $\Delta\Phi_v$ is given by, (Leyendekkers 1974)

$$\Delta\Phi_v = -0.287 + 0.888I^{\frac{1}{2}} - 0.491I \quad (4.4.10)$$

so that

$$I\partial\Delta\Phi_v/\partial I = 0.444I^{\frac{1}{2}} - 0.491I \quad (4.4.11)$$

At S = 35‰ (I = 0.723 molkg^{-1}), $I\partial\Delta\Phi_v/\partial I$ = 0.02 mlmol^{-1}.

The corresponding value of $\Sigma f_i(I/W)(\partial\Delta\Phi_v/\partial m_i)_{m_j}$ (with $I/W = m_s$) is -0.08, which is reasonably consistent, within the experimental uncertainties.

TABLE 4.7 The partial molal value* of $\Delta\Phi_v$, the excess mean apparent molal volume of seasalt. ($ml mol^{-1}$)

S = 35‰ 25^oC 1 bar

Anion group	Q/ν_-[a]	$\bar{s}^o$ cal deg^{-1} mol^{-1}	Anion group	Q/ν_-	$\bar{s}^o$ cal deg^{-1} mol^{-1}
OH^-	(+0.85)	-2.52	$H_2PO_4^-$	(+0.96)	21.3
Cl^-	-0.23	13.2	$HAsO_4^{2-}$	(-0.22)	0.9
Br^-	(+0.69)	19.29	$H_2AsO_4^-$	(+1.22)	28
I^-	(+1.28)	26.14	$HCOO^-$	(+1.02)	21.9
SO_4^{2-}	-0.69	4.1	HCO_3^-	+1.10	22.7
HSO_4^-	(+0.944)	30.52	CO_3^{2-}	+8.68	-12.7
NO_3^-	-0.23	35	$H_2BO_3^-$	(-0.78)	7.3
HPO_4^{2-}	(+4.57)	-8.6	$NaSO_4^-$	+1.3	(26)

* averaged per mole of anion

a $Q = m_s(\partial\Delta\Phi_v/\partial m_i)_{m_j}$

bracketed values estimated from entropy correlation (Fig.4.2)

viz. $Q/\nu_- = -0.268\bar{s}^o + 0.026(\bar{s}^o)^2 - 5.31\times10^{-4}(\bar{s}^o)^3$, eq (4.4.9).

Values of $m_s(\partial\Delta\Phi_v/\partial m_i)_{m_j}$ (or Q), for various anion groups, are listed in Table 4.7 (for S = 35‰, 25°C).

Estimates of $\bar{v}_i^{sw}$ (at S = 35‰, 25°C) for a number of electrolytes in seawater are given in Table 4.8, together with experimental values, and estimates from the "free" ion values given by Millero (1969). The corresponding values in the binary solutions, $\bar{v}_i$, at the same ionic strength ($0.723 mol kg^{-1}$) are shown for comparison. In many cases, the value of $\bar{v}_i^{sw} - \bar{v}_i$ is only 1-2$ml mol^{-1}$ (or less, for simple 1:1 electrolytes). This might be expected from eq 4.4.7 which shows that the pmv change depends on

TABLE 4.8 The partial molal volumes of electrolytes in seawater

S = 35‰ 25°C 1-atm

Salt	$\bar{V}_i^{SW}$ exp.[a]	calc.[b]	Millero[c]	$\bar{V}_i$
$AgNO_3$	-	30.73	30.5	31.36
$BaCl_2$	-	28.06	27.0	27.02
$CaCl_2$	22.00 ± 0.28	22.76 (22.80)	22.0	22.29
$CaCO_3$	-5.4 ± 2.0	-4.0	-18.0	-13.13
$Ca(HCO_3)_2$	38.35 ± 0.76	39.23	32.2	36.79
$Ca(NO_3)_2$	45.13 ± 0.64	45.39	44.4	43.94
$CaSO_4$	5.24 ± 0.61	5.18	-0.2	5.66
$CdCl_2$	-	28.68	20.0	29.63
CsBr	-	49.66	48.1	48.58
CsCl	-	41.65	41.2	41.88
CsI	-	60.95	59.7	59.72
HCl	-	19.55	20.1	18.90
HOH	-	-0.91	-1.6	-1.29
KBr	-	37.24	35.8	36.16
KCl	29.20 ± 0.14	29.11	28.90	29.45
K_2CO_3	30.3 ± 0.8	30.2	17.8	21.94
K_2CrO_4	-	44.83	41.4	45.58
KF	-	10.89	10.1	10.8
$KHCO_3$	37.38 ± 0.3	36.81	34.0	35.96
KI	-	48.55	47.4	47.31
KNO_3	40.76 ± 0.21	40.74	40.1	41.07
K_2SO_4	41.63 ± 0.37	41.61	35.6	44.31
$LaCl_3$	-	24.47	21.0	22.09
LiBr	-	27.93	26.3	25.54
LiCl	-	18.93	19.4	18.87
LiI	-	38.09	37.9	36.55
LiOH	-	-1.87	-2.3	-2.26
$MgCl_2$	19.62 ± 0.21	19.22 (19.18)	19.6	18.66
$Mg(HCO_3)_2$	35.98 ± 0.67	35.46	29.8	32.61
$Mg(NO_3)_2$	42.75 ± 0.52	42.55	42.0	41.67
$MgSO_4$	2.86 ± 0.48	2.96 (2.95)	-2.6	2.90

TABLE 4.8 (continued)

Salt	$\bar{v}_i^{sw}$ exp.[a]	calc.[b]	Millero[c]	$\bar{v}_i$
NaBr	-	26.72	25.7	25.66
NaCl	18.90 $\pm$ 0.08	18.85 (18.84)	18.8	19.04
Na_2CO_3	11.0 $\pm$ 2	9.8	-2.4	1.45
$NaHCO_3$	27.07 $\pm$ 0.27	27.05	23.9	26.56
NaI	-	38.11	37.3	36.79
$NaNO_3$	30.46 $\pm$ 0.17	30.46	30.0	30.73
NaOH	-	-1.59	-2.9	-1.48
Na_2SO_4	21.03 $\pm$ 0.14	18.93 20.72[d]	15.4	20.13
NH_4Cl	-	37.87	37.8	37.80
RbBr	-	42.3	40.9	41.27
RbCl	-	34.41	34.0	34.65
RbI	-	53.54	52.5	52.33

a Duedall and Weyl (1967), Duedall (1968)

b eqs 4.4.8 and 4.4.4 (bracketed values) and Table 4.7

c Millero (1969) sum of "free" ion values

d using Q/ν_- of $NaSO_4^-$

the ϕ_v and Φ_v gradient differences, and on the mixing terms which are generally small in value, Fig. 4.2, eqs4.4.9, 4.4.10.

It is of interest to look at the difference $\bar{v}_i^{sw}-\bar{v}_i$ for a component, i, in a little more detail.

4.41 The volume change on transfer from the binary solution to seawater, at constant ionic strength

From eq 4.4.7 the volume of transfer is given by

$$\Delta\bar{v}_i^{tr,I} = (\bar{v}_i^{sw} - \bar{v}_i) = -I\partial\phi_i/\partial I$$
$$+(w_i/W)(I\partial\Phi_v/\partial I - I\partial\Delta\Phi_v/\partial I)$$
$$+ \Delta\Phi_v + m_s(\partial\Delta\Phi_v/\partial m_i)_{m_j} \qquad (4.41.1)$$

The partial ionic strength terms can be expressed in terms of the partial molal volume of water in the binary solution and in seawater.

From eq 4.2.6 (since $V_w^o = M_w/d_w$, and with subscript 1 now replaced by w).

$$I\partial\phi_i/\partial I = w_i 55.51 M_w/d_w I - w_i 55.51 \bar{V}_w^i/I \qquad (4.41.2)$$

with M_w, d_w, $\bar{V}_w^i$ representing the molecular weight, density and partial molal volume of water (in the presence of the solute i), respectively.

($w_i = \frac{1}{2}z_+z_-\nu_i$, as noted previously)

Similarly, with $W = \Sigma_i f_i w_i$, $I\partial\Phi_v/\partial I = (55.51W/I)(M_w/d_w - \bar{V}_w^{sw})$ (4.41.3)

Differentiating eq 4.4.5 (i.e. $\Delta\Phi_v = \Phi_v - \Sigma_i f_i \phi_i$) w.r.t. I, at constant temperature and pressure, and constant f_i values, gives

$$I\partial\Delta\Phi_v/\partial I = I\partial\Phi_v/\partial I - \Sigma_i f_i I\partial\phi_i/\partial I \qquad (4.41.4)$$

Then, since $\Sigma_i f_i I\partial\phi_i/\partial I = (55.51W/I)(M_w/d_w - \Sigma_i(I_i/I)\bar{V}_w^i)$

(from eq 4.41.2 and using $f_i w_i = I_i/m_s = I_i W/I$) eq 4.41.4 becomes

$$I\partial\Delta\Phi_v/\partial I = (55.51W/I)(\Sigma_i (I_i/I)\bar{V}_w^i - \bar{V}_w^{sw}) \qquad (4.41.5)$$

Thus, the concentration gradient of Φ_v, the mean apparent molal volume of seasalt, is proportional to the change in the partial molal volume of water on addition of the seasalt. The concentration gradient of $\Delta\Phi_v$, the excess mean apparent molal volume, emamv, of seasalt, is proportional to the difference between the summed pmv's of water in the binary solutions (weighted by I_i/I) and the pmv of water in seawater.

For a given charge system and ionic strength

$$I\partial\phi_i/\partial I = \text{const.} + b_v I \qquad (4.41.6)$$

from eq 4.2.18. So that, under the same conditions,

$$\left[\Delta\bar{v}_i^{tr,I} - m_s(\partial\Delta\Phi_v/\partial m_i)_{m_j}\right],$$ represented hereafter

by $\Delta_i^{tr,I}$, should be simply related to b_v, the "deviation" constant of the Redlich equation. This is illustrated in Fig. 4.3 for 1:1 charge systems. The b_v value for KNO_3 is predicted as $0.2 cm^3 l.mol^{-2}$, which agrees with the predicted value in Table 4.4, although there is no simple correlation of

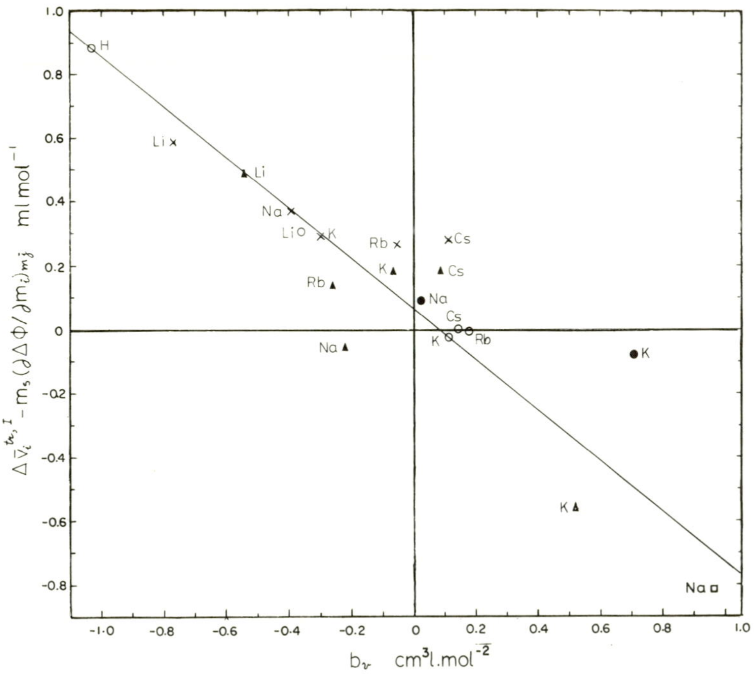

FIGURE 4.3

The volume of transfer at constant ionic strength (minus the partial molal emamv of seasalt, $m_s(\partial\Delta\Phi_v/\partial m_i)_{m_j}$) versus the deviation function of the Redlich equation (4.2.18).

Anion	Cl^-	Br^-	I^-	F^-	NO_3^-	OH^-
Symbol	○	▲	×	△	●	□

$\phi^o_{v_i}$ and $\Delta_i^{tr,I}$ for the 1:1 electrolytes in Fig. 4.3.

Since the effect of the salt is to increase the internal pressure of the solution, the concentration gradient might be expected to be correlated with the limiting pressure gradient, i.e. ϕ^o_K. This is illustrated in Fig. 4.4 where $(1/w_i)(\Delta_i^{tr,I}-\Delta\Phi_v)$, at I = 0.723molkg^{-1}, (or $-I\partial\phi_i/\partial I$ + constant, see eq 4.41.1) is plotted against $\phi^o_{K_i}/w_i$ The deviations for the

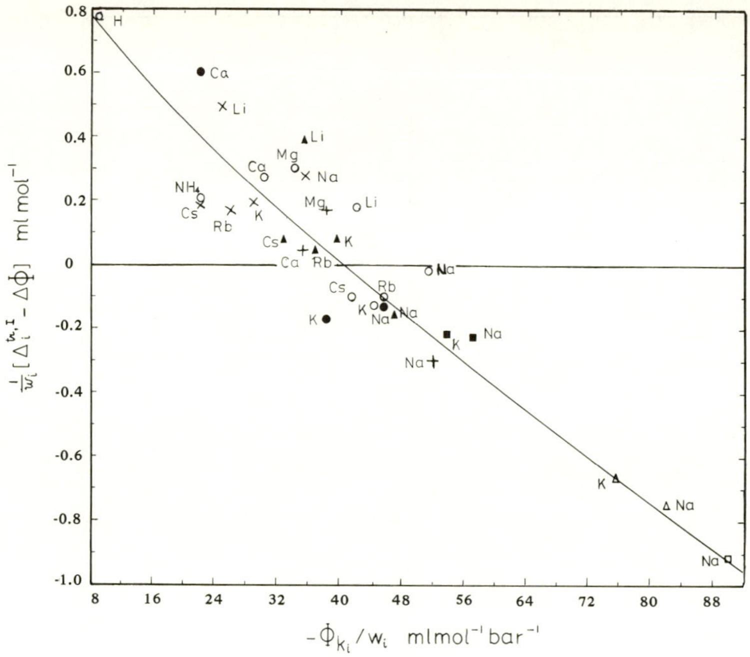

FIGURE 4.4

The concentration gradient of ϕ_{V_i} (plus a constant) derived from eq 4.41.1

$(\Delta_i^{tr,I} = (\Delta\bar{v}_i^{tr,I} - m_s(\partial\Delta\Phi_v/\partial m_i)_{m_j})$

versus the pressure gradient of ϕ_{V_i}

Anion	Cl^-	Br^-	I^-	F^-	NO_3^-	OH^-	SO_4^{2-}	CO_3^{2-}
Symbol	○	▲	×	△	●	□	+	■

averaged curve are generally less than 0.2mlmol^{-1} (i.e. within the experimental errors). The value of $\Delta\Phi_v$ at 25°, I = 0.723molkg^{-1}, was 0.11 mlmol^{-1} (from eq 4.4.10).

These relationships are somewhat limited by lack of data. On the other hand, values of $\phi_{v_i}^o$ are readily available (or can be estimated from the ionic values - Appendix 4, Table 4A.1). A more generally useful relationship has been developed by comparing the pmv's of electrolytes at infinite dilution in water and in seawater; this will now be briefly outlined.

4.42 Volume of transfer of electrolytes from water to seawater.

It has been found, Leyendekkers (1974), that the experimental values of $\bar{v}_i^{sw} - \phi_{v_i}^o$ (or $\Delta\bar{v}_i^{tr}$) can be correlated with the standard ionic entropy of the anion of the salt, i. For oxyanions, at 25°C, S = 35‰, 1 atm

$$\Delta\bar{v}_i^{tr}/\nu_- = 11.4 - 0.38\bar{s}_-^o + 0.0034(\bar{s}_-^o)^2 \quad (4.42.1)$$

to ± 0.02 ml mol^{-1} of the average value for the anion group.

TABLE 4.9 The partial molal volume changes of electrolytes on transfer from water to seawater.

$$\Delta\bar{v}_i^{tr} = (\bar{v}_i^{sw} - \bar{v}_i) \text{ ml mol}^{-1}$$

S = 35‰ 25°C, 1-atm

Anion group	$\Delta\bar{v}_i^{tr}/\nu_-$	Anion group	$\Delta\bar{v}_i^{tr}/\nu_-$
OH-	(5.1)	HPO_4^{2-}	(14.9)
Cl-	2.4	$H_2PO_4^-$	(4.9)
ClO_4^-	(1.3)	$HAsO_4^{2-}$	(11.1)
Br-	(3.5)	BrO_3^-	(1.8)
I-	(3.2)	HCOO-	(4.7)
SO_4^{2-}	9.8	HCO_3^-	4.3
HSO_4^-	(3.0)	CO_3^{2-}	16.8
NO_3^-	2.3	MnO_4^-	(1.2)
PO_4^{3-}	40.4	$H_2BO_3^-$	(8.8)
CN-	4.2	CrO_4^{2-}	(7.7)

Bracketed values predicted from estimates of $\bar{v}_i^{sw}$ (Table 4.8 or from eq 4.4.8).

N.B. For Li+ and NH_4^+ salts, subtract 0.6 per mole of cation
For acids subtract 1.0 per mole of H+

The values of $\bar{s}^o_-$ are listed in Appendix 3, Table 3A.1. ν_- represents the number of moles of the anion from one mole of fully dissociated electrolyte. Values of $\Delta\bar{v}_i^{tr}$ for a number of anion groups are given in Table 4.9. The bracketed values are based on estimates of $\bar{v}_i^{sw}$ from eq 4.4.8 and Table 4.7. The pmv of the salt, i, is estimated from

$$\bar{v}_i^{sw} = \phi_{v_i}^o + \Delta\bar{v}_i^{tr} \qquad (4.42.2)$$

($\phi_{v_i}^o$ is the limiting value of ϕ_{v_i} - Table 4A.1)

The lithium, hydrogen and ammonium salts have negative deviations from the mean value of $\Delta\bar{v}_i^{tr}/\nu_-$ for each anion group. This shows that these cations partly neutralise the effect of the anion. When estimating $\bar{v}_i^{sw}$ for salts of these cations, the contributions to $\Delta\bar{v}_i^{tr}/\nu_-$ should be taken into account (-0.6 for Li+ and NH_4+ and - 1.0 for H+ in ml per mole of cation). The value of $\Delta\bar{v}_i^{tr}$, like $(\partial\Delta\Phi_v/\partial m_i)_{m_j}$, appears to depend largely on anion-solvent interactions. These may be described as changes in the contributions of the water-water molecule interactions to the total internal pressure, changes in the electrostatic ordering of water about the cations, or changes in structure.

Another property of the ion that reflects the interactions with the solvent is the limiting mobility λ^o_{ion}. At infinite dilution the ion will migrate (under the influence of an applied electric field) independently of the other ions, so that its size and structure (whether compatible with the solvent or not) must be important factors in its mobility. As well, λ^o data for a large range of ions are readily available. In Figure 4.5 values of $\Delta\bar{v}_i^{tr}/\nu_-z_-$ have been plotted against the corresponding value of λ^o/z_-^2 for the anion. This curve can be used to supplement and check the entropy relationship. The value for λ^o_{OH-}, 197.6, is much higher than that predicted by the curve, viz. 83 $ohm^{-1}cm^2$. However, the mobility of H_3O^+ is about 90 $ohm^{-1}cm^2$ and the mobility of H+ and OH- are abnormal, having

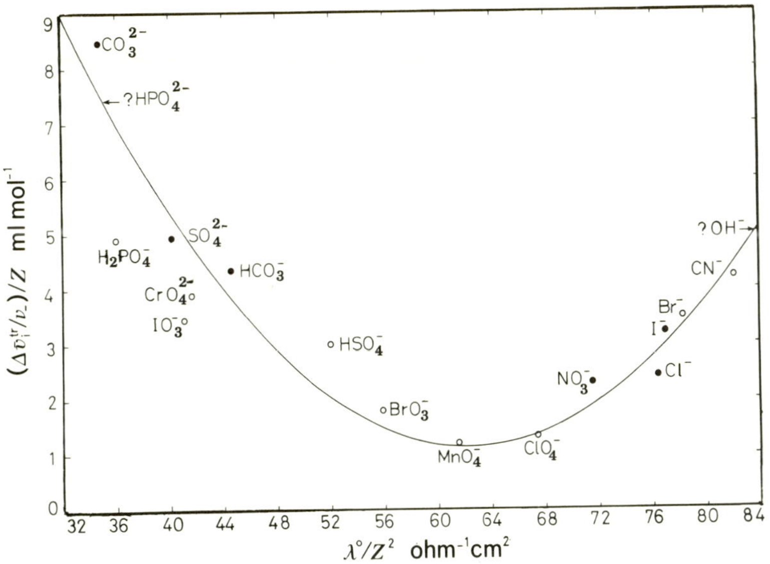

FIGURE 4.5

Volume of transfer function $(\Delta\bar{v}_i^{tr}/\nu_- z_-)$ versus the limiting ionic equivalent-conductivity (λ^o/z_-^2). Line represents the eq $y = 34.2 - 1.06x + 8.48\times10^{-3}x^2$ ($\pm$ 0.5 mlmol^{-1}), eq 4.42.3

● experimental

○ predicted from eq 4.42.1

---

been interpreted as being due to a proton-transfer process (Robinson and Stokes 1959, p121).

It does not appear, on the evidence of the Figure 4.5 and eq 4.42.1 that an ion such as PO_4^{3-} (1/3 λ^o = 92.8ohm^{-1}cm^2) is likely to fit very easily into a seawater environment. The $\Delta\bar{v}_i^{tr}/\nu_-$ predicted from Figure 4.5 is about 30mlmol^{-1} and eq 4.42.1 predicts an even higher value (around 40mlmol^{-1} - Leyendekkers 1974). This is, of course, what is expected from the pH.

The curve in Figure 4.5 is given by

$$(\Delta\bar{v}_i^{tr}/\nu_-)/z_- = 34.2 - 1.06(\lambda^o/z_-^2) + 8.48\times10^{-3}(\lambda^o/z_-^2)^2$$

$$(\text{with } \sigma = \pm0.5\text{ml mol}^{-1}) \quad (4.42.3)$$

As for the transfer volume at constant ionic strength, $\Delta\bar{v}_i^{tr}$ may be expressed in terms of the partial molal volume of the solvent, water. Since, from eq 4.4.8 and eq 4.2.21 (omitting δ')

$$\bar{v}_i^{sw} = \phi_i^o + 2I\partial\phi_i/\partial I - t_iI + (w_i/W)(I\partial\Phi_v/\partial I - I\partial\Delta\Phi_v/\partial I) + \delta \quad (4.42.4)$$

where $\delta = \Delta\phi_v + m_s(\partial\Delta\phi_v/\partial m_i)_{m_j}$

and since $2I\left[\frac{\partial\phi_i}{\partial I}\right] = s_iI^{\frac{1}{2}} + 2t_iI$

from eqs 4.42.4, 4.2.6 and 4.41.3, 4.41.4,

with $n_w = 55.51$. (The subscript v of ϕ_{v_i} is omitted here for convenience)

$$\Delta\bar{v}_i^{tr} = \bar{v}_i^{sw} - \phi_i^o = (2w_in_w/I)(M_w/d_w - \bar{V}_w^i) - t_iI + (n_ww_i/I)(M_w/d_w - \Sigma_k(I_k/I)\bar{V}_w^k) + \delta \quad (4.42.5)$$

with $\underline{k}$ the general representative of the seawater components, so that

$$\Delta\bar{v}_i^{tr} = (n_ww_i/I)(3M_w/d_w - 2\bar{V}_w^i - \Sigma_k(I_k/I)\bar{V}_w^k) - t_iI + \delta \quad (4.42.6)$$

(with $t_i \simeq b_{v_i}/w_i$).

Thus, apart from $m_s(\partial\Delta\Phi_v/\partial m_i)m_j$, which seems most sensitive to the anion of the salt, the value of $\Delta\bar{v}_i^{tr}$ will largely depend on (1) the pm volume changes of the solvent due to the dissolution of the salt $\underline{i}$, and the averaged effect of all the seawater salts on water, and (2) the deviation terms in the binary solution. Terms other than t_i may be included (i.e. δ' cannot be ignored). ϕ_{MgSO_4}, for example, is given by (see Table 4A.4 , Appendix 4).

$$\phi_{MgSO_4} = \phi^o_{MgSO_4} + sI^{\frac{1}{2}} + tI + r_1I^{3/2} + r_2I^2 \quad (4.42.7)$$

so that the additional term $-(2r_1I^{3/2}+3r_2I^2)$ must be added to $\Delta\bar{v}_i^{tr}$. The deviation term for $MgSO_4$ (at $I=0.723$ molkg^{-1}, or S = 35‰, 25°C) is about 8 compared with -0.8 $ml\,mol^{-1}$ for $m_s(\partial\Delta\Phi_v/\partial m_i)_{m_j}$, whilst for simple monovalent salts it is negligible.

So far, the emphasis has been on the pmv's of the components of the solution (the water and salts). Non-thermodynamic assumptions have generally to be made when the corresponding ionic properties are used. However, as shown by Millero (1969, 1971, 1974) consideration of the ionic volumes can provide useful insights into solution structure and equilibria in multicomponent electrolyte solutions such as seawater. It is now intended therefore to outline some relationships and ideas on ionic volumes that should be useful in the study of seawater.

4.5 The partial molal volumes of ions.

At infinite dilution the pmv of an electrolyte, *i*, is identical with the amv, i.e.

$$\bar{V}_i^o = \phi_{V_i}^o$$

and ion-solvent interactions are dominant. Since the principle of additivity applies

$$\bar{V}_i^o = \phi_{V_i}^o = \nu_+\bar{V}_+^o + \nu_-\bar{V}_-^o \qquad (4.5.1)$$

and if one of the ionic volumes is known, by measurement (Zana and Yeager 1966, 1967), or by some reasonable convention, then the values of the rest follow. The relationship of the ionic volumes to other properties of the ion, such as mobility, size, mass, etc., can then be studied. A large amount of work has been done along these lines and the interested reader is referred to the review by Millero (1971) and the other references listed at the end of this Chapter.

The overall results of these studies indicate that, for simple monovalent ions, using the symbols commonly employed,

1. $\bar{v}^o_{H+}$ in water at 25°C is about $-5.0\ \text{mlmol}^{-1}$; the experimental value is -5.4mlmol^{-1}, derived from ultrasonic vibration potentials.

2. $$\bar{v}^o_{ion} = \bar{v}^o_{int} + \bar{v}^o_{elect} = Ar^3 - Bz^2/r \qquad (4.5.2)$$

where int represents "intrinsic", elect represents "electrostrictive", r is the radius of the ion in solution, z the charge and A and B are constants. The value of r is most simply taken as the crystal radius, but the value of Ar^3 does not then correspond to $2.52r^3$ (Table 4.10) the crystal volume. The difference between $\bar{v}^o_{int}$ and $\bar{v}^o_{cryst}$ is usually accounted for by assuming some positive disorder effect. Such an effect has been allowed for generally in two ways

1. Modification of r on a theoretical or empirical basis, e.g. by adding a constant, a, to r (i.e. assuming that the increase of the radius in solution is the same for all ions), or modifying A by making a = const.r (i.e. assuming that the increase is proportional to the radius).
2. Separating Ar^3 into two components, the crystal volume $2.52r^3$ and $\bar{v}^o_{disord}$ or $A'r^2$ where r is the crystal radius.

The various theories and results are outlined fully by Millero (1971). However, it is of interest to compare values of Ar^3 derived from the various interpretations with what might be considered $\bar{v}^o_{int}$ in very concentrated solutions, i.e. as the solution-to-crystal transition point is approached.

When a crystalline solid is expanded from its normal equilibrium volume, the cohesive forces reach a maximum at the "critical disruptive volume", V* which corresponds to the transition from solid to liquid. Scott (1931) has estimated V* for most of the alkali halides, so that a

TABLE 4.10 Comparison of estimates of $\bar{v}^o_{int}$ of ions (summed) in aqueous salt solutions 25°C 1-atm

Salt	V*	ϕ*	$\bar{v}^o_{cryst}$ [x]	a	b	c
LiCl	32.4	25.3	15.5	27.6	26.9	37.0
LiBr	39.1	30.0	19.2	34.3	32.3	43.2
LiI	51.2	39.2	25.9	46.3	42.8	54.0
NaCl	41.6	26.9	17.2	30.5	30.3	41.6
NaBr	49.0	31.5	20.8	37.2	35.7	47.9
NaI	61.1	40.6	27.6	49.2	46.1	58.7
KF	37.2	24.0	12.3	21.9	23.7	29.8
KCl	56.6	36.3	20.9	37.3	36.8	45.4
KBr	64.7	41.3	24.6	43.9	42.2	51.6
KI	78.0	51.0	31.3	55.9	52.6	62.4
RbCl	63.8	40.7	23.1	41.3	40.3	54.2
RbBr	72.7	46.3	26.9	47.9	45.7	60.5
RbI	86.7	55.9	33.5	59.9	56.2	71.2
CsCl	73.4	47.2	27.1	48.4	46.4	61.4
CsBr	82.6	52.8	30.8	55.1	51.8	67.7
CsI	97.5	62.7	37.6	67.0	62.3	78.5

x $\bar{v}^o_{cryst} = (4\Pi N/3)r^3 x10^{-24} = 2.52r^3$

a $\bar{v}^o_{int} = (1.213)^3 2.52r^3 = 4.50r^3$ (Mukerjee 1961)

b $\bar{v}^o_{int} = 2.52r^3 + 3.15r^2$ (Conway *et al* 1965)

c $\bar{v}^o_{int} = 2.52(r + 0.55)^3$ (Glueckauf 1965)

maximum value of the amv, ϕ^*_v, can be calculated from eq 4.2.10, viz.

$$\phi^*_v = \phi^o_v + s_v(1000/V^*)^{1/2} \quad (4.5.3)$$

assuming that V* corresponds to the hypothetical maximum concentration in the supersaturated salt solution. Table 4.10 lists values of ϕ^*_v (Harned and Owen 1958, Table 8-5-2) and $\bar{v}^o_{int}$ calculated by summing the ionic values, calculated on the basis of various theories, as indicated.

The closest agreement to ϕ_v* is given by the set of values (column b) derived from the theory of Conway, *et al* (1965) who estimated the disordered volume as $3.15r^2$. The radius of the ion in solution is taken as the crystal radius in this theory. This value of the radius, on present evidence, seems the most appropriate one (Millero 1971). Stokes and

Robinson (1957) have measured the void space for the packing of spheres and hence estimated that $\bar{v}^o_{int} = 4.35r^3$. This gives values of $\bar{v}^o_{int}$ close to those of column b in Table 4.10. On the assumption that the values in column b represent the best estimates of $\bar{v}^o_{int}$ then a value for A (eq 4.5.2) around 4.3-4.4 seems appropriate from various theoretical approaches. The empirical value at 25°C estimated by Curthoys and Mathieson (1970) for monovalent ions (plus Mg, Sr and La halides) was 4.46 (Pauling radii, as used in above calculations); see also Table 4.11. The second term of eq 4.5.2 accounts for the electrostriction partial molal volume of the ion, the component due to the compression of the solvent by the ion and proportional to z^2/r. The value for the constant B, derived on a semi-empirical basis, will depend on the form chosen for the first term of the equation and the interpretation of r. For aqueous solutions empirical B estimates range from 8.0 to 11.0 and theoretical ones from 4-6 (Millero 1971).

$\bar{v}^o_{elect}$ has the same form as the electrostatic entropy of solvation $\Delta\bar{s}$, which represents quantitatively the profound effect that an electrical charge has on the interaction of a solute with water. It seems reasonable to suppose that $\bar{v}^o_{elect}$ would be proportional to $\Delta\bar{s}$ and that its temperature variation would be parallel. This would also apply if $\bar{v}^o_{elect}$ is made up of a constant term and a structural term (below 100°C) as assumed by Curthoys and Mathieson (1970). The entropy term is represented by (Pitzer and Brewer, 1961, p523).

$$\Delta\bar{s} = (166\times10^3(\partial D/\partial T)/D^2)\,\frac{z^2}{r_0} \quad \text{caldeg}^{-1}$$
$$= B'z^2/r_o \qquad (4.5.4)$$

derived by differentiating the Born equation for the free energy of a charged sphere in a continuous ideal dielectric medium. (See Chapter 3). Where, D is the dielectric constant and r_o the radius of the spherical

TABLE 4.11 Comparison of the electrostatic entropy of solvation and $\bar{V}^{o}_{elect}$ at various temperatures

Temp ^{o}C	$-\frac{10^6}{D}$ $(\partial D/\partial T)^*$ deg^{-1}	D	B' eq 4.5.4 Å $caldeg^{-1}$	B^X_G (eq 4.5.2) Å $mlmol^{-1}$	B^X_P (eq 4.5.2) Å $mlmol^{-1}$	A^X_G (eq 4.5.2) $(Å)^{-3}$	A^X_P $mlmol^{-1}$
0	4657	88.15	8.8	8.96	10.18	3.97	4.23
5	4641	86.12	8.9	8.74	10.08	4.06	4.32
15	4609	82.23	9.3	8.71	9.94	4.21	4.48
25	4579	78.54	9.7	8.66	9.90	4.30	4.58
25**				7.97	8.54	4.30	4.46
35	4553	75.04	10.1	8.78	10.03	4.41	4.68
45	4531	71.70	10.5	8.90	10.16	4.48	4.75
55	4516	68.53	10.9	9.10	10.43	4.53	4.81
65	4510	65.51	11.4	9.32	10.37	4.57	4.79

* (Harned and Owen 1958, p161, Table 5-1-3)
X Coefficients of eq 4.5.2 (Curthoys and Mathieson 1970)
** data for the halides of rubidium, cesium, magnesium, strontium and lanthanum included

boundary between the ion and the dielectric. Since water molecules are comparable in size to ions and the electrically positive side can be expected to be quite different from the negative side, the crystal radii can hardly be expected to be the appropriate value for r_o. However, if r_o is proportional to the crystal radius, which seems to be the case, for the simple monovalent ions, then the difference can be absorbed into B' just as it was for the Ar^3 term. B' might therefore be useful in terms of an indicator of solvent differences, especially since the temperature data are more readily available for different solvents than the corresponding pressure data.

Values of B' are compared with the semi-empirical values of B derived by fitting very accurate data for the chlorides, bromides, the iodides of sodium, potassium, calcium and barium to eq4.5.2 (Curthoys and Mathieson 1970). Both the Pauling and Goldschmidt radii were used, designated by suffixes P and G, respectively, in Table 4.11. Values of A are also shown.

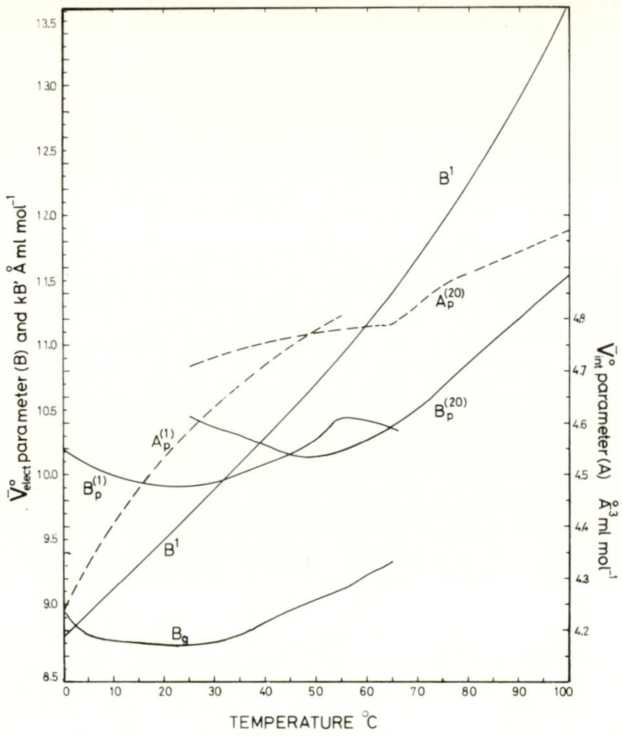

FIGURE 4.6

Parameters of $\bar{v}^o = Ar^3 + Bz^2/r$ (eq 4.5.2), and $B' = (166x10^3(\partial D/\partial T)/D^2)$ Åcaldeg^{-1} see eq 4.5.4. $k = 1$ degmlmol^{-1}cal^{-1} A and B values from Curthoys and Mathieson (1970).

The value of ϕ^* is practically independent of temperature, and the value of $\bar{v}^o_{int}$ (the intrinsic volume of the ion) would be expected to be constant with temperature. However, the empirical value of A changes with temperature. This suggests that A is absorbing some of the electrostriction term, possibly portion of the structural term, $\bar{v}^o_{st}$ defined by

$$\bar{v}^o_{elect} = B''z^2/r + \bar{v}^o_{st} \tag{4.5.5}$$

This conclusion is supported by the temperature trends of A which closely parallel those of $\Delta\bar{s}$, Figure 4.6. Values at 20 bars are included to extend the range to 100°C. At these higher temperatures (and pressure) the A and B versus temperature curves are similar. This is not a feature of the lower temperature curves. The actual values of B and B' are quite close, and within the range of the estimates for the two radii (Pauling and Goldschmidt). Such comparisons might be useful for seawater solutions, and the corresponding entropy term for seawater will be considered in the next section.

In the above outline, the aim has been to emphasize the concepts of $\bar{v}^{o}_{int}$ and $\bar{v}^{o}_{elect}$ as reasonable starting points for interpretation. For purposes of utility, eq 4.5.2 may be considered as a useful semi-empirical equation.

So far, only simple monovalent ions have been considered, although eq 4.5.2 seems appropriate for higher charge systems, e.g. see Table 4.11, especially if the first term is treated as a two component term containing $\bar{v}^{o}_{disord}$ as a positive contribution. Inclusion of the higher charge systems results in a decrease of B (at 25°C to 8-8.5). Millero (1969) assumes values of A and B of $4.48(\text{Å})^{-3}\text{ml mol}^{-1}$ and $8.0\ \text{Å}\ \text{ml mol}^{-1}$ respectively for pure water. Apart from the above reasons for variations in A and B it is probable that these parameters should be different for anions and cations. $\bar{v}^{o}_{elect}$ is probably larger for anions than for cations of the same size and charge, and the effect of temperature is different (Millero 1971).

4.51 Seawater solutions.

Because of its simple form, eq 4.5.2 is especially useful for summarising the effects of various solvents, so that predictions of $\bar{v}^{o}_{i}$ for electrolyte $\underline{i}$, dissolved in such solvents, can be made. Seawater has often been treated as a special solvent, say for S= 45‰, mixed with water. The pmv of electrolytes in seawater could then be considered as $\bar{v}_i^o$ for a particular salinity, and be compared with $\phi_{v_i}^{\ o}$. This device is, however, not very satisfactory theoretically, since some of the comparative interactions involved (i.e. comparing different electrolytes) are between the ions. It can be seen from eq 4.4.8 and using

$$\phi_{v_i} = \phi_{v_i}^{o} + s_v^L (I/w_i)^{\frac{1}{2}} + (I/w_i) b_v \qquad (4.2.18)*$$

that, for the electrolyte, i, at constant I,

$$(1/w_i)(\bar{v}_i^{sw} - \phi_{v_i}^{o}) = s_v^L I^{\frac{1}{2}}/w_i^{3/2} + (I/w_i)(b_v/w_i) + \delta/w_i + E \qquad (4.51.1)$$

where E is constant for a given concentration.

with $\delta = \Delta\Phi_v + m_s(\partial\Delta\Phi_v/\partial m_i)_{m_j}$

$= \Delta\Phi_v + Q$

and, $E = (1/W)(I\partial\Phi_v/\partial I - I\partial\Delta\Phi_v/\partial I) \qquad (4.51.2)$

or, $$\Delta\bar{v}_i^{tr}/w_i = kI^{\frac{1}{2}} + (I/w_i)(b_v/w_i) + \delta/w_i + \text{constant} \qquad (4.51.3)$$

Values of $k(=s_v^L/W^{3/2})$, b_v, and $m_s(\partial\Delta\Phi_v/\partial m_i)_{m_j}$ are given in Tables 4.2, 4.1, and 4.7 (average values) respectively.

Values of the various terms, for some seawater components, are listed in Table 4.12. The value of the constant, E, (eq 4.51.2) (I = 0.723molkg^{-1}) is 0.74mlmol^{-1}. The last column of Table 4.12 indicates that the deviations from this value are not very large. This illustrates the consistency between the various relationships and the data.

In terms of eq 4.5.2, since

$$\bar{v}_i^o = A(\nu_+ r_+^3 + \nu_- r_-^3) - B(\nu_+ z_+^2/r_+ + \nu_- z_-^2/r_-) \qquad (4.51.4)$$

and assuming the equation for $\bar{v}_i^{o\ sw}$ has the same form, with A_{sw} and B_{sw} replacing A and B, then

$$\Delta\bar{v}_i^{tr} = -(B_{sw}-B)(\nu_+ z_+^2/r_+ + \nu_- z_-^2/r_-) \qquad (4.51.5)$$

if the intrinsic volume is unchanged. In practice, a small deviation term

TABLE 4.12 Component terms of the volume of transfer $\Delta\bar{v}_i^{tr}$ (S = 35‰, I = 0.723molkg^{-1}) 25°C 1-atm mlmol^{-1}

Salt	$\Delta\bar{v}_i^{tr}$	$w_i k I^{1/2}$	$\frac{Ib_v}{w_i}$	$Q+\Delta\Phi_v$	$\frac{1}{w_i}(\Delta\bar{v}_i^{tr}-(a+b+c))$ **
		a	b	c	
NaCl	2.283	1.588	0.014	-0.057	0.7
$CaCl_2$	4.142	4.765	-0.304	-1.087	0.3
	4.35 *		-1.496*		0.7
$MgCl_2$	4.963	4.765	-1.518	0.073	0.6
	5.600*		-1.352*		0.7
Na_2SO_4	9.610	4.765	0.714	1.513	0.9

** should equal $(1/W)(I\partial\Phi_v/\partial I - I\partial\Delta\Phi_v/\partial I)$, i.e. 0.74

$\Delta\bar{v}_i^{tr} = (\bar{v}_i^{sw}-\phi_{v_i}^{o})$ Tables 4.3 and 4.8, * Tables 4.1a and 4.8

a Table 4.2, $k = (s_v^L/w_i^{3/2})$

b Table 4.1, * Table 4.1a

c Leyendekkers (1974), see eq 4.4.10, Table 4.7

arising from changes in $\bar{v}^o_{disord}$ and/or $\bar{v}^o_{st}$ (the structural contribution) might be expected. The ion-ion interations represented by the Harned Coefficients have been shown to be linear functions of the conventional ionic entropy for many systems (Leyendekkers 1971). The value of Q is also related to this quantity, as is $\Delta\bar{v}_i^{tr}$ (Section 4.4). The conventional ionic entropy, on the other hand, is a function of z^2/r, viz. for simple ions

$$\bar{s}^o_{298} = -5.5z + 3/2R\ln M + C_1 - C_2z^2/r$$

C_1 and C_2 are empirical constants, M is the atomic mass, and r Pauling's univalent ion radius. See Chapter 3 (Section 3.72 and Tables 3.4 and 3.5). It is not surprising therefore that a correlation between $\Delta\bar{v}_i^{tr}$ and z^2/r exists.

Millero (1969) has developed an equation for predicting the pmv of ions in seawater on the basis of eq 4.5.2. He first established that the volume of the hydrogen ion was -3.7mlmol^{-1} (cf. -5mlmol^{-1} in water).

This allowed the pmv of the salts (experimental values) to be split into the ion contributions in a reasonable way. The $\bar{v}^{o\,sw}_{ion}$ values were then fitted to eq 4.5.2. A was found to be $4.58\pm0.02(\text{Å})^{-3}\text{mlmol}^{-1}$ and B = $7.5\pm0.1\text{Å mlmol}^{-1}$ at 25°C, S = 35.1‰. He also derived the semi-empirical equation

$$\Delta\bar{v}^{otr}_{ion} = 0.37_2 z^2/r + 0.83_3 \qquad (4.51.6)$$

Millero also considered the effect of changing the salinity, i.e. the solvent, over the range 30-40‰, the final equation for 25°C is

$$\Delta\bar{v}^{otr}_{ion} = 0.0105\underline{S}(z^2/r) + 0.0237\underline{S} \qquad (4.51.7)$$

and

$$\Delta\bar{v}^{otr}_{i} = a((nz^2/r)_+ + (nz^2/r)_-) + (n_+ + n_-)b \qquad (4.51.8)$$

where n represents the number of ions, a = $0.0105\underline{S}$ and b = $0.0237\underline{S}$. The temperature effect will be considered in Chapter 5.

The value of E, in eq 4.51.3 at 25°C, S = 35‰ was 0.74 which corresponds very well to the value of 0.83 in eq 4.51.6. The value of E, at 25°C, as a function of concentration is given by

$$E = (1/W)(I\partial\Phi_V/\partial I - I\partial\Delta\Phi_V/\partial I)$$
$$= (1/W)(1.155I^{\frac{1}{2}} + 0.04I) \qquad (4.51.9)$$

from Leyendekkers (1974). This is equivalent to

$0.0214\underline{S}$ approximately, for S = 30-40‰,

which is consistent with eq 4.51.7 (i.e. w_ix E is nearly the same as the last term of eq 4.51.8).

As noted previously, strong ionic ± interactions, such as ionpairing, may be interpreted as interactions between water and the anion (i.e. the electrostatic ordering of water is disrupted by the interactions). This accounts for the equivalence between $\Delta\bar{v}^{o\,sw}_{elect}$ and the ion interaction terms of eq 4.51.3.

It is of interest to compare the first term of eq 4.51.6 with the

TABLE 4.13 Dielectric constants * of aqueous electrolyte solutions
Coefficient of the equation

$$D_i = 78.54 - 2A_i\underline{c}$$

$25^{o}C$, concentration range 0.5-2.0 mol $\ell.^{-1}$

Electrolyte i	A_i ℓmol^{-1}	Electrolyte i	A_i ℓmol^{-1}
$BaCl_2$	14	$MgCl_2$	15
HCl	10	NaCl	5.5
KCl	5	NaF	6
KF	6.5	NaI	7.5
KI	8	NaOH	10.5
$LaCl_3$	22	Na_2SO_4	11
LiCl	7	RbCl	5

* Parsons (1959)

change in the entropy of solvation in going from water to seawater. The dielectric constant of water is decreased in the presence of an ionic solute. The effects of a number of salts are listed in Table 4.13, for 25°C. The effect of temperature on the coefficient A_d is apparently not known. However, an estimate could be made from $B_{sw}-B$ assuming that the difference in the two parameters is the same as, or close to, the difference in the entropy coefficient.

Since,

$$D_{sw} = D - 2A_d I \qquad (4.51.10)$$

where A_d is the averaged value of NaCl, $MgCl_2$ and Na_2SO_4, weighted according to the relative compositions, i.e.

$$A_d = 0.83A_{NaCl} + (0.11/3)A_{MgCl_2} + (0.06/3)A_{Na_2SO_4}$$

Thus, $$\partial D_{sw}/\partial T = \partial D/\partial T - 2I\partial A_d/\partial T$$

and $B_{sw}-B = 166x10^3((\partial D_{sw}/\partial T)/D_{sw}^2 - (\partial D/\partial T)/D^2)$

$= 0.106B' - 431.6\ \partial A_d/\partial T$ (4.51.11)

If $B_{sw}-B = -0.4,\quad \partial A_d/\partial T = 3x10^{-3}(kgdeg^{-1}mol^{-1})$

Equation 4.51.6 strictly only applies to simple monovalent ions. However, as for binary solutions, the relationship makes a useful reference to assess the behaviour of more complex ions and to interpret the solution state. The measured value of the pmv of an electrolyte in seawater is the sum of the fractional contributions of all the species involving the ions of the particular electrolyte (eq 4.1.3). If the electrolyte is fully dissociated, then the ionic volumes can be estimated unambiguously from the experimental results (assuming, of course that the estimate of $\bar{v}_{H+}$ in seawater is close to the true value). In this case, the volumes are well represented by eq 4.51.6.

If we assume that, in general, ions obey eq 4.51.6 then we can obtain the values for any ions provided we know the appropriate value for r. However, in order to estimate the pmv of the salt, further assumptions and estimates, mainly concerning ion-pairing, have to be made. In other words, the ionic values derived from eq 4.51.6 are those for the "free" ions, i.e. the values to be expected when dissociation is complete.

The values of ionic volumes estimated on the basis of different values for $\bar{v}^o_{H+}$ in water and seawater are listed in Appendix 4. The different sets of assumptions and interpretations give widely different results and this limitation should always be kept in mind when using ionic volumes. It is of interest to note that Bramhall (1963) showed that $\bar{v}^o_{ion} - \bar{v}^o_{cryst}$ is linearly related to $(\underset{\sim}{B} - \bar{v}^o_{cryst}/400)$ where $\underset{\sim}{B}$ is the viscosity coefficient of the ion, which characterises its short range solvating properties. The values of $\bar{v}^o_{ion}$ were based on $\bar{v}^o_{H+} = 0.2 mlmol^{-1}$ and the correlation does not apply otherwise.

4.52 Estimate of ionic volumes in seawater from $\Delta\bar{v}_i^{tr}$

If it is assumed that the change in volume of an electrolyte in going from water to seawater is due to the anion, this simplifies the estimates of the ionic values. In addition, for "average" cations (e.g. Na^+, K^+, Ca^{2+}, Mg^{2+}, Sr^{2+} and other simple divalent cations), this assumption is compatible with the data (within the limits of the experimental errors). The values given in Appendix 4, Table 4A.2 (convention L) have been compiled on this assumption, and with $\bar{v}^o_{H+}$ taken as the experimental value of $-5.4 ml mol^{-1}$ (at 25°C).

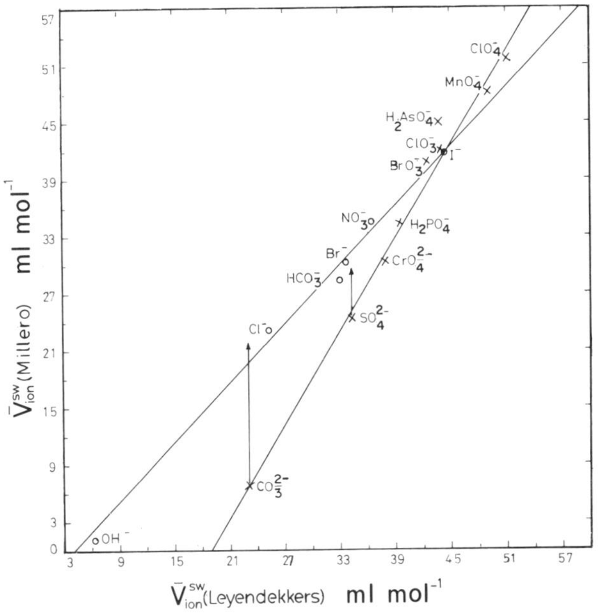

FIGURE 4.7

Comparison of the partial molal volumes of anions in seawater using the M and L conventions (Table 4A.2)
The arrows indicate the change in the M values when ion-pairing is taken into account.

In Figure 4.7 the values of $\bar{v}^{sw}_{anion}$ estimated with the L convention are compared with the corresponding values on the M scale. The values for the M scale are assumed to be for the unassociated ions. Ions such as CO_3^{2-} and SO_4^{2-} and most of the other oxyanions fall on the same line. These ions form ion-pairs and an additional volume increment has to be added to account for this on the M scale. The values estimated by Millero (1969) and Millero and Berner (1972) for SO_4^{2-} and CO_3^{2-} respectively are shown by the arrow tips. The agreement with the L scale is quite good. Figure 4.7 has been used as a guide to estimate ionic values on the L scale. These estimates are given in Table 4A.2 (bracketed values).

To estimate the values of $\bar{v}^{sw}_{anion}$ at other salinities, it is only necessary to add $\Delta\bar{v}^{tr}_i$ to ϕ^o_v (anion). The method of calculating $\Delta\bar{v}^{tr}_i$ has been outlined in Section 4.42 and 4.51 (see eqs4.42.4 and 4.51.3). The changes in $\Delta\bar{v}^{tr}_i$ for a change in salinity of 24.6-48‰ are shown in Table 4.14. for a few electrolytes to illustrate the trends.

4.53 Hydration numbers from partial molal volumes.

Since hydration numbers are useful for estimating the activity coefficients of electrolytes and ions, it is of interest to briefly note some methods of estimating hydration numbers from data on partial molal volumes. For more extensive information, see Millero's (1971) review on "The molal volumes of electrolytes" and the references therein, especially those of J. Padova. See, also, Millero, Ward, et al (1974) and Millero and Masterton (1974).

Stokes and Robinson (1957) have estimated hydration numbers, n, from

$$n = (4.35r^3 - \bar{v}^o_{ion})/18 \qquad (4.53.1)$$

i.e. $(\bar{v}^o_{int} - \bar{v}^o_{ion})/18$ or $\bar{v}^o_{elect}/18$. As noted previously, $4.35r^3$ probably gives good estimates of $\bar{v}^o_{int}$ so that n should be a reasonable estimate of the actual hydration number for dilute solutions.

TABLE 4.14 The effect of concentration on the volume of transfer of electrolytes from water to seawater solutions.
25°C 1 bar

	$\bar{v}_i^{sw}$	$\bar{v}_i$	$\Delta\bar{v}^{tr}$ *	S‰	Increase
$CaCl_2$	21.77	17.858	3.912	24.6	
	22.28		4.422	48	0.51
KCl	28.97	26.5	2.47	24.6	
	29.48		2.98	48	0.51
KNO_3	40.41	38.18	2.23	24.6	
	41.19		3.01	48	0.78
$MgCl_2$	19.22	14.657	4.563	24.6	
	20.11		5.453	48	0.89
$MgSO_4$	1.73	-7.157	8.887	24.6	
	4.26		11.417	48	2.63
NaCl	18.54	16.617	1.923	24.6	
	19.32		2.703	48	0.78
Na_2SO_4	19.60	11.420	8.18	24.6	
	22.79		11.37	48	3.19

* See eq 4.42.4 and 4.51.3.

Padova (1964) derived equations giving the solvation number, n, of a salt as a function of concentration, e.g.

$$n = n^o(1 - (s_v/(v_{in}-\phi_v^o))c^{1/2}) \qquad (4.53.2)$$

with n^o the solvation number at infinite dilution of the salt and v_{in} is approximately the molar volume in the crystal. This equation was valid up to saturation where, in most cases, n corresponded to the number of solvent molecules in the crystallohydrate of the salt. Also the equation

$$\phi_v^o = v_{in} - (-\phi_k^o)(s_v/s_k) \qquad (4.53.3)$$

was derived, from which v_{in} can be calculated provided the necessary parameters are available. The values of v_{in} from this equation

TABLE 4.15 Comparison of crystal molar volumes of salts and v_{in} calculated from eq 4.53.3

Salt	v_c *	v_{in} **	Salt	v_c *	v_{in} **
LiCl	20.5	22.7	KI	53.2	52.4
LiBr	25.1	29.3	RbCl	43.2	40.1
LiI	33.0	39.1	RbBr	49.4	46.7
NaF	15.1	14.1	RbI	59.8	56.5
NaCl	27.0	27.1	CsCl	42.3	46.2
NaBr	32.1	33.7	CsBr	47.9	52.8
NaI	40.9	43.5	CsI	57.6	62.6
NaOH	18.8	18.0[X] (12.8)[XX]	HCl		20.7
			Na_2SO_4		52.7[X]
KF	23.4	23.1			
KCl	37.5	36.0			
KBr	43.3	42.6	Na_2CO_3		32.9

* Handbook of Chemistry and Physics (Chemical Rubber Publishing Company, Cleveland, Ohio 1960-61).

** Parsons (1959)

X McDevit and Long (1952)

XX Marks (1960)

are generally within 10-15% of the anhydrous salt volume. It should be noted that values obtained for v_{in} are independent of the solvation model, and of the electrostriction values. Table 4.15 lists some representative values. The solvation numbers may therefore be obtained from

$$n = n^o(1- (s_k/(-\phi_k^o))c^{1/2}) \tag{4.53.4}$$

Another equation from Padova (1964) which might be useful is,

$$n = -\phi_k d_w/\beta_w M_w \tag{4.53.5}$$

where β_w is the isothermal compressibility, d_w the density, and M_w the molecular mass of water.

Millero, Ward, <u>et al</u> (1974) derived a similar relationship by

TABLE 4.15a Hydration numbers for ions calculated from volume and compressibility data

for 25°C unless otherwise indicated.

Cations

Ion	Temp.	a	b
Ag(+)			2
Ba(2+)			8
Ca(2+)			8
Cd(2+)			5
Cs(+)			1
Er3+)			12
K(+)			2
La(3+)			11
Li(+)			2
Mg(2+)	$0^{o}C$	8.4	
	15	7.7	
	25	7.7	8
	30	7.9	
	45	7.8	
Na(+)	$0^{o}C$	5.4	
	15	4.3	
	25	3.7	3
	30	3.4	
	45	3.2	
Nd(3+)			12
NH_4(+)			0
Rb(+)			2
Tl(+)			2
Yb(3+)			13
$MgSO_4^o$ (ion pair)			
	$0^{o}C$	15.9	
	15	15.9	
	25	15.2	
	30	14.9	
	45	14.3	

Anions

Ion	Temp.	a	b
Br(-)			2
BrO_3(-)			3
Cl(-)	$0^{o}C$	2.8	
	15	2.3	
	25	2.0	3
	30	1.8	
	45	1.7	
ClO_3(-)			2
ClO_2(-)			1
CNS(-)			0
CH_3COO(-)			4
CO_3(2-)			15
CrO_4(2-)			13
F(-)			7
HCO_3(-)			2
I(-)			0
IO_3(-)			7
NO_3(-)			1
OH(-)			8
SO_4(2-)	$0^{o}C$	10.0	
	15	9.6	
	25	8.6	11
	30	8.1	
	45	7.9	

a Millero, Ward et.al. (1974)

b Padova (1964)

differentiating the following equation w.r.t. pressure,

$$\bar{V}^o_{elect} = n(\bar{V}^o_E - \bar{V}^o_B) \qquad (4.53.6)$$

where $\bar{V}^o_E$, $\bar{V}^o_B$ are the molar volumes of the electrostricted and bulk water molecules, respectively. This equation is, of course, the basis of eq 4.53.1. The intrinsic volume, $\bar{V}^o_{int}$, and the hydration number, n, were assumed to be independent of pressure. The equation so derived was

$$n = -\phi^o_k(\text{ion})/\beta^o_B\bar{V}^o_B \qquad (4.53.7)$$

where B refers to the bulk water.

Values of n calculated from this equation for a number of temperatures are listed in Table 4.15a. These values are apparently in good agreement with values from other types of equations and data (Millero, Ward, et al 1974). The ionic hydration numbers, n^o, given by Padova, for 25°C are also listed.

The n values for the salts are obtained from the ionic sums, viz. $n(\text{salt}) = (\nu_+ n_+ + \nu_- n_-)$. These values agree (to 0.2 or better) with those calculated directly from eq 4.53.7, using the ϕ^o_k values for the salts.

4.6 Seasalt

The approach adopted for dealing with the "solute" in seawater is outlined in general terms in Chapter 1. For the partial molal volume, and its temperature and pressure derivatives, this approach may be summarised.

1. For convenience, seawater may be considered as a mixed electrolyte solution according to a suitable "recipe". Two recipes, of the many possibilities, are shown in Table 4.16. The first one, A, has been used as a basis for a number of studies (Leyendekkers 1973, 1974).

2. The contribution of each of the electrolytes, *i*, to the volume of seawater is summed to give the total effect of the solute, i.e. $\Sigma_i m_i \bar{V}^{sw}_i$, where m_i represents the number of moles of *i* and $\bar{V}^{sw}_i$ the partial molal

TABLE 4.16 Seawater as a mixture of electrolytes, Recipes for the combination of the constituents, the ions, of seawater.

S = 35‰

SOLUTION A*

Electrolyte	$\underline{m}_i$ molkg^{-1} water	f_i $\underline{m}_i/\Sigma_i\underline{m}_i$	ν_i	"Molecular weights"
NaCl	0.424310	0.796440	2	
Na_2SO_4	0.029245	0.054895	3	
$NaHCO_3$	0.002418	0.004539	2	$M_t = \Sigma_i f_i M_i$
NaF	0.000074	0.000139	2	= 68.4148 g
KCl	0.009412	0.017666	2	
KBr	0.000854	0.001602	2	$M_t' = 68.0811$ g
$MgCl_2$	0.055211	0.103630	3	(corrected M_t)**
$CaCl_2$	0.010707	0.020098	3	
$SrCl_2$	0.000093	0.000173	3	
H_3BO_3	0.000436	0.000818	2	

SOLUTION B

Electrolyte	$\underline{m}_i$ molkg^{-1} water	f_i $\underline{m}_i/\Sigma_i\underline{m}_i$	ν_i	"Molecular weights"
NaCl	0.485218	0.863370	2	
$MgSO_4$	0.029245	0.052037	2	
$Mg(HCO_3)_2$	0.001209	0.002151	3	$M_t = \Sigma_i f_i M_i$
NaF	0.000074	0.000132	2	= 64.8547 g
KCl	0.009412	0.016747	2	
KBr	0.000854	0.001520	2	
$MgCl_2$	0.024757	0.044051	3	$M_t' = 64.5358$ g
$CaCl_2$	0.010707	0.019051	3	(corrected M_t)**
$SrCl_2$	0.000093	0.000165	3	
H_3BO_3	0.000436	0.000776	2	

* based on the artificial seawater recipe of Kester et al. (1967).

** allows for the difference between the total salt per kg of seawater (S*) and salinity (S).

volume of $\underline{i}$ in the mixture. To remain consistent with binary solution data, a volume of seawater is chosen that contains 1 kg of water, so that the moles of ions or electrolytes are simply given by the familiar unit of molality, $\underline{m}$. In other words, the solvent is taken as water, the same as for a single electrolyte solution.

For convenience, assuming the ratios of the ions remain effectively constant with changing ionic strength or salinity, the sum may be expressed

$$\Sigma m_i \bar{v}_i^{sw} = (\Sigma m_i)\ \Sigma(f_i \bar{v}_i^{sw})$$
$$= m_s \bar{V}_s \qquad (4.6.1)$$

where $m_s = \Sigma m_i$ and $\bar{V}_s = \Sigma(f_i \bar{v}_i^{sw})$ with $f_i = m_i/m_s$

The quantity $\bar{V}_s$, thus represents the collective contribution of the salts per mole of salt, i.e. averaged over the salts of the chosen recipe. $\bar{V}_s$ will depend on the recipe chosen. However, for any two recipes, A and B, say.

$$\underline{m}_s^A \bar{V}_s^A = \underline{m}_s^B \bar{V}_s^B \qquad (4.6.2)$$

i.e. the different $\bar{V}_s$ will be proportional to each other and also to the average volume effect per mole of ion, $\bar{V}_s^I$, given by

$$\Sigma \underline{m}_i \bar{v}_i^{sw} / \Sigma m_{\underline{ion}} \qquad (4.6.3)$$

where ion represents any ion of seawater. For example, for solution A, for any component, i,

$$(\underline{m}_- + \underline{m}_+)_i = (\Sigma \underline{m}_{ion})_i = \nu_i \underline{m}_i = \nu_i f_i \underline{m}_s$$

$$\Sigma(\Sigma \underline{m}_{ion})_i = \underline{m}_s \Sigma \nu_i f_i$$
$$\bar{V}_s^I = \bar{V}_s^A / \Sigma \nu_i f_i \qquad (4.6.4)$$

Table 4.17 shows the values of $\Sigma\nu_i f_i$ for the two recipess. As $\Sigma \underline{m}_{ion}$ is independent of any recipe chosen, $\bar{V}_s^I$, is a more general quantity. Another general quantity is $\bar{V}_s/M_t'$ representing $\Sigma \underline{m}_i \bar{v}_i^{sw}/S(1-c)$. However, all the quantities are only averaged values for a solution containing 1 kg of water. For oceanographers, a more convenient weighting term might be "per gram of seawater". To convert to this unit, $\underline{m}_s \bar{V}_s$ is simply divided by the mass of seawater in grams. This is given by $(\underline{m}_s M_t + 1000)$ or $(1.0049S/(1-c) + 1000)$; the mass of water is 1000 g and the mass of salt is $\underline{m}_s M_t$ or $1.0049S/(1-c)$ g.

For purposes of estimating $\Sigma f_i \bar{v}_i^{sw}$, recipe A is chosen as convenient, so

TABLE 4.17 Conversion factors for obtaining various "averages" of the volume contribution of the solute ($\Sigma \underline{m}_i \bar{v}_i^{sw}/\bar{w}$) in a seawater solution (containing 1 kg water)

Designation	Weight, $\bar{w}$	General value	$\bar{w}$ Recipes, Table 4.16 Value for S=35‰ A	B
per mole of salts	$\Sigma \underline{m}_i = \underline{m}_s$	$\underline{m}_s = S/(1-c)M_t'$	0.532760	0.562005
per mole of ions	$\Sigma \underline{m}_{ion}$	$\underline{m}_s \Sigma \nu_i f_i =$ $\Sigma \nu_i \underline{m}_i$	$2.178796\underline{m}_s^A$ = 1.160776	$2.065418\underline{m}_s^B$ = 1.160776
per gram seawater	(mass of water + mass of salts) in g	$(10^3 + \underline{m}_s M_t)$ or $(10^3 + 1.0049S/(1-c))$	1036.447	1036.447
per "corrected" salt mass*	$S/(1-c)$	$\underline{m}_s M_t'$	36.2694	36.2694

* allows for the difference between total salt per kg seawater (S*) and salinity (S).

that the seasalt terms $\bar{V}_s$ and Φ_v in the following equations were estimated on the basis of this recipe. To convert to $\Sigma \underline{m}_i \bar{v}_i^{sw}$ the values of $\bar{V}_s^A$ are simply multiplied by $S/(1-c)68.0811$ or $\underline{m}_s^A$. All the other averages or means are obtained by dividing by the appropriate "weight", $\bar{w}$, Table 4.17.

The quantity $\underline{m}_s \bar{V}_s$ and its temperature and pressure derivatives, may be obtained (i) from estimates based on theory and/or binary solution data, (ii) from direct measurements of $\bar{v}_i^{sw}$ for the seawater components, or (iii) from measurements (direct or indirect) of seawater properties such as density, expansibility and compressibility. By comparing the values derived from the different sources, information can be obtained, not only of value for the interpretation of solution processes, but also for assessing the relative accuracy of basic oceanographic data.

Usually the results will be presented as $\bar{V}_s^A$, $\Sigma \underline{m}_i \bar{v}_i^{sw}$, or as the average per gram of seawater, $\bar{V}_s^{sw}$. Conversion to any other average (or mean) can readily be made by referring to Table 4.17.

4.61 The mean partial molal volume of seasalt.

The generalised form of eq 4.3.4 is

$$(\Sigma_i m_i)\Phi_V = \Sigma_i m_i \phi_{v_i} + (\Sigma_i m_i)\Delta\Phi_V \qquad (4.3.4)^*$$

For seawater $\underline{i}$ represents the major electrolyte components (e.g. recipe A, Table 4.16) and Φ_V is the mean apparent molal volume of seasalt. In terms of the density of seawater, d_{sw} and of pure water, d_w,

$$\Phi_V = 10^3(d_w - d_{sw})/m_s d_{sw} d_w + M_t/d_{sw} \qquad (4.61.1)$$

analogous to eq 4.2.8

with $m_s = \Sigma_i m_i$, $M_t = \Sigma_i f_i M_i$ (Table 4.16)

In terms of salinity, with $c = 10^{-3}S(‰)$.

$$\Phi_V = \frac{(1-c)}{c} M_t'(d_w - d_{sw})/d_{sw} d_w + M_t/d_{sw}$$

$$= \frac{(1-c)}{c} M_t'(v_{sw} - v_w) + M_t v_{sw} \qquad (4.61.2)$$

with v representing the specific volume.

Values of Φ_V calculated from seawater density data (Dorsey 1940, Table 108) have been fitted to the corresponding values of $I^{1/2}$ and S, Appendix 4, Table 4A.7. As might be expected from the binary solution results, the $I^{1/2}$ function gives the best fit. For the range 16-38‰ a linear fit to $I^{1/2}$ gives deviations of less than $0.1 ml mol^{-1}$ whereas a quadratic in S is required for a comparative fit.

In general, Φ_V is given by

$$\Phi_V = a_o + a_1 I^{1/2} + a_2 I \qquad (4.61.3)$$

with a standard deviation of $\pm 0.05 ml mol^{-1}$. It should be noted that eq 4.72.5$\underline{a}$ has been fitted over a smaller concentration range than the corresponding equation given in Appendix 4, Table 4A.7.

The mean partial molal volume of seasalt, mpmv, or $\bar{V}_s$, is thus given by

$$\bar{V}_s = \Phi_v + I\partial\Phi_v/\partial I \qquad (4.61.4)$$

or
$$\bar{V}_s = a_o + (\tfrac{3}{2})a_1 I^{\frac{1}{2}} + 2a_2 I \qquad (4.61.5)$$

These relationships enable Φ_v and $\bar{V}_s$ to be calculated directly from the density data. It is more convenient to calculate these quantities from an equation of state. Nevertheless, eq 4.61.4 and 4.61.5 will be useful in a comparative sense, to check on any distortions introduced by the equation of state.

Simple equations for $\bar{V}_s$ and Φ_v may be derived from the Tait-Gibson equation (which is dealt with in detail in Section 6.3). The effect of the solute in compressing the water and reducing its volume is illustrated in Figures 4.1 and 4.9.

The effect may be accounted for by assuming that the solution is subject to a pressure $(P+P_e)$ where P is the atmospheric and P_e the effective pressure due to the ionized solute (Tammann's hypothesis 1907). The value of P_e for a given solution, estimated from a single compression over any convenient pressure range, is directly related to the partial molal volume of the solute, for, at any pressure P bars,

$$\bar{V}_s = M_t\Psi_t - 434.3C(\partial(B+P_e)/\partial m_s)_T/(B+P_e+P) \qquad (4.61.6)$$

where Ψ_t is the specific volume of the pure solute (as a liquid), C and B are positive constants characterising the solvent. With water as solvent, $C = 0.315v_w^{(1)}$, $v_w^{(1)}$ being the specific volume $(1/d_w^{(1)})$ of pure water at 1 bar and at the temperature of the solution. P_e and B are also functions of temperature and are taken to be independent of pressure. An outline of the derivation of this equation will be given in the next section.

Values of $(B+P_e)$ or B*, have been derived from various seawater data sets by Li (1967), as functions of salinity (‰) and temperature (°C). The

B* function used here is from Ekman's (1908) compressibility data, viz.

$$B^* = (2670.8+6.8966S) + (19.39-0.0703S)t - 0.223t^2 \tag{4.61.7}$$

so that, at 25°C

$$B^* = 3016.175 + 5.1391S \tag{4.61.8}$$

The quantity Ψ_t is effectively independent of concentration for the range of interest, say 0.1 to 1 molal (ionic strength). For the Ekman (1908) (or E) data the value is given by

$$M_t\Psi_t = 37.09 - 0.1498t \tag{4.61.9}$$

Equation 4.61.6 may be expressed in terms of salinity,

$$\bar{V}_s = M_t\Psi_t - (434.3CM_t'(1-c)^2/(B^*+1))\partial B^*/\partial S \tag{4.61.10}$$

with, from eq 4.61.7

$$\partial B^*/\partial S = 6.8966 - 0.0703t \tag{4.61.11}$$

Wilson and Bradley (1968) measured the relative volumes of seawater samples at temperatures from 0-40°C and under pressures up to nearly 1000 bars. They fitted the results to a "Tumlirz" form of equation of state, to obtain for the specific volume, v_{sw}, of seawater.

$$v_{sw} = 0.70200 + Q'/R' \tag{4.61.12}$$

with $Q' = 1752.73 + 11.01t - 0.0639t^2 - 3.9986S - 0.0107tS$

and $R' = p + 5880.9 + 37.592t - 0.34395t^2 + 2.2524S$

so that $Q'_{25} = 1988.0425 - 4.2661S$

and $R'_{25} = p + 6605.7313 + 2.2524S$

The pressure, p, is in bars, t in °C, S in ‰ and v in cm^3g^{-1}.

These results can be used to calculate $\bar{V}_s$, most conveniently, as follows

Equation 4.61.1 may be expressed

$$\Phi_v = 10^3(v_{sw}-v_w)/m_s + M_tv_{sw} \tag{4.61.13}$$

so that $\bar{V}_s = \Phi_v + m_s\, \partial\Phi_v/\partial m_s$ (4.61.14)

$= (m_s M_t + 10^3)\partial v_{sw}/\partial m_s + M_t v_{sw}$

and since $m_s = S/(1-c)M_t'$, $\partial m_s = \partial S/(1-c)^2 M_t'$

this gives $\bar{V}_s = (SM_t + 10^3(1-c)M_t')(1-c)\ \partial v_{sw}/\partial S$

$+ M_t v_{sw}$ (4.61.15)

The values of v_{sw} and $\partial v_{sw}/\partial S$ are obtained from eq 4.61.12. The raw WB data (Wilson and Bradley 1968) may also be used directly in eq 4.61.2 from which $\bar{V}_s$ may be estimated via eq 4.61.4.

More recently, Wang and Millero (1973) have developed an equation of state using sound velocity measurements of Wilson (1959,1960), on water and seawater. An outline of the derivation of this equation is given in Section 6.6 where the use of acoustic data, for calculating various thermodynamic quantities, is discussed. Here, in the present section, a summary of the equation is given together with the coefficients, listed in Table 4.18.

The specific volume of seawater, $v_{sw}^{(p)}$, at an applied pressure of p' (at atmospheric pressure p' is zero), is given by

$$v_{sw}^{(1)} - v_{sw}^{(p)} = p'v_w^{(1)}/(B_{sw} + A_{sw_1}p' + A_{sw_2}p'^2) \quad (4.61.16)$$

where p represents the absolute pressure, as usual, and is equivalent to (p'+1), and

$$v_{sw}^{(1)} = 1.000027/(1.0+\sigma_t 10^{-3})\ (cm^3 g^{-1}) \quad (4.61.17)$$

with $\sigma_t = A_t + B_t\sigma_o + C_t\sigma_o^2$ (4.61.18)

The coefficients are functions of temperature and/or salinity (Table 4.18). The corresponding equation for water is given in Section 6.72.

Equation 4.61.16 has been used to calculate Φ_v and $\bar{V}_s$, from eqs 4.61.13 and 4.61.15.

TABLE 4.18 Coefficients and functions for the equation of state of Wang and Millero, eq 4.61.16

Term	Function
σ_o	$-9.344586324\text{x}10^{-2}+8.14876576925\text{x}10^{-1}S$ $-4.824961403\text{x}10^{-4}S^2 + 6.767861356\text{x}10^{-6}S^3$
A_t	$(t+67.26)^{-1}(4.5316842620t - 0.5459391107t^2$ $- 1.9824839871\text{x}10^{-3}t^3$ $-1.438030609\text{x}10^{-7}t^4)$
B_t	$1.0 - 4.7867\text{x}10^{-3}t + 9.8185\text{x}10^{-5}t^2 -1.0843\text{x}10^{-6}t^3$
C_t	$1.8030\text{x}10^{-5}t -8.164\text{x}10^{-7}t^2 + 1.667\text{x}10^{-8}t^3$
$v_w^{(1)}$	$(0.9998396+18.224944\text{x}10^{-3}t -7.922210\text{x}10^{-6}t^2$ $-55.44846\text{x}10^{-9}t^3 + 149.7562\text{x}10^{-12}t^4$ $- 393.2952\text{x}10^{-15}t^5)^{-1}(1 + 18.159725\text{x}10^{-3}t)$ cm^3g^{-1}
B_{sw}	$B_w^o + S(7.270886\text{x}10^{1} - 4.77264\text{x}10^{-1}t$ $+1.53360\text{x}10^{-2}t^2 -5.24748\text{x}10^{-4}t^3$ $+ 6.59438\text{x}10^{-6}t^4 + 0.109644(S-35))$ bar
B_w^o	$(1.0 + 2.165928\text{x}10^{-2}t)10^6\text{x}(50.88630 + 0.7171582t$ $+ 0.7819867\text{x}10^{-3}t^2 + 31.62214\text{x}10^{-6}t^3$ $-0.1323594\text{x}10^{-6}t^4 + 0.6345750\text{x}10^{-9}t^5)^{-1}$
A_{sw_1}	$3.4823- 4.937\text{x}10^{-3}t + 1.763\text{x}10^{-4}t^2$ $+ 6.2\text{x}10^{-3}(S-35)$
A_{sw_2}	$4.028\text{x}10^{-5} - 2.776\text{x}10^{-6}t + 1.049\text{x}10^{-7}t^2$ $- 7.3\text{x}10^{-7}(S-35)$

Direct experimental values of $\bar{V}_s$ are available for the salinity range 30-40‰ and it is of interest to compare these with the values calculated from the above equations. Table 4.19 lists the values in the form of the deviations from the direct experimental values. The latter were obtained by summing the contributions of the individual seawater components, i.e.

$$\bar{V}_s = \Sigma_i f_i \bar{v}_i^{sw} \qquad (4.61.19)$$

TABLE 4.19 The mean partial molal volume of seasalt at 25°C, 1 bar
Solution A[a] $ml\,mol^{-1}$(salts)

S‰	I $mol\,kg^{-1}$	$\bar{V}_s^A$ (exp)	$\bar{V}_s^A$ (exp) - $\bar{V}_s^A$ (calc)						
			Bin	E	WM	WB(a)	WB(b)	NK	D
5	0.10020	18.203	+0.733	+0.477	+0.624	+0.364	+1.100	+0.803	+0.803
10	0.20141	18.399	+0.464	+0.387	+0.328	+0.392	+0.631	+0.499	+0.389
15	0.30365	18.593	+0.297	+0.301	+0.101	+0.419	+0.453	+0.293	+0.163
20	0.40694	18.789	+0.186	+0.222	-0.057	+0.450	+0.319	+0.159	+0.019
25	0.51128	18.983	+0.106	+0.146	-0.150	+0.480	+0.263	+0.083	-0.057
30	0.61670	19.178	+0.051	+0.076	-0.178	+0.513	+0.268	-0.052	-0.102
35	0.72321	19.373	+0.014	+0.011	-0.144	+0.548	+0.323	-0.187	-0.137
40	0.83083	19.568	-0.010	-0.049	-0.050	+0.584	+0.408	-0.232	-0.132

a see Table 4.16
Experimental values from eq 4.61.19
Bin Binary solution data (Lee 1966 and Parsons 1959)
E Ekman (1908) eqs 4.61.7, 4.61.9, and 4.61.10.
WM Wang and Millero (1973) eqs 4.61.15, 4.61.16
WB Wilson and Bradley (1968)
(a) Smoothed data, Tumlirz form of equation (4.61.12)
(b) Raw data, eqs 4.61.13 and 4.61.4
NK Newton and Kennedy (1965) eqs 4.61.13 and 4.61.4
D Dorsey (1940, Table 108) eqs 4.61.1 and 4.61.4

where f_i is given in Table 4.16 (solution A). Using the experimental results of Duedall (1968).

$$\bar{V}_s^A = 15.76 + 16.36\times10^{-2}t - 0.304\times10^{-2}t^2 + (1.075 + 0.112t)\times10^{-2}S + RES^{(A)} \qquad (4.61.20)$$

covering the electrolytes NaCl, Na_2SO_4, $NaHCO_3$, KCl, $MgCl_2$ and $CaCl_2$; RES represents the contribution of $SrCl_2$, NaF, and KBr. The values of $\bar{v}_i^{sw}$ for these latter salts were estimated from eq 4.4.8, using ϕ_{v_i} equations from Appendix 4 (Table 4A.3).

$RES^{(A)}$ varied from 0.059 to 0.063 and $RES^{(B)}$ from 0.056 to 0.060 $ml\,mol^{-1}$ (salts), for the salinity range 5-40‰. The value of $RES/\Sigma\nu_i f_i$ varied from 0.027 to 0.029 over the same salinity range.

Since the experimental values only cover the salinity range 30-40‰ the extrapolated values at lower salinities are not reliable. The results indicate that these values are too high. Over the experimental range the values of D, E and WM are all in good agreement with the experimental values. The NK values are slightly too high at S≥ 35‰. The large deviations of the WB set indicate some systematic error in the volume measurements probably emphasized in the salinity gradient. The smoothing, via the Tumlirz form of equation (4.61.12) accentuates the deviations. Two points are of interest in regard to these deviations:

(1) The salinity values were derived via conductivity measurements, except at S≤ 20‰ (Wilson, personal communication). Under the experimental conditions, changes in conductivity occur, as noted in similar cases (Newton and Kennedy 1965). There does not seem to be any conclusive evidence that such changes (apart from evaporative processes) are directly related to salinity changes (i.e. the normal compositional seawater salinity). Natural compositional deviations of the seawater samples could be a factor, but there is no other evidence to support this. In any case, it is suggested that, in any set of measurements involving salinity as an important variable, checks on the conductivity measurements should be made via chemical titration of selected samples. In addition, compositional analysis would indicate whether the samples were representative of "average" seawater. Alternatively, artificial seawater samples could be used as a check.

(2) The Tumlirz form of equation appears to distort some of the derivatives of the specific volume. The salinity derivative, in this case, appears to be subject to a "flattening" effect. The temperature derivative also appears to be affected (Bradshaw and Schleicher 1970 and Caldwell and Tucker 1970).

Values of $\bar{V}_s$ were also calculated from $\Sigma f_i \bar{v}_i$, where $\bar{v}_i$ represents the pmv

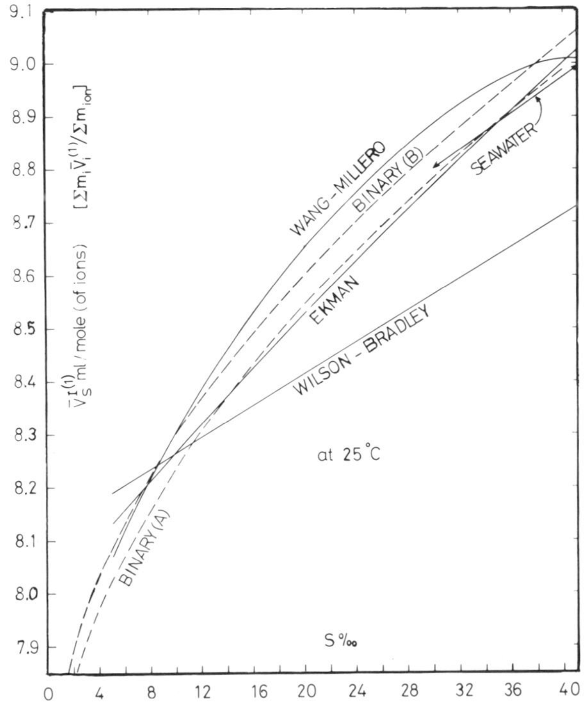

FIGURE 4.8

The partial molal volume of seasalt as a function of salinity, 1 bar pressure 25°C, in mlmol^{-1} (of ions).

Estimates made from seawater P-V-T-S data:

Ekman (1908) eq 4.61.10
Wang and Millero (1973) eqs 4.61.16 and 4.61.15
Wilson and Bradley (1968) eqs 4.61.12 and 4.61.15
Binary and seawater curves estimated from the pmv's of the components, eq 4.61.19
Binary: eq 4.61.21
Seawater:eq 4.61.20 (only covers the range 30-40‰)

---

of the seawater component in the binary solution at the same ionic strength as the seawater solution. These values are compared with the others in Figure 4.8, using the ionic, rather than the salt, mean, i.e. $\Sigma m_i \bar{v}_i^{sw}/\Sigma m_{ion}$, (or $\bar{V}_s^A/2.178796$ - See Table 4.17)

Over the range 30-40‰, the values of Ekman give the closest agreement with

TABLE 4.20 Experimental [a] values of the mean partial molal volume [b] of "seasalt" in various units (1 bar)

(a) As a function of salinity, 25°C

S‰	Solution A	Solution B	Solution A or B[c]			Ionic Strength[d]
	$ml\,mol^{-1}$ (salts)		$ml\,mol^{-1}$ (ions)	ml per kg water	ml per g seawater	$mol\,kg^{-1}$ water
	$\bar{V}_s^A$	$\bar{V}_s^B$	$\bar{V}_s^I$	$m_s\bar{V}_s$	$\bar{V}_s^{sw} \times 10^3$	
21.69	18.856	17.878	8.657	6.141	6.008	0.44209
30.52	19.199	18.203	8.814	8.879	8.607	0.62773
35.00	19.373	18.369	8.894	10.323	9.960	0.72321
36.13	19.418	18.410	8.915	10.692	10.304	0.74744
41.03	19.608	18.591	9.002	12.325	11.817	0.85314

(b) As a function of temperature at S = 35‰

t^oC					
0	(16.190)	(15.350)	(7.433)	(8.621)	(8.317)
5	(17.130)	(16.241)	(7.864)	(9.123)	(8.802)
10	17.919	16.989	8.226	9.545	9.210
15	18.556	17.593	8.518	9.885	9.538
20	19.040	18.052	8.741	10.144	9.788
25	19.374	18.369	8.894	10.323	9.960
30	19.554	18.539	8.977	10.419	10.052

a from the values for the individual components, Duedall and Weyl (1967), Duedall (1968)

b $\bar{V}_s = \Sigma m_i \bar{v}_i^{sw}/\bar{w}$, where $\bar{w}$ is the averaging quantity - see Table 4.17

c maximum deviation between A and B values is 0.01%

d $\frac{1}{2} m_s \Sigma f_i (z_+ z_-)_i \nu_i = 0.01994S/(1-c)$

bracketed values less reliable - see Section 5.3 Table 5.6.

the experimental results. Except at the highest salinities, the WM values are too high. The WB curve is in very poor agreement with the other results. The binary solution values obviously give a very good estimate of the directly measured values. Binary solution data, therefore, might be expected to give a useful guide at other temperatures and pressures, especially data of A solution salts.

The experimental values of $\bar{V}_s$, given in various units, are summarised in Table 4.20. The changes with temperature are also given, although the temperature effect will be discussed in Chapter 5.

For 25°C, the binary solution data (solution A electrolytes) gives

$$\bar{V}_s^A(\text{bin}) = 16.367 + 3.46672I^{\frac{1}{2}} + 0.06162I \qquad (4.61.21)$$

and (for solution B electrolytes)

$$\bar{V}_s^B(\text{bin}) = 15.517 + 4.16376I^{\frac{1}{2}} - 1.836457I + 1.60013I^{3/2} - 0.476138I^2 \qquad (4.61.22)$$

Note that the limiting values do not refer to the same quantities. However, for $\bar{V}_s^I$ (the mean value per mole of ions - see Table 4.17) the two equations give the same limiting value (within 0.001), viz

$$\bar{V}_s^I = \bar{V}_s^A/2.178796 = \bar{V}_s^B/2.065418$$

so that $$\bar{V}_s^I(\text{bin})_{I\to 0} = 7.512 \ \text{mlmol}^{-1}$$

Eqs 4.61.21 and 4.61.22 were derived on the basis of data from Lee (1966) and Parsons (1959).

4.62 The Mean apparent molal volume of seasalt, mamv and the volume of mixing term $\Delta\Phi_v$.

The quantity Φ_v (or mamv), has already been defined by eqs (4.3.4*), 4.61.1, 4.61.2, 4.61.13 where it was used to derive values for the (mean) pmv of seasalt. While not a thermodynamic quantity Φ_v is a useful practical function from which $\Delta\Phi_v$, eq (4.3.4)*, can be estimated using binary solution data. For 25°C, solution A, $\Delta\Phi_v$ may be represented by (Leyendekkers 1974),

$$\Delta\Phi_v^A = -0.287 + 0.888I^{\frac{1}{2}} - 0.491I$$

to $\pm 0.03\text{mlmol}^{-1}$ for $S \geq 10‰$ (4.62.1)

For the binary solution, from eq 4.61.21 and using eqs 4.61.3 and 4.61.5, at 25°C

$$\Phi_v^A(\text{bin}) = 16.367 + 2.31182I^{\frac{1}{2}} + 0.0308I \qquad (4.62.2)$$

(All of these equations can be converted to salinity functions via, $I = 0.01994S/(1-10^{-3}S)$)

As the Tait-Gibson equation (Ekman's data) gives reliable results for $\bar{V}_s$, the corresponding values for Φ_v, as a function of salinity, temperature and pressure may be calculated from this equation. In this section, only the equations for 1-bar will be given, but these are easily extended to higher pressures (Chapter 6). The specific volume of seawater, v_{sw} may be expressed,

$$v_{sw} = x_w\Psi_w + x_t\Psi_t \tag{4.62.3}$$

where x_w, x_t are the number of grams of solvent and solute, respectively, per gram of solution. Ψ_t has been defined above (eqs 4.61.6, 4.61.9) and Ψ_w is the apparent specific volume of the solvent in solution, i.e. the water in seawater.

since $x_w = 10^3/(m_sM_t+10^3);\quad x_t = m_sM_t/(10^3+m_sM_t)$

and with $Q = 10^3/m_sM_t$,

$$v_{sw} = (Q\Psi_w + \Psi_t)/(1+Q) \tag{4.62.4}$$

The addition of the ionic solute seasalt (as a liquid) to one gram of pure water causes an increase in the effective pressure and a volume contraction of $(v_w-\Psi_w)$, where (Harned and Owen, 1958, p381).

$$v_w - \Psi_w = C\log(B^*+1)/(B+1) \tag{4.62.5}$$

so that

$$\Psi_w = v_w - C\log(B^*+1)/(B+1) \tag{4.62.6}$$

combining eqs 4.62.4, 4.61.13 gives

$$\Phi_v = M_tQ(\Psi_w - v_w) + M_t\Psi_t \tag{4.62.7}$$

and using eq 4.62.6,

$$\Phi_v = M_t\Psi_t - M_tQC\log(B^*+1)/(B+1)$$

or

$$\Phi_v = M_t\Psi_t - 315v_w^{(1)}M_t'\frac{(1-c)}{S}\log\frac{(B^*+1)}{(B+1)} \tag{4.62.8}$$

with $B = B^*(S=0)$, since the Ψ terms were derived on this basis. B^* and $M_t\Psi_t$ are given by eqs 4.61.7 and 4.61.9, respectively. $v_w^{(1)}$ by eq 5.71.1 (Chapter 5).

TABLE 4.21 The mean apparent molal volume of seasalt, Φ_v^A (solution A) $\mathrm{ml\,mol^{-1}}$

$25^{o}C$ 1 bar

S‰	I $\mathrm{molkg^{-1}}$	Binary[a]	Ekman	Dorsey	Wang-Millero	Wilson-Bradley
10.00	0.20141	17.410	17.726	17.362	17.351	17.506
15.00	0.30365	17.650	17.870	17.711	17.666	17.703
21.69	0.44209	17.918	18.060	18.021	18.000	17.885
25.00	0.51128	18.035	18.153	18.155	18.141	17.962
30.72	0.63197	18.225	18.313	18.354	18.356	18.083
35.00	0.72321	18.354	18.431	18.493	18.494	18.168
36.13	0.74744	18.388	18.462	18.520	18.528	18.190
41.55	0.86442	18.543	18.611	18.689	18.673	18.292

a from eq 4.62.2 References for other data sets given in Table 4.19

The pmv can casily be obtained from eq 4.61.4 by differentiating with respect to S, viz.

$$(1-c)S\partial\Phi_v/\partial S = (315v_w^{(1)}M_t'(1-c)/S)\ \log\frac{(B^*+1)}{(B+1)}$$

$$- (315v_w^{(1)}M_t'x0.4343(1-c)^2\partial B^*/\partial S)/(B^*+1)$$

so that $$\bar{V}_s = \Phi_v + (1-c)S\partial\Phi_v/\partial S \qquad (4.62.9)$$

becomes

$$\bar{V}_s = M_t\Psi_t - 434.3x0.315v_w^{(1)}M_t'(1-c)^2(\partial B^*/\partial S)/(B^*+1)$$

which is eq 4.61.10.

As for $\bar{V}_s$, values of Φ_v calculated from the various equations and data sets, including binary solution data, are shown in Table 4.21.

The results follow the trends to be expected from $\bar{V}_s$. However, the values of WB show good agreement with the binary results up to a salinity of 25‰ which indicates that the specific volume data is quite reliable in this range; the mixing term should be small in this region.

The salinity values of WB at 20‰ and below are probably more reliable than the higher values, as outlined above, which would explain the results in Table 4.21. On the other hand, Ekman's values are too high at the lower salinities. This is because Ψ_t does not remain constant at the lower salinities (see Figure 6.11). In addition, Ekman's experimental range did not extend to these low salinities.

4.7 The apparent specific volume, asv, and the partial molal volume, pmv, of water in seawater.

4.71 Asv. As noted above, the addition of the ionic solute, seasalt, (as a liquid) to one gram of pure water causes an increase in the effective pressure. This increase in pressure causes a volume contraction of $v_w^{(p)} - \Psi_w^{(p)}$, where $v_w^{(p)}$ is the specific volume of pure water at a pressure of p bars and $\Psi_w^{(p)}$ is the apparent specific volume of water in the solution, at the same pressure. The more general form of eq 4.62.5 is

$$v_w^{(p)} - \Psi_w^{(p)} = C\log\frac{(B^*+p)}{(B+p)} \qquad (4.71.1)$$

with $C = 0.315v_w^{(1)}$

At 1 bar $$v_w^{(1)} - \Psi_w^{(1)} = C\log\frac{(B+p_e+1)}{(B+1)} \qquad (4.62.5)^*$$

since $B+p_e = B^*$, p_e being the effective pressure due to the ionized solute. The total pressure is thus (p_e+1) bars .

The change in the volume of pure water subject to the same pressure change, 1 bar to (p_e+1)bars, is, by the Tait equation,

$$v_w^{(1)} - v_w^{(p_e+1)} = C\log\frac{(B+p_e+1)}{(B+1)} \qquad (4.71.2)$$

In other words, the value of $\Psi_w^{(1)}$, the apparent specific volume of water in seawater, should be identical with $v_w^{(p_e+1)}$.

Values of p_e corresponding to a given temperature and salinity are given by

$$p_e(\text{bars}) = B^*-B^*_{S=0} = 6.8966S - 0.0703St \qquad (4.71.3)$$

(as derived by Li (1967) from Ekman's 1908 bulk compression data). p_e is independent of pressure. This equation is probably not reliable above 30°C.

TABLE 4.22 Comparison of the volume contractions of water due to applied and solute pressures.

	0°C			25°C		
S‰	P_e (bars)	$\Psi_w^{(1)}$ (mlg^{-1})	$v_w^{(p_e+1)}$ (mlg^{-1})	P_e (bars)	$\Psi_w^{(1)}$ (mlg^{-1})	$v_w^{(p_e+1)}$ (mlg^{-1})
0	0.0		1.000132	0.0		1.002933
5	34.483	0.998378	0.998378	25.696	1.001770	1.001772
10	68.966	0.996646	0.996645	51.391	1.000616	1.000619
20	137.932	0.993245	0.993244	102.782	0.998337	0.998341
30	206.898	0.989928	0.989928	154.173	0.996096	0.996097
35	241.381	0.988298	0.988298	179.869	0.994988	0.994988
40	275.864	0.986688	0.986689	205.564	0.993890	0.993887

P_e from eq 4.71.3; Ψ_w(at 1 bar) eqs 4.62.5, 4.71.3; v_w(at 1 bar) from Kell (1970). eq 5.71.1 and Table 5.18; v_w(at (p_e+1) bars), calculated from eq of Fine and Millero (1973), see Sections 6.72 and 6.8 and eq 6.63.5

Table 4.22 lists values of p_e at 0°C and 25°C for a range of salinities. The corresponding values of $v_w^{(p_e+1)}$ were calculated from the P-V-T equation for pure water given by Fine and Millero (1973) (based on sound velocity data, see Section 6.72, eq 6.63.5). These are compared with the values of $\Psi_w^{(1)}$ calculated from eqs 4.62.5, 4.71.3 and values of $v_w^{(1)}$ from Kell (1970), converted to mlg^{-1}, see Table 5.18 and eq 5.71.1. The agreement indicates that the assumptions of Gibson (and Tammann) are consistent with the experimental results. The specific volumes are probably accurate to at least 30ppm (the two sets of data, Ekman's and the acoustic are shown to give results consistent with each other, See Section 6.8).

Figure 4.9 shows the values of $\Psi_w^{(1)}$, for a range of salinities, as a function of temperature, together with the specific volume of pure water. The effect of the salt decreases with increasing temperature, as already indicated by eq 4.71.3. At 0°C the decrease in volume of the water at a salinity of 40‰ (I = 0.8308 mol/kg water) is 13.42 ml per kg of water, whereas at 30°C the corresponding decrease is only 8.36 ml (ca 62%).

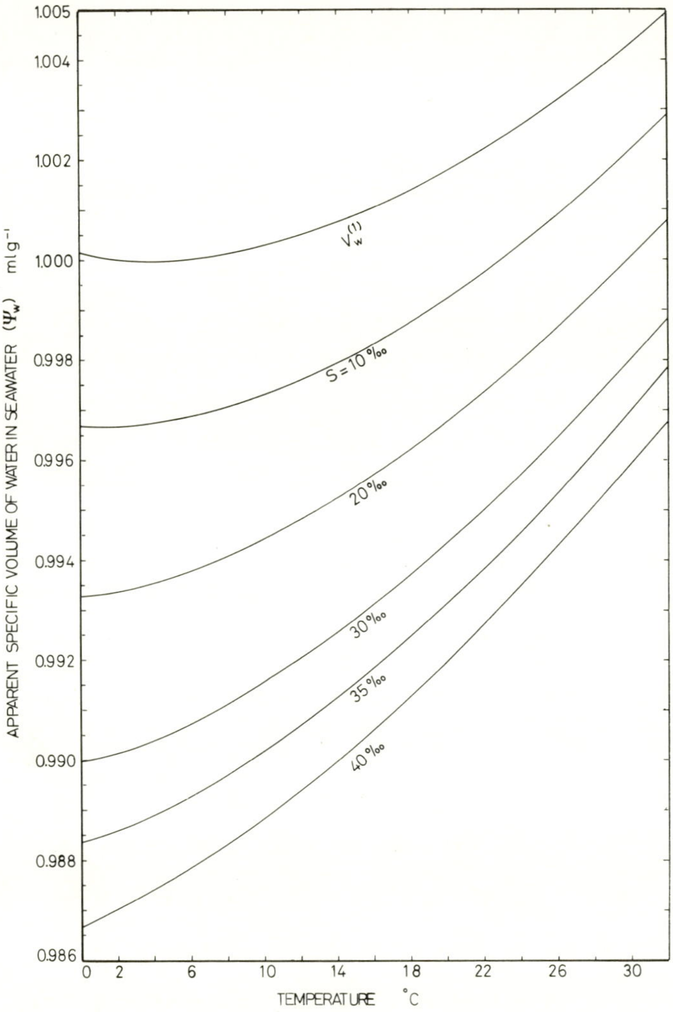

FIGURE 4.9

The reduction in the volume of 1 g of water due to the compression by seasalt, as a function of temperature, at 1 bar pressure.

4.72 Pmv.

The contribution of the solute, $m_s\bar{V}_s$ to the seawater volume has been given in Section 4.61 so that the pmv of the water, $\bar{V}_w$, is obtained from

$$55.51\bar{V}_w = V - m_s\bar{V}_s \qquad (4.72.1)$$

where V is the volume of seawater containing m_s moles of salts, i.e. containing 1 kg or 55.51 moles of water.

For any pressure, p,

$$55.51\bar{V}_w^{(p)} = v_{sw}^{(p)}w_s - m_s\bar{V}_s^{(p)} \qquad (4.72.2)$$

where w_s is the mass of seawater containing 1 kg of water, and $v_{sw}^{(p)}$ is the specific volume of seawater.

The first term on the R.H.S. may be replaced by the apparent specific volumes of the water, Ψ_w, and of seasalt, Ψ_t, (eq 4.62.3)

$$w_s v_{sw}^{(p)} = 10^3\Psi_w^{(p)} + m_s M_t \Psi_t^{(p)}$$

and the second term by

$$m_s\bar{V}_s^{(p)} = m_s M_t \Psi_t - Av_w^{(1)} \left[I(\partial B^*/\partial I)\,/(B^*+p)\right]$$

from eq 4.61.6 with A = 434.3x0.315

so that eq 4.72.2 becomes

$$55.51\bar{V}_w^{(p)} = 10^3\Psi_w^{(p)} + (Av_w^{(1)} I\partial B^*/\partial I)/(B^*+p) \qquad (4.72.3)$$

Combined with eq 4.71.1 this gives

$$\bar{V}_w^{(p)} = (v_w^{(1)}/0.05551) \left[1-0.315\log\frac{(B^*+p)}{(B+1)} + (10^{-3}AI\partial B^*/\partial I)/(B^*+p)\right] \qquad (4.72.4)$$

$$\text{with } B^* = 2670.8 + 6.8966S + (19.39-0.0703S)t - 0.223t^2 \qquad (4.61.7)^*$$

$$\text{and } I\partial B^*/\partial I = S(1-10^{-3}S)\ \partial B^*/\partial S$$

$$(\text{since } I = 0.01994S/(1-10^{-3}S)).$$

Values of $\bar{V}_w^{(1)}$ calculated from eq 4.72.4 are listed in Table 4.23 as a function of salinity, at 25°C. The changes are only very small. This is because the two R.H.S. terms of eq 4.62.4 compensate each other as illustrated in Figure 4.10. The apparent specific volume of the water

TABLE 4.23 The partial molal volume of water in seawater at 25°C as a function of salinity.

S‰	I $molkg^{-1}$	$\bar{V}_w^{(1)}$ $mlmol^{-1}$ eq 4.72.4	eq 4.72.7
0	0	18.0681	18.0681
5	0.100201	18.0674	18.0675
10	0.201414	18.0668	18.0664
15	0.303655	18.0659	18.0651
20	0.406939	18.0646	18.0636
25	0.511282	18.0630	18.0619
30	0.616701	18.0610	18.0601
35	0.723212	18.0587	18.0582
40	0.830833	18.0561	18.0562.

Ψ_w decreases as the salt concentration (or the internal pressure) increases, whereas the second term increases with concentration.

The pmv of water in seawater may also be expressed in the same form as eq 4.2.5 or 4.2.6.

Since $w_s v_{sw}^{(p)} = m_s \Phi_v^{(p)} + 10^3 v_w^{(p)}$

(from eq 4.61.1 cf eq 4.2.9),

and $m_s(\bar{V}_s - \Phi_v)^{(p)} = \frac{I^2}{W}(\partial\Phi_v/\partial I)^{(p)}$

(cf. eq 4.2.4)

with $W = \frac{1}{2}\Sigma_i(z_+z_-)_i\nu_i f_i$, $f_i = m_i/m_s$ and $m_s W = I$,

eq 4.72.2 may be written

$$\bar{V}_w^{(p)} = 10^3 v_w^{(p)}/55.51 - (I^2/55.51W)(\partial\Phi_v/\partial I)^{(p)} \quad (4.72.5)$$

For 25°C, 1 bar, the mean apparent molal volume of seasalt (solution A) is given by eq 4.61.3, viz.

$$\Phi_v^A = 16.080 + 3.198I^{\frac{1}{2}} - 0.451I \quad (4.72.5a)$$

with $0.2 \leq I \leq 0.8$,

based on Knudsen et al's seawater density data (Dorsey 1940, Table 108), so that, with W = 1.3576,

$$(I^2\partial\Phi_v/\partial I)/55.51W = 0.0212180I^{3/2} - 5.9846x10^{-3}I^2 \quad (4.72.6)$$

and eq 4.72.5 becomes, with $v_w^{(1)}$ (25°C) = $1.002933mlg^{-1}$

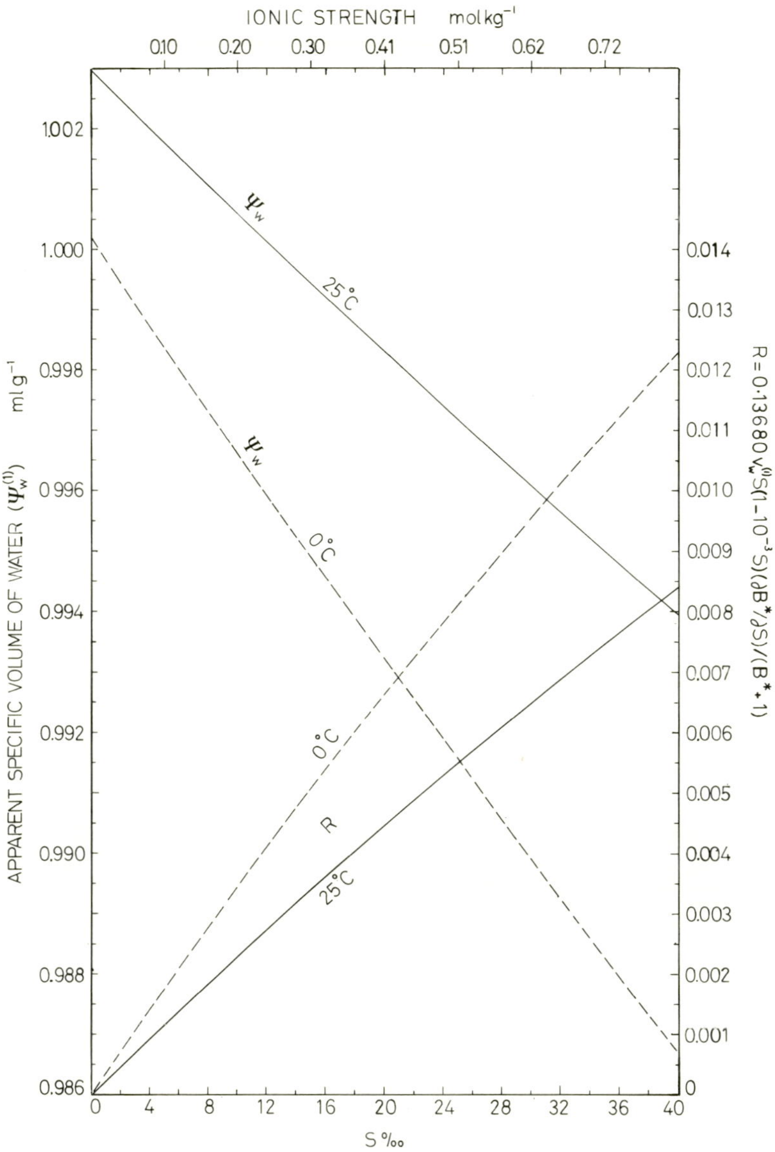

FIGURE 4.10

Components of the partial molal volume of water in seawater, as a function of salinity at 0°C and 25°C, 1 bar pressure. From eq 4.72.3.

$$\bar{V}^{(1)}_{w,25^\circ C} = 18.06811 - 0.0212180 I^{3/2} + 5.9846\text{x}10^{-3} I^2 \qquad (4.72.7)$$

Values of $\bar{V}^{(1)}_w$ from eq 4.72.7 are shown in Table 4.23.

The agreement with those from eq 4.72.4 is good, being generally better than 0.005% (except S = 20,25‰).

For seawater solutions deviating from the "average" composition, a useful approximation of the second term of eq 4.72.5 is obtained from

$(1/55.51)\Sigma_i m_i I\partial\phi_i/\partial I$, where $I\partial\phi_i/\partial I$ (i.e. $(\bar{v}_i - \phi_{v_i})$)

refers to the binary solution of i at the same ionic strength as the seawater solution.

REFERENCES Chapter 4.

Bradshaw A. and K.E. Schleicher 1970.

Direct measurement of thermal expansion of sea water under pressure.
Deep-Sea Res. 17: 691-706.

Caldwell D.R. and B.E. Tucker 1970.

Determination of thermal expansion of sea-water by observing onset of convection.
Deep-Sea Res. 17: 707-719.

Conway B.E., R.E. Verrall, and J.E. Desnoyers 1965.

Specificity in ionic hydration and the evaluation of individual ionic properties.
Z.Phys.Chem.(Leipzig) 230: 157-178.

Curthoys G. and J.G. Mathieson 1970.

Partial molal volumes of ions.
Trans.Faraday Soc. 66(1): 43-50.

Dorsey N.E. 1940.

"Properties of ordinary water-substance", Reinhold, New York. 673pp.

Chapter 4.

Duedall I.W. 1968.

Partial molal volumes of 16 salts in seawater.
Environ.Sci.Technol., 2: 706-707.

Duedall I.W. and P.K. Weyl 1967.

The partial equivalent volumes of salts in seawater.
Limnol.Oceanogr. 12(1): 52-59.

Ekman V.W. 1908.

Die Zusammendrueckbarkeit des Meerwassers.
Publ.Circonst.perm.Explor.Mer, 43: 1-47.

Fine R.A. and F.J. Millero 1973.

The compressibility of water as a function of temperature and pressure.
J.Chem.Phys. 59(10): 5229-36.

Fortier J.-L., P.-A. Leduc, and J.E. Desnoyers 1974.

Thermodynamic properties of alkali halides. II. Enthalpies of dilution and heat capacities in water at 25°C.
J.Solution Chem. 3(4): 323-349.

Glueckauf E. 1965.

Molar volumes of ions.
Trans.Faraday Soc. 61: 914-921.

Harned H.S. and B.B. Owen 1958.

"The Physical chemistry of electrolytic solutions", 3rd ed, ACS monograph, Reinhold, New York, 803pp.

Kell G.S. 1970.

Isothermal compressibility of liquid water at 1 atm.
J.Chem.Eng.Data 15(1): 119-122.

Chapter 4.

Kester D.R., I.W. Duedall, D.N.Connors, and R.M. Pytkowicz 1967.

Preparation of artificial seawater.
Limnol.Oceanogr. 12: 176-179.

Lee J. 1966.

The apparent and partial molal volumes of electrolytes in water and in aqueous sodium-chloride solution.
Thesis, Yale Univ., Ann Arbor, Mich., Univ.Microfilms, 66-4906, Abstr. B,27,131.

Leyendekkers J.V. 1971.

Thermodynamics of mixed electrolyte solutions.
Ionic entropy correlations and volume fraction statistics.
J.Phys.Chem. 75(7): 946-956.

Leyendekkers J.V. 1973.

The chemical potentials of seawater components.
Mar.Chem. 1(2): 75-88.

Leyendekkers J.V. 1974.

Prediction of the partial molal volumes of electrolytes in seawater.
Mar.Chem. 2(2): 89-110.

Li Y.-H. 1967.

Equation of state of water and seawater.
J.Geophys.Res. 72(10): 2665-2678.

Marks G.W. 1960.

Dependence of acoustic velocity in some aqueous hydroxide solutions on temperature and concentration.
J.Acoustic.Soc. America. 32: 327-335.

Masterton W.L., H.Welles, J.H.Knox, and F.J.Millero 1974.

Volume change of ion pair formation: rubidium nitrate and thallium(I) nitrate in water.
J. Solution Chem. 3(2): 91-102.

Chapter 4.

McDevit W.F. and F.A. Long 1952.

The activity coefficient of benzene in aqueous salt solutions.
J.Amer.Chem.Soc. 74: 1773-1777

Millero F.J. 1969.

The partial molal volumes of ions in seawater.
Limnol.Oceanogr. 14: 376-385.

Millero F.J. 1971.

The molal volume of electrolytes.
Chem. Rev. 71(2): 147-176.

Millero F.J. 1971a.

Effect of pressure on sulfate ion association in seawater.
Geochim.Cosmochim.Acta, 35(11): 1089-1098-

Millero F.J. 1974.

Sea water as a multicomponent electrolyte solution, in "The Sea - ideas and observations". E. Goldberg, editor, Wiley & Sons, 5.

Millero F.J. and R.A. Berner 1972.

Effect of pressure on carbonate equilibria in seawater.
Geochim.Cosmochim.Acta, 36(1): 92-98.

Millero F.J. and W.L. Masterton 1974.

Volume change for the formation of magnesium sulfate ion pairs at various temperatures.
J.Phys.Chem. 78(13): 1287-1294.

Millero F.J., G.K. Ward, F.K. Lepple, and E.V. Hoff 1974.

Isothermal compressibility of aqueous sodium chloride, magnesium chloride, sodium sulfate, and magnesium sulfate solutions from 0 to 45° at 1 atm.
J.Phys.Chem. 78(16): 1636-1643.

Chapter 4.

Mukerjee P. 1961.

On ion-solvent interactions. Part I. Partial molal volumes of ions in aqueous solution.
J.Phys.Chem. 65: 740-744.

Mukerjee P. 1961.

On ion-solvent interactions. Part II. Internal pressure and electrostriction of aqueous solutions of electrolytes.
J.Phys.Chem. 65: 744-746.

Newton M.S. and G.C. Kennedy 1965.

P-V-T-S relations of sea water.
J. Marine Res. 23(2): 88-103.

Padova J. 1964.

Solvation approach to ion solvent interaction.
J.Chem.Phys. 40(3): 691-694.

Parsons R. 1959.

"Handbook of electrochemical constants", Butterworths, London, 113pp.

Perron G., J.E.Desnoyer, and F.J. Millero 1974.

Apparent molal volumes and heat capacities of alkaline earth chlorides in water at 25°C.
Canadian J.Chem. 52(22): 3738-3741.

Pitzer K.S. and L. Brewer 1961.

revised edition of "Thermodynamics" by G.N. Lewis and M. Randall. McGraw-Hill, New York, N.Y., 723pp.

Powell R.E. and W.M. Latimer 1951.

The entropy of aqueous solutes.
J.Chem.Phys. 19: 1139-1141.

Chapter 4.

Redlich O. and D.M. Meyer 1964.

The molal volume of electrolytes.
Chem.Rev. 64: 221-227.

Robinson R.A., and R.H. Stokes 1959.

"Electrolyte solutions" 2nd ed(revised),
Butterworths, London 571pp.

Scott A.F. 1931.

The apparent volumes of salts in solution II.
The problem of their interpretation.
J.Phys.Chem. 35: 3379-3396.

Stokes R.H. and R.A. Robinson 1957.

The application of volume fraction statistics to the calculation of activities of hydrated electrolytes.
Trans. Faraday Soc. 53: 301-304.

Tammann G. 1907.

"Über die Beziehungen zwischen den inneren Kräften und Eigenschaften der Lösungen", p36, Voss, Leipzig.

Wang Dong-Ping and F.J. Millero 1973.

Precise representation of the P-V-T properties of water and seawater determined from sound speeds.
J.Geophys.Res. 78(3): 7122-7128.

Ward G.K. and F.J. Millero 1974.

The effect of pressure on the ionization of boric acid in aqueous solutions from molal-volume data.
J.Solution Chem. 3(6): 417-430.

Wilson W. 1959.

Speed of sound in distilled water as a function of temperature and pressure.
J.Acoust.Soc. Am. 31: 1067-1072.

Chapter 4.

Wilson W. 1960.

Speed of sound in sea water as a function of T, P, and salinity.
J.Acoust.Soc.Am. 38: 641-644.

Wilson W. and D. Bradley 1968.

Specific volume of seawater as a function of temperature, pressure, and salinity.
Deep-Sea Res. 15: 335-363.

Young T.F. and M.B. Smith 1954.

Thermodynamic properties of mixtures of electrolytes in aqueous solutions.
J.Phys.Chem. 58: 716-724.

Zana R. and E. Yeager 1966.

Determination of ionic partial molal volumes from ionic vibration potentials.
J.Phys.Chem. 70: 954.

Zana R. and E. Yeager 1967.

Ultrasonic vibration potentials and their use in the determination of ionic partial molal volumes.
J.Phys.Chem. 71: 521-536.

APPENDIX 4.

Table 4A.1

Conventional partial molal volumes of ions at infinite dilution in water

Based on $\bar{v}^{o}_{H+} = -5.1 - 0.008t - 1.7x10^{-4}t^2$

$cm^3 mol^{-1}$

0°C

H+	Li+	Na+	K+	Rb+	Cs+	Ag+	NH_4^+	Me_4N+	Et_4N+
-5.1	-5.55	-8.61	2.07	7.21	14.58	-7.24	12.37	83.49	142.37

n-Pr_4N+	n-Bu_4N+	Ph_4As+
207.43	266.0	286.02

Mg^{2+}	Ca^{2+}	Sr^{2+}	Ba^{2+}
-26.91	-30.04	-30.97	-25.99

La^{3+}
-56.01

F-	Cl-	Br-	I-	OH-	NO_3^-	HCO_3^-	SCN^-	$PhSO_3^-$	MnO_4^-
2.89	21.55	28.16	38.67	-1.7	31.7	(27.1)	36.4	98.9	43.6

ClO_4^-
45.8

CO_3^{2-}	SO_4^{2-}	CrO_4^{2-}
0.4	21.3	23.8

Table 4A.1 (ctd)

Conventional partial molal volumes of ions at infinite dilution in water

Based on $\bar{v}^{o}_{H+} = -5.1 \ -0.008t - 1.7x10^{-4}t^{2}$

$cm^{3}mol^{-1}$

$25^{o}C$

H^+	Li^+	Na^+	K^+	Rb^+	Cs^+	Ag^+	Tl^+	NH_4^+	Me_4N^+
-5.4	-6.28	-6.61	3.62	8.67	15.94	-6.1	5.2	12.46	84.17

Et_4N^+	n-Pr_4N^+	n-Bu_4N^+	Ph_4As^+	n-Am_4N^+	$MeNH_3^+$	$EtNH_3^+$	n-$PrNH_3^+$
143.72	209.04	270.26	295.25	333.8	30.71	47.54	64.04

n-$BuNH_3^+$	n-$PenNH_3^+$	n-$HexNH_3^+$	n-$HepNH_3^+$	n-$OctNH_3^+$	$(HOEt)_4N^+$
80.1	96.04	111.93	127.83	143.75	146.6

$(Bu_3N(CH_3)_8NBu_3)^{2+}$
523.1

Be^{2+}	Mg^{2+}	Ni^{2+}	Co^{2+}	Zn^{2+}	Fe^{2+}	Mn^{2+}	Cu^{2+}	Cd^{2+}
-22.8	-31.97	-34.8	-34.8	-32.4	-35.5	-28.5	-38.56	-30.8

Ca^{2+}	Hg^{2+}	Sr^{2+}	Pb^{2+}	Ba^{2+}
-38.65	-30.1	-28.96	-26.3	-23.27

Al^{3+}	Fe^{3+}	Cr^{3+}	Yb^{3+}	Er^{3+}	Ho^{3+}	Dy^{3+}	Tb^{3+}	Gd^{3+}
-58.4	-59.9	-55.7	-60.42	-59.6	-57.96	-57.03	-56.44	-56.61

Sm^{3+}	Nd^{3+}	Pr^{3+}	La^{3+}	Th^{4+}
-58.53	-59.51	-58.73	-55.30	-69.7

Table 4A.1 (ctd)

$cm^3 mol^{-1}$

25°C

F^-	Cl^-	Br^-	I^-	OH^-	Ph_4B^-	ReO_4^-	OCN^-	$SeCN^-$	BF_4^-
4.24	23.23	30.11	41.62	1.36	283.02	53.58	31.52	55.08	49.58

SO_3F^-	SO_3NH_2	NO_3^-	NO_2^-	SCN^-	CHO_2^-	$CH_3CO_2^-$	$CH_3CH_2CO_2^-$	$CH(CH_2)_2CO_2^-$
53.33	46.89	34.40	31.6	41.1	31.67	45.86	59.4	75.80

$PhSO_3^-$	HCO_3^-	MnO_4^-	ClO_3^-	BrO_3^-	IO_3^-	ClO_4^-	HSO_4^-	$HSeO_4^-$
114.3	28.8	47.9	42.06	40.7	30.7	49.52	41.07	36.5

$H_2PO_4^-$	$H_2AsO_4^-$	$p\text{-}CH_3PhSO_3^-$	PhO^-
34.5	40.6	125.0	74.1

S^{2-}	SO_4^{2-}	SeO_4^{2-}	CO_3^{2-}	CrO_4^{2-}	WO_4^{2-}	MoO_4^{2-}	$PtCl_6^{2-}$	HPO_4^{2-}
2.6	24.78	31.8	6.5	30.5	36.5	39.7	160.8	18.5

$S_2O_3^{2-}$	$C_2O_4^{2-}$	$Cr_2O_7^{2-}$	SO_3^{2-}	$Fe(CN)_6^{2-}$
44.8	26.8	83.8	19.7	84.8

AsO_4^{3-}	$Fe(CN)_6^{3-}$
0.6	137.0

Table 4A.2

Conventional partial molal volumes of ions in seawater at S = 35‰.

25°C M convention:* $\bar{v}^{sw}_{H+} = -3.2$, and $\bar{v}^{sw}_{ion} = \bar{v}^{o}_{ion} + 0.37_2 z^2/r$

$+ 0.83_3$

with $\bar{v}^{o}_{ion}$ based on $\bar{v}^{o}_{H+} = -4.5$

L convention:** $\bar{v}^{o}_{H+} = -5.4$, $\bar{v}^{sw}_{H+} = -6.4$, and

$(\bar{v}^{sw}_{i} - \bar{v}^{o}_{i}) / \nu_{-} = \bar{v}^{sw}_{anion} - \bar{v}^{o}_{anion}$

$ml\,mol^{-1}$

	H^+	Li^+	Na^+	Ag^+	K^+	Tl^+	Rb^+	Cs^+	NH_4^+
M	-3.2	-3.9	-4.5	-4.0	5.6	7.2	10.7	17.9	14.5
L	-6.4	-6.8	-6.6	-6.1	3.6	5.2	8.7	15.9	12.0

	Be^{2+}	Mg^{2+}	Ni^{2+}	Co^{2+}	Zn^{2+}	Fe^{2+}	Mn^{2+}	Cu^{2+}	Cd^{2+}
M	-25.0	-27.0	-30.1	-30.2	-29.8	-32.3	-20.0	-32.9	-26.6
L	-22.8	-32.0	-34.8	-34.8	-32.4	-35.5	-28.5	-38.6	-30.8

	Ca^{2+}	Hg^{2+}	Sr^{2+}	Pb^{2+}	Ba^{2+}
M	-24.6	-26.5	-24.5	-22.4	-19.6
L	-38.7	-30.1	-29.0	-26.3	-23.3

	Al^{3+}	Fe^{3+}	Cr^{3+}	Yb^{3+}	Er^{3+}	Ho^{3+}	Py^{3+}	Tb^{3+}	Gd^{3+}
M	-47.5	-40.9	-47.4	-53.0	-51.5	-51.0	-50.1	-49.6	-49.8
L	-58.4	-60.0	-55.7	-60.4	-59.6	-58.0	-57.0	-56.4	-56.6

	Sm^{3+}	Nd^{3+}	Pr^{3+}	La^{3+}	Th^{4+}
M	-52.6	-52.9	-52.1	-48.9	
L	-58.5	-59.5	-58.7	-55.3	-69.7

	F^-	Cl^-	Br^-	I^-	OH^-	SH^-	CN^-	N_3^-	SCN^-	$HCOO^-$
M	-4.5	23.3	30.2	41.8	1.6	24.7	26.4	30.5	41.1	31.1
L	7.3	25.6	33.6	44.8	6.5	(28)	(30)	(33)	(44)	(38)

	CH_3COO^-	NO_3^-	NO_2^-	HCO_3^-	MnO_4^-	ClO_3^-	BrO_3^-	IO_3^-
M	45.9	34.5	30.5	28.4	48.2	42.0	40.8	29.9
L		36.7	(38)	33.1	49.1	43.9	42.5	(38)

Table 4A.2 (ctd)

	ClO_4^-	HSO_4^-	VO_3^-	$H_2PO_4^-$	$HSeO_4^-$	$H_2AsO_4^-$	ϕSO_3^-	
M	51.6	34.0	32.5	34.6	36.5	44.9	114.4	
L	50.9	44.03	(38.5)	39.35	(41)	44.0		
	S^{2-}	SO_4^{2-}	SeO_4^{2-}	CO_3^{2-}	$PtCl_6^{2-}$	HPO_4^{2-}	CrO_4^{2-}	WO_4^{2-}
M	1.4	24.4	31.5	6.6	155.7	18.6	30.2	36.2
L	(6.3)	34.6		23.3		33.4	38.2	(41)
	MoO_4^{2-}	AsO_4^{3-}						
M	39.4	-0.4						
L	(43)	(18.5)						

* Millero (1969)
** Leyendekkers (1974)
Bracketed values estimated from Fig. 4.7 (assuming L and M conventions give proportional ionic values)

For Li and NH_4 salts and acids, each mole of cation is assumed to contribute -0.6 (salts) and -1.0 (acids) to $(\bar{v}_i^{sw}-\bar{v}_i^{o})$.

TABLE 4A.3

25°C

Coefficients for the equations

$$\phi_v = a_o + a_1 I^{1/2} + a_2 I \qquad \text{mlmol}^{-1}$$

$$\bar{v} = a_o + (3/2) a_1 I^{1/2} + 2a_2 I$$

with I in mol/kg water

Salt Ref.	ϕ_v^o s_v molar eq 4.2.10	a_o	a_1	a_2	std dev. $\pm\sigma$	Ionic strength range
$AgNO_3$	28.01	28.027	2.5571	-	0.004	
*1	2.61	28.004	2.6466	-0.0721	0.0005	
$BaCl_2$	23.6	23.630	2.7360	-	0.007	
*1	4.83	23.588	2.8269	-0.0423	0.0009	
			See Table 4.1a also			
$B(OH)_3$		39.22	-	0.105	0.001 (av.dev)	0-0.5m
*2						
$CaCl_2$	18.25	18.282	3.4079	-	0.008	
*1	6.00	18.237	3.5063	-0.0459	0.001	
*3		17.858	3.611	-0.117	0.05 (std.err)	
			See Table 4.1a also			
$CdAc_2$	71.35	71.864	13.1579	-	0.1	
*1	24.25	71.151	14.6883	-0.6940	0.02	0.151-3.142
$CdCl_2$	23.24	23.300	4.9854	-	0.02	
*1	8.82	23.214	5.1776	-0.0894	0.002	0.151-2.948
HCl	18.07	18.074	0.9375	-	0.0009	
*1	0.95	18.069	0.9576	-0.0163	0.0001	0.05-0.97
KBr	33.73	33.745	1.8936	-	0.004	
*1	1.939	33.725	1.9702	-0.0616	0.0004	0.05-0.986
KCl	26.52	26.534	2.2824	-	0.003	
*1	2.327	26.515	2.3575	-0.0606	0.0004	0.05-0.98
*4		26.85	1.802	0.135	0.005 (av.dev)	0 - 1
K_2CrO_4	37.1	37.225	6.5437	-	0.03	
*1	11.7	37.049	6.9258	-0.1766	0.004	0.151-2.996
KF	6.60	6.607	3.3249	-	0.002	
	3.35	6.597	3.3651	-0.0326	0.0003	0.05-0.962
$KHCO_3$	32.7	32.720	2.5398	-	0.005	
*1	2.6	32.693	2.6420	-0.0822	0.0005	0.05-0.986

TABLE 4A.3 (ctd) 25°C

Salt / Ref.	ϕ_v^o / s_v	a_o	a_1	a_2	std dev $\pm\ \sigma$	Ionic strength range
	molar eq 4.2.10					
KI	45.36	45.376	1.5087	-	0.004	0.05-0.997
*1	1.556	45.355	1.5888	-0.0641		
KNO_3	38.18	38.200	2.2393	-	0.005	
*1	2.3	38.173	2.3424	-0.0827	0.0005	
KOH	2.9	2.907	4.3248	-	0.002	0.05-0.959
*1	4.35	2.896	4.3655	-0.0331	0.0003	
K_2SO_4	32.28	32.481	10.1203	-	0.05	0.151-3.001
*1	18.11	32.192	10.7498	-0.2907	0.007	
		See Table 4A.4 also				
$LaCl_3$	16.0	16.073	4.7572	-	0.02	0.3-5.9
*1	11.87	15.966	4.9227	-0.0545	0.003	
LiBr	24.08	24.086	1.1393	-	0.002	0.05-0.97
*1	1.159	24.078	1.1718	-0.0262	0.0002	
LiCl	17.00	17.006	1.4690	-	0.001	0.05-0.97
*1	1.488	16.998	1.4998	-0.0250	0.0002	
LiOH	-6.0	-6.002	3.0014	-	0.0003	0.05-0.95
*1	3.00	-6.000	2.9943	+0.0058	0.0000	
$MgCl_2$	15.3	15.318	2.4266	-	0.004	0.151-2.912
*1	4.26	15.293	2.4826	-0.0262	0.0006	
*3		14.657	3.117	0.026	0.02	0.05-1.3
		See Tables 4A.4 and 4.1a also				
$Mg(HCO_3)_2$		27.649	4.5776	-0.1391	additivity	
$MgSO_4$	see Table 4A.4					
NH_4Cl	35.98	35.992	1.4145	-	0.003	0.05-0.988
*1	1.45	35.976	1.4743	-0.0481	0.0003	
NH_4NO_3	47.56	47.570	0.9395	-	0.002	0.05-0.999
*1	0.97	47.557	0.9910	-0.0412	0.00002	
NaAc	40.10	40.117	1.8481	-	0.004	0.05-0.992
*1	1.9	40.094	1.9361	-0.0706	0.0004	
$NaB(OH)_4^x$		20.63	1.868	4.708	0.04	0-1
*2						
NaBR	23.51	23.520	1.7301	-	0.002	0.05-0.988
*1	1.76	23.507	1.7800	-0.0401	0.0003	
*3		23.558	1.649	-	0.03	0.05-1
Na_2CO_3	-6.74	-6.727	6.4984	-	0.004	0.15-2.87
*1	11.30	-6.753	6.5542	-0.0263	0.001	

TABLE 4A.3 (ctd) 25°C

Salt Ref.	ϕ_v^o s_v molar eq 4.2.10	a_o	a_1	a_2	std dev ± σ	Ionic strength range
NaCl	16.40	16.409	2.1254	-	0.002	0.05-0.97
*1	2.153	16.397	2.1707	-0.0367	0.0002	
*3		16.617	1.850	0.05	<u>0.03</u>	0.16-3.4
*4		16.62	1.871	-0.041	<u>0.011</u>	0-1
NaF		-3.52	3.178	-0.009	additivity	
$NaHCO_3$		22.57	2.455	-0.058	*	
NaOH	-6.7	-6.70	4.18	-	0.0003	0.05-0.950
*1	4.18					
Na_2SO_4	11.52	11.583	6.9104	-	0.02	0.151-2.923
*1	12.16	11.489	7.1189	-0.0973	0.003	
*3		11.420	7.009	-0.0144	<u>0.04</u>	
*4		11.62	6.617	+0.15	<u>0.02</u>	0-1
$RbNO_3^x$ *5		43.07	2.46	0.15	<u>0.02</u>	0-0.41
$SrCl_2$	17.9	17.960	5.6111	-	0.02	0.151-2.935
	9.9	17.873	5.8024	-0.0891	0.002	

See Table 4.1a also

$TlNO_3$ *5 $\quad \phi_v = 39.5 + 10.83c^{1/2} - 25.46c + 26.0c^{3/2} \pm \underline{0.03}$ (0.03-0.25M)

*1 Parsons (1959)
*2 Ward and Millero (1974)
*3 Lee (1966)
*4 Millero and Masterton (1974), *5 Masterton, Welles, Knox and Millero (1974)
X molar scale

std dev. underlined values are deviations from experimental values, otherwise the ± σ indicates the fit for the values from the original Masson equation (column 2 constants)

TABLE 4A.4

Coefficients for the equation

$$\phi_v = \phi_v^o + a_1 m^{\frac{1}{2}} + a_2 m + a_3 m^{3/2} + a_4 m^2 \text{ mlmol}^{-1}$$

with m in mol/kg water

25°C

Salt	ϕ_v^o	a_1	a_2	a_3	a_4	av.dev.
K_2SO_4	31.99 32.00*	16.493	-20.205	35.520	-21.403	0.022
$MgCl_2$	14.49	8.633	-13.305	18.592	- 8.156	0.053
$MgSO_4$	-7.18	35.44	-73.72	98.400	-48.800	0.052
Na_2CO_3*	-6.13	9.692	1.77			
$NaHCO_3$*	23.45	1.865	1.32			
NaOH*	-5.20	1.865	1.12			
Na_2SO_4*	11.76	9.692	2.70			

0°C

Salt	ϕ_v^o	a_1	a_2	a_3	a_4	av.dev.
K_2SO_4	23.79	3.467	83.526	-204.836	164.316	0.011
$MgSO_4$	-13.06	33.528	-43.528	41.845	-15.203	0.025
NaCl	12.90	-0.053	6.153	-3.610	-	0.06
Na_2SO_4	2.33	0.964	57.195	-76.722	34.822	0.054
KCl	23.63	1.407	1.235	-	-	0.005
$MgCl_2$	10.42	8.132	-0.820	-	-	0.019

From the compilation of Millero and Masterton (1974)
* Perron, Desnoyers and Millero (1975): a_1 is limiting slope for these values, i.e. $1.868(\frac{1}{2}\Sigma\nu_{ion} z^2_{ion})^{3/2} d_o^{\frac{1}{2}}$ at 25°C

TABLE 4A.5 The apparent molal volumes of aqueous boric acid and sodium borate for the temperature range 0-50°C

Parameters for the Redlich equation

$$\phi_V = \phi_V^o + s_V^L c^{1/2} + b_V c \qquad \mathrm{ml\,mol^{-1}}$$

	$t\,^oC$	0	15	25	50
$NaB(OH)_4$	ϕ_V^o	12.39	17.89	20.63	23.51
	s_V^L	1.444	1.697	1.868	2.333
	b_V	8.850	5.981	4.708	4.092
	av dev.	0.06	0.04	0.04	0.06

$$\phi_V^o = 12.38 + 0.434t - 4.229\times10^{-3}t^2 \pm 0.02$$

$$b_V = 8.856 - 0.235t + 2.794\times10^{-3}t^2 \pm 0.01$$

Parameters for the equation

$$\phi_V = \phi_V^o + a_2 m \qquad \mathrm{ml\,mol^{-1}}$$

	$t\,^oC$	0	15	25	50
$B(OH)_3$	ϕ_V^o	36.56	38.44	39.22	41.04
	a_2	0.09_5	0.09_5	0.10_5	0.38_5
	av dev.	0.001	0.000	0.001	0.005

$$\phi_V^o = 36.56 + 0.13t - 8.1\times10^{-4}t^2 \pm 0.05$$

Data from Ward and Millero (1974)

TABLE 4A.6 Coefficients for the equation

$$\phi_v = \phi_v^o + s_v c^{\frac{1}{2}}$$

0°C 1-atm

Salt	ϕ_v^o	s_v	Salt	ϕ_v^o	s_v
CsBr	41.95	3.046	LiI	32.50	1.425
CsCl	35.23	3.293	NaBr	18.86	2.976
CsI	52.74	2.885	NaCl	12.36	3.240
KBr	29.38	3.219	NaI	29.34	2.886
KCl	23.00	3.291	RbBr	34.61	3.112
KI	40.05	3.133	RbCl	28.11	3.287
LiBr	21.94	1.673	RbI	45.15	3.158
LiCl	15.30	1.990			

From the compilation of Parsons (1959)

TABLE 4A.7 The Mean Apparent Molal Volume of Seasalt

(Solution A, Table 4.16) at 1 bar

The equation recommended is

$$\Phi_v^A = M_t^A\psi_t - 315v_w^{(1)}M_t'^A\frac{(1-c)}{S}\log\frac{(B^*+1)}{(B+1)} \quad \text{ml/mol of salts}$$

with $M_t'^A = 68.0811$ g (4.62.8)*

$$M_t^A\psi_t = 37.09 - 0.1498t \quad \text{g} \qquad (4.61.9)*$$

$$B^* = 2670.8 + 6.8966S + (19.39-0.0703S)t - 0.223t^2 \quad \text{bar} \qquad (4.61.7)*$$

$B = B^*(S=0)$

$v_w^{(1)}$ is given by eq 5.71.1 and Table 5.18

Then $I\partial\Phi_v/\partial I = (1-c)S\frac{\partial\Phi_v}{\partial S}$ is given by eq 4.62.9, see Table 5.3 also.

Note that $\left[\frac{S}{(1-c)M_t'}\right]\Phi_v$ is independent of the recipe and gives the total solute effect for a seawater solution containing 1 kg of water. For interest, Φ_v^A has been calculated directly from the specific volume data compiled by Dorsey (1940, Table 108), from the data of Knudsen _et al_, and fitted to $I^{1/2}$ mol/kg and the salinity, S‰. The coefficients, for temperatures, 0,10 and 25°C are listed hereunder. See Table 4.21 also.

TABLE 4A.7 (ctd)

Coefficients for the equations

$$\Phi_v^A = b_o + b_1 S + b_2 S^2$$

$$\Phi_v^A = c_o + c_1 I^{1/2}$$

$$\Phi_v^A = d_o + d_1 I^{1/2} + d_2 I$$

ml/ mol of salts at 1-bar, concentration range

S= 1.8-41.6‰ or I= 0.04-0.86 mol/kg water

t^oC	n	0	1	2	± st. dev.
25	b_n	16.21	0.11691	-1.4531×10^{-3}	0.1
	c_n	15.82	3.203	-	0.1
	d_n	15.09	6.082	-2.454	0.05
10	b_n	14.23	0.12794	-1.5186×10^{-3}	0.1
	c_n	13.74	3.673	-	0.1
	d_n	13.05	6.398	-2.323	0.06
0	b_n	11.99	0.14314	-1.6477×10^{-3}	0.1
	c_n	11.40	4.228	-	0.1
	d_n	10.72	6.933	-2.305	0.06

The density data of Knudsen et al as given by Dorsey (1940, Table 108) were used to calculate Φ_v from eq 4.61.1. The values were then fitted to the above equations via a least-squares analysis.

REFERENCES Appendix 4

Dorsey N.E. 1940.

"Properties of ordinary water-substance", Reinhold, New York. 673pp.

Lee J. 1966.

The apparent and partial molal volumes of electrolytes in water and in aqueous sodium-chloride solution. Thesis, Yale Univ., Ann Arbor, Mich., Univ.Microfilms, 66-4906, Abstr. B,27,131.

Leyendekkers J.V. 1974.

Prediction of the partial molal volumes of electrolytes in seawater. Mar.Chem. 2(2): 89-110.

Masterton W.L., H. Welles, J.H. Knox, and F.J. Millero 1974.

Volume change of ion pair formation: rubidium nitrate and thallium (I) nitrate in water. J.Sol.Chem. 3(2): 91-102.

Millero F.J. 1969.

The partial molal volumes of ions in seawater. Limnol.Oceanogr. 14: 376-385.

Millero F.J. and W.L. Masterton 1974.

Volume change for the formation of magnesium sulfate ion pairs at various temperatures. J.Phys.Chem. 78(13): 1287-1294.

Parson R. 1959.

"Handbook of electrochemical constants", Butterworths, London, 113pp.

Perron G., J.E. Desnoyers, and F.J. Millero 1975.

Apparent molal volumes and heat capacities of some sulfates and carbonates in water at 25°C. Can.J.Chem. 53: 1134-1138.

Ward G.K. and F.J. Millero 1974.

The effect of pressure on the ionization of boric acid in aqueous solutions from molal-volume data. J.Sol.Chem. 3(6): 417-430.

CHAPTER 5

EXPANSIBILITY

The Temperature Derivative of the Partial Molal Volume

5.1 BASIC EQUATIONS.

The change of the pmv of a component, i, with temperature, at constant pressure, is called the partial molal thermal expansibility $\bar{E}_i$.

$$\bar{E}_i \equiv (\partial\bar{v}_i/\partial T)_p \qquad (5.1.1)$$

This differential is of interest, also, because of its relationship to the partial molal entropy, viz.

$$\bar{E}_i \equiv -(\partial\bar{s}_i/\partial p)_T \qquad \text{(See Chapter 3)}$$

Similarly, the apparent molal expansibility of the solute is defined by

$$\phi_E = (\partial\phi_v/\partial T)_p \qquad (5.1.2)$$

Since the molality is independent of temperature the following equations are simply obtained from the appropriate equations of Chapter 4 by differentiating with respect to temperature.

From eq 4.2.2

$$\phi_E = ((\partial V/\partial T)_p - 55.51(\partial V_w^o/\partial T)_p)/\underline{m} \qquad (5.1.3)$$

The coefficient of thermal expansion of the solution, α, and the coefficient of thermal expansion of the pure solvent, α_w, are given by

$$\alpha \equiv (1/V)(\partial V/\partial T)_p = -(1/d)(\partial d/\partial T)_p$$

$$\alpha_w \equiv (1/V_w^o)(\partial V_w^o/\partial T)_p = -(1/d_w)(\partial d_w/\partial T)_p$$

where V is the volume of the solution, d the density. The subscript w indicates the solvent (water).

Equation 3 becomes

$$\phi_E = (\alpha V - \alpha_w 55.51 V_w^o)/m \qquad (5.1.4)$$

and from eq 4.2.8

$$\phi_E = (10^3/mdd_w)(d_w\,\alpha - d\alpha_w) + M_i\,\alpha/d \qquad (5.1.5)$$

$$= (10^3/md_w)(\alpha - \alpha_w) + \alpha\phi_v \qquad (5.1.5a)$$

From eqs 4.2.4,4.2.6,4.2.7, by differentiation w.r.t. temperature,

$$\bar{E}_i = \phi_E + I\partial\phi_E/\partial I \qquad (5.1.6)$$

$$\bar{E}_w = \alpha_w V_w^o - (I^2/w_i(55.51))\,\partial\phi_E/\partial I \qquad (5.1.7)$$

(where $\alpha_w \bar{V}_w^o = \bar{E}_w^o$, and $w_i = \frac{1}{2}z_+z_-\nu_i$)

$$\bar{E}_i - \phi_E = 55.51(\alpha_w V_w^o - \bar{E}_i)/m \qquad (5.1.8)$$

5.11 Volume concentration scale. As for ϕ_v, ϕ_E data for binary solutions are usually in terms of the molar scale. For convenience, the corresponding equations for this scale are listed.

Differentiating the Masson eq 4.2.10 gives

$$\phi_E = (\partial\phi_v^o/\partial T)_p + ((\partial s_v/\partial T)_p - \alpha s_v/2)\,c^{\frac{1}{2}}$$

since $\alpha = -(1/c)(\partial c/\partial T)_p$ (5.11.1)

thus, $\phi_E = \phi_E^o + s_E c^{\frac{1}{2}}$ (5.11.1a)

where $s_E = (\partial s_v/\partial T)_p - \frac{1}{2}\alpha s_v$ (5.11.1b)

At moderate concentrations $s_E \approx (\partial s_v/\partial T)_p - \frac{1}{2}\alpha_w s_v$ (5.11.1c)

Another relationship that could be useful is

$$(\partial s_v/\partial T)_p \simeq -(V^*/10^3)^{\frac{1}{2}}\,\phi_E^o \qquad (5.11.1d)$$

(Where V* is the "critical disruptive volume", Section 4.5)

The coefficient of thermal expansion of the solution may be expressed in terms of its characteristic parameters, viz. (See Harned and Owen 1958).

$$\alpha = \alpha_w + (\phi_E^o - \alpha_w \phi_v^o)10^{-3}c$$

$$+ (s_E - \alpha_w s_v)10^{-3}c^{3/2} \qquad (5.11.2)$$

Values of these parameters for a number of electrolytes are listed in Table 5.1, where $A = (\phi_E^o - \alpha_w \phi_v^o)10^{-3}$ and $B = (s_E - \alpha_w s_v)10^{-3}$

TABLE 5.1 Coefficients for equations 5.11.1a and 5.11.2, 25°C, 1-atm

Salt	$10^2\phi_E^o$	$10^2 s_E$	$10^2\phi_E$ I = 0.7	$10^4 A$	$10^4 B$
HCl	3.4	-0.02	3.4	0.29	-0.004
KCl	8.50	-2.20	6.7	0.78	-0.226
LiCl	2.5	-0.65	2.0	0.21	-0.069
LiOH[x]	4.5	-1.8	3.0	0.47	-0.19
NaCl	9.30	-2.10	7.5	0.89	-0.216
NaOH	11.7	-3.5	8.8	1.19	-0.36
Na_2SO_4	22.19	-9.33	14.38	2.19	-0.966

x less reliable data (Harned and Owen 1958).

$A = (\phi_E^o - \alpha_w \phi_v^o)10^{-3}$; $B = (S_E - \alpha_w S_v)10^{-3}$

For the pme's on the molar scale the $I\partial\phi_E/\partial I$ term of eq 5.1.6 is replaced by $(1000-c\phi_v)/\{2000+c^{3/2}\partial\phi_v/\partial c^{1/2}\}\ c^{1/2}\partial\phi_E/\partial c^{1/2}$ and for eq 5.1.7 $(I^2/w_i 55.51)\ \partial\phi_E/\partial I$ is replaced by $(M_w/d_w)\ \{1/(2000+c^{3/2}\partial\phi_v/\partial c^{1/2})\}\ c^{3/2}(\partial\phi_E/\partial c^{1/2})$

Interconversion between the c and m scales is conveniently made via eq 4.2.20, viz. $m = c/d_w(1-10^{-3}c\phi_v)$ (4.2.20*)

or $c = md_w/(1+m\phi_v d_w 10^{-3})$

5.2 The partial molal expansibilities of ions

The temperature effect on the pmv of an ion in pure water cannot be simply predicted from the charge and radius because of the differences between the expansibility of cations and anions of the same size

(Millero 1968). In any case, such values would be conventional because of the thermodynamic limitations. It is convenient to tabulate $\bar{E}^o_{Ion}$ rather than the individual salts; a procedure already adopted for the pmv's. The values of the salts can easily be derived by summing the ionic values e.g.

$$\bar{E}^o_{CaCl_2} = \bar{E}^o_{Ca^{2+}} + 2\bar{E}^o_{Cl}$$

The available data have been summarised in Appendix 5A.1c.

Millero (1969) has suggested that the pme of salts in seawater can be obtained from

$$\bar{E}_i = \bar{E}^o_i - 0.035 \qquad (5.2.1)$$

This equation is based on the experimental data of Duedall and Weyl (1967) and Duedall (1968). However, there is evidence to suggest that the values of $\bar{v}^{sw}_i$ (exp) are too low at 0°C (Section 5.3) and this would affect the reliability of eq 5.2.1. Nevertheless, it should be useful as a first approximation. The quantity -0.035 corresponds to $\partial\Delta\bar{v}^{tr}_i/\partial t$, which averages -0.08, -0.01 and -0.02 per deg for sulfates, nitrates and chlorides respectively, over the temperature range 0-25°C (as outlined in the next Section, see, e.g., Table 5.5)

5.3 Prediction of the pmv's of salts in seawater at any temperature.

When $\bar{E}^o_i$ data are available, eq 5.2.1 can be used to estimate $\bar{E}^{sw}_i$. The pmv can then be estimated from the value at 25°C, via

$$\bar{v}^{sw}_{i,t^oC} = \bar{v}^{sw}_{i,25^oC} + \bar{E}^{sw}_i(t - 25) \qquad (5.3.1)$$

where t is any temperature over the range 0-30°C.

As illustrated in Chapter 4, there are a number of ways in which the pmv's of salts in seawater can be estimated directly. For temperatures other than 25°C, however, the binary solution data are rather limited, Tables 5.2, 5.2a.

TABLE 5.2 Volume properties at 0°C $ml\,mol^{-1}$ ($I=0.723 mol\,kg^{-1}$, S=35‰ 1 bar pressure)

Electrolyte	$\phi^o_{v_i}$	$\bar{v}_i$	ϕ_{v_i}	$(\phi_{v_i} + \frac{w_i I}{W}(\frac{\partial \Phi_v}{\partial I}))$	$\bar{v}_i^{sw}$ (exp)
$BaCl_2$	17.1	23.28	22.6	26.3	
$CaCl_2$	13.1		16.6	20.3	18.29
CsCl	35.23		38.0	39.2	
KCl	23.00	27.2	25.8	27.0	26.63
	23.63	27.21	25.72	26.93	
LiCl	15.30		17.0	18.2	
$MgCl_2$	10.42	16.01	14.21	17.93	16.66
NaCl	12.36	16.8	15.1	16.3	15.95
	12.90	16.2	15.1		
RbCl	28.11		30.9	32.1	
CsBr	41.95		44.5	45.7	
KBr	29.38		32.1	33.3	
LiBr	21.94		23.3	24.5	
NaBr	18.86		21.4	22.6	
RbBr	34.61		37.2	38.4	
CsI	52.74		55.2	56.4	
KI	40.05		42.7	43.9	
LiI	32.50		33.7	34.9	
NaI	29.34		31.8	33.0	
RbI	45.15		47.8	49.0	
K_2SO_4	23.79	34.65	30.93	34.65	35.41
$MgSO_4$	-13.06	-0.87	-3.96	1.00	-1.16
Na_2SO_4	2.33	14.0	9.53	13.3	14.07
	(4.1)	(16.3)	(12.9)		

ϕ^o_v, $\bar{v}$ and ϕ_v (Parsons 1959), underlined values from data listed in Table 5.2a, $\bar{v}^{sw}$(exp) from Duedall (1968), bracketed values for Na_2SO_4 estimated from $\phi_v(0°C) = \phi_v(25°C) + \phi_E^{25°}(t-25)$ with ϕ_E from Table 5A.1a.

TABLE 5.2a Coefficients* for the equations,

$$\phi_v(0°C) = \phi_v^o + Am^{1/2} + Bm + Cm^{3/2} + Dm^2$$

$$\bar{v}(0°C) = \phi_v^o + \frac{3}{2} Am^{1/2} + 2Bm + \frac{5}{2} Cm^{3/2} + 3Dm^2$$

Salt	ϕ_v^o	A	B	C	D	Average deviation
KCl	23.63	1.407	1.235	-	-	0.005
K_2SO_4	23.79	3.467	83.526	-204.836	164.316	0.011
$MgCl_2$	10.42	8.132	-0.820	-	-	0.019
$MgSO_4$	-13.06	33.528	-43.528	41.845	-15.203	0.025
NaCl	12.90	-0.053	6.153	-3.610	-	0.06
Na_2SO_4	2.33	0.964	57.195	-76.722	34.822	0.054

* Millero and Masterton (1974)

The available data have been used to calculate ϕ_{v_i} at an ionic strength equivalent to S = 35‰, i.e. 0.723, molkg^{-1}, and at 0°C. The pmv of the salt in seawater was then estimated from

$$\bar{v}_i^{sw} = \phi_{v_i} + w_i(I/W)\partial\Phi_v/\partial I + M \qquad (4.4.8)^*$$

where M, the mixing term (see eq 4.4.8) has been neglected. The gradient, $(I/W)\partial\Phi_v/\partial I$, or $((1-c)S/W)\partial\Phi_v/\partial S$, was estimated from the Tait-Gibson equation, (see eq 4.62.9)

$$((1-c)S/W)\partial\Phi_v/\partial S = Av_w^{(1)}(1-c)\ \{(1/S)\log\frac{(B^*+1)}{(B+1)} - (0.434(1-c)\partial B^*/\partial S)/(B^*+1)\} \qquad (5.3.2)$$

where A = 315x68.0811/1.3576, B* is given by eq 4.61.7 and $v_w^{(1)}$ by eq 5.72.6 and Table 5.18 Section 5.7. Values of this gradient (Ekman's (1908) data) are listed in Table 5.3 for salinities 30‰, 35‰, 40‰ and 34‰, for the temperature range 0-25°C.

TABLE 5.3 The concentration gradient of the mean apparent molal volume of seasalt eq 5.3.2

$((1-c)S/W)\partial\Phi_v/\partial S \equiv (I/W)\partial\Phi_v/\partial I$

$ml\,mol^{-1}$

S‰ / t°C	30	34	35	40
0	1.08	1.21	1.24	1.39
5	0.95	1.06	1.09	1.22
10	0.84	0.94	0.97	1.09
15	0.75	0.84	0.86	0.97
20	0.67	0.75	0.77	0.86
25	0.60	0.67	0.69	0.77

Data of Ekman (1908) as given by Li (1967)

Values of $\bar{v}^{sw}_{i,0^oC}$ calculated in this way are shown in Table 5.4 together with the experimental values. The value of $\bar{V}_s$ ($\Sigma f_i \bar{v}_i^{sw}$) at 0°C, from direct measurement, is low relative to the value calculated from the most reliable seawater density data (Ekman (1908), and Millero (1973), Dorsey's (1940) Table 108 - Section 5.5). It could therefore be assumed that the $\bar{v}_i^{sw}$(exp) values at this temperature are too low. The results in Table 5.4 do not contradict this assumption but are somewhat inconclusive, as the mixing term is not known. The volume properties of calcium chloride seem to be anomalous in binary solutions, relative to the alkali halides and barium chloride (Dunn 1968). This salt had a larger Q term at 25°C than the other chlorides (Chapter 4). This also occurs at 0°C. Further experimental data for $\bar{v}_i^{sw}$ and $\phi^o_{v_i}$, in the low temperature region (0-10°C) would be of value, especially since this is the temperature area of most interest in ocean-

TABLE 5.4 The partial molal volume change on transfer from water to seawater at $0^{o}C$, $ml\,mol^{-1}$ (S = 35‰, 1 bar)

Electrolyte		$\phi^{o}_{v_i}$ *	$\bar{v}^{sw}_i$ exp	$\bar{v}^{sw}_i$ calc	$\Delta\bar{v}^{tr}_i/\nu$ exp		$\Delta\bar{v}^{tr}_i/\nu$ calc
$CaCl_2$		13.05	18.29	20.3	2.6		3.6
CsCl		36.13		39.2			3.1
		(35.23)					(4.0)
KCl	23.63	23.63	26.63	27.0	3.0		3.4
		(23.00)			(3.6)		(4.0)
$MgCl_2$	10.42	11.09	16.66	17.2	2.8		
NaCl	12.90	12.95	15.95	16.3	3.0		3.3
		(12.36)			(3.6)		(3.9)
RbCl		28.76		32.1			3.3
		(28.11)					(4.0)
$Ca(HCO_3)_2$		24.2	31.93		3.9		
$KHCO_3$		29.2	33.45		4.3		
$Mg(HCO_3)_2$		22.2	30.32		4.1		
$NaHCO_3$		18.5	22.77		4.3		
$Ca(NO_3)_2$		33.4	38.0		2.3		
KNO_3		33.8	36.49		2.7		
$Mg(NO_3)_2$		31.4	36.37		2.5		
$NaNO_3$		23.1	25.80		2.7		
$CaSO_4$	(-11.3)	-8.7	0.50		9.2		
K_2SO_4	23.79	25.5	35.41		9.9	11.62	
$MgSO_4$	-13.06	-10.7	-1.16		9.5	11.90	
Na_2SO_4	2.33	4.1	14.07		10.0	11.74	
CsBr		42.74		45.7			3.0
		(41.95)					(3.7)
KBr		30.23		33.3			3.1
		(29.38)					(3.9)
NaBr		19.55		22.6			3.0
		(18.86)					(3.7)
RbBr		35.37		38.4			3.0
		(34.61)					(3.8)
CsI		53.25		56.4			3.1
		(52.74)					(3.7)
KI		40.74		43.9			3.2
		(40.05)					(3.8)
NaI		30.06		33.0			2.9
		(29.34)					(3.7)
RbI		45.88		49.0			3.1
		(45.15)					(3.8)

* Table 4A.1 (summed ionic values), bracketed values (Parsons 1959) underlined values (Millero and Masterton 1974, and references therein) exp Duedall (1968)

ographic studies of systems under pressure. If the $\bar{v}_i^{sw}$ (exp) values at 0^oC are too low, then this would limit the usefulness of eq 5.2.1 which is based on these results.

The values for the volume of transfer ($\Delta\bar{v}_i^{tr}$) given previously (Leyendekkers 1974) will also be affected. The $\Delta\bar{v}_i^{tr}$ method of prediction is particularly useful since the only binary solution data required is $\phi^o_{v_i}$. Data for $\phi^o_{v_i}$ are generally available for most temperatures, or can be predicted more readily than other volume data (Table 4A.1) The pmv is then estimated from

$$\bar{v}_i^{sw} = \phi^o_{v_i} + \Delta\bar{v}_i^{tr} \qquad (4.42.2)*$$

Values of $\Delta\bar{v}_i^{tr}$ for 0^oC estimated from the experimental $\bar{v}_i^{sw}$ values and from the calculated values (mixing term neglected), are compared in Table 5.4 according to the anion group. The mean values for 0^oC and 25^oC are listed in Table 5.5.

TABLE 5.5 Mean $\Delta\bar{v}_i^{tr}/\nu-$ for anion groups (S = 35‰, 1 bar)

Anion	0^oC^x		25^oC^{xx}
	exp	calc	exp
Cl-	2.9	3.3	2.4
Br-		3.0	
I-		3.1	
HCO_3^-	4.2		4.3
SO_4^{2-}	9.7 (11.8)*		9.8
NO_3^-	2.6		2.3
CO_3^{2-}			16.8

x Table 5.4 , only unbracketed values used

xx Table 4.9

* From ϕ^o_v values of Millero and Masterton (1974), probably more reliable than summed ionic values.

This comparison was only made for the chlorides as the mixing term is not likely to be large, and there are adequate binary solution data, for these salts. The difference in $\Delta\bar{v}^{tr}$ for chlorides is within the uncertainties of ϕ_v^o, so that the previous conclusion (Leyendekkers 1974) that this quantity ($\Delta\bar{v}^{tr}$) is not very sensitive to temperature (over the range 0-25°C) is reasonable for chlorides. Probably this is valid for other strong electrolytes. The sulfate value increases by 2 $ml mol^{-1}$, however, at 0°C and the change in the carbonate value is likely to be larger.

The discrepancy between the $\bar{V}_s$(exp) and $\bar{V}_s$ from seawater density data at 0°C (see next section), appears to derive from low experimental values for NaCl and Na_2SO_4 and possibly $MgCl_2$. The discrepancy is only 0.3-0.4 $ml mol^{-1}$, however, and in general the results suggest that the $\Delta\bar{v}^{tr}$ equation should give useful estimates for many salts, even if the 25°C value of $\Delta\bar{v}^{tr}$ is used. Carbonates and other oxy-anion salts will probably be exceptions.

5.4 Seawater solutions

As in previous chapters, seawater is considered as a mixed electrolyte solution, and the recipe of Kester et al (solution A, Table 4.16) is used as the basis for the calculations. That is, the properties of the salts NaCl, Na_2SO_4, $NaHCO_3$, $MgCl_2$, $CaCl_2$ $SrCl_2$, NaBr and KF, in the binary solutions, at the same ionic strength as seawater, will be used as a reference for seawater properties. Wherever possible, the predicted or measured values of the component properties in seawater will be used as well. The properties of the solution can be predicted from the summed contributions of the components and comparison made with directly measured values, as was done for the partial molal volume at one bar pressure (Chapter 4).

It is convenient to define the mean partial molal expansibility $\bar{E}_s$, and the mean apparent molal expansibility, Φ_E, as the average value per mole of salt (solution A) as was done for $\bar{V}_s$. These quantities are proportional to the mean value per ion, $\bar{E}_s^I$ and Φ_E^I, and for a given S‰, to $\bar{E}_s^{sw}$ and Φ_E^{sw}, the mean values per gram of seawater. The relationships are:

$$\bar{E}_s^A = \Sigma_i f_i \bar{E}_i^A \quad \text{mldeg}^{-1}\text{mol}^{-1}\text{(salt) solution A}$$

where $\underline{i}$ refers to a component of solution A and $f_i = m_i/\Sigma m_i$

$$\bar{E}_s^I = \bar{E}_s^A/2.178796 \quad \text{mldeg}^{-1}\text{mol}^{-1}\text{(ion)}$$

$$\bar{E}_s^{sw} = \bar{E}_s^A(S/68.0811(10^3+0.0049S)) \ \text{mldeg}^{-1}\text{g(seawater)}$$

$$\simeq 10^{-3}\bar{E}_s^A S/68.0811$$

where S represents the salinity as usual.

The value of $\bar{V}_s$ at any temperature may be obtained directly as it was for 25°C, or from

$$\bar{V}_{s,t} = \bar{V}_{s,25} + \bar{E}_s(t - 25) \qquad (5.4.1)$$

where $\bar{E}_s$ is the average value over the temperature range. Before considering the expansibility of seasalt, the direct calculation of $\bar{V}_s$, for a range of temperatures will be briefly considered.

5.5 The mean partial molal volume of seasalt over the temperature range 0–30°C.

The equations for estimating the mean pmv of seasalt over a range of temperatures and salinities have been given in Chapter 4 (eqs 4.61.6, 4.61.15) The values at 25°C were calculated from direct experimental values (Duedall, 1968) and from seawater density and compressibility data. The same calculations for 0°C are given in Table 5.6.

At S = 35‰, the estimates for the WB data (unsmoothed) give better agreement with the direct experimental value (eq 4.61.20). This is in contrast to the results for 25°C and for lower salinities. A comparison of the experimental $\bar{V}_s$ with estimates from the E and WM data at 5°C intervals (Table 5.7) suggests that the experimental value at 0°C, S = 35‰, is probably too low. This was suggested by the results for the estimates of

TABLE 5.6 The mean partial molal volume of seasalt* $ml\,mol^{-1}$(salt) at 0°C, 1 bar

S(‰)	D	E	WM	EXP	NK	WB	WB'
1.84	12.825	13.238	13.430			13.293	12.233
10.86	14.266	14.198	14.323			14.480	14.182
21.69	15.466	15.297	15.539			15.229	15.366
30.52	16.260	16.141	16.192	16.09	16.08	15.569	15.969
30.72	16.276	16.159	16.204			15.577	15.979
34.99	16.626	16.553	16.441	16.14	16.70	15.740	16.202
36.13	16.719	16.657	16.496			15.783	16.258
41.03	17.099	17.095	16.694		17.23	15.967	16.467
41.55	17.141	17.141	16.712			15.987	16.489

* Mixture of salts according to Solution A (Table 4.16).
D Dorsey (Table 108, 1940), E Ekman (1908), WM Wang Millero (1973), EXP (Duedall 1968), NK Newton and Kennedy (1965), WB Wilson and Bradley (1966), smoothed data, WB' raw data.

TABLE 5.7 The mean partial molal volume of seasalt* $ml\,mol^{-1}$ (salt) at 1 bar, as a function of temperature

Temp °C	S = 30‰ a	S = 30‰ b	S = 30‰ c	S = 35‰ a	S = 35‰ b	S = 35‰ c
0	16.09	16.09	16.16	16.14	16.55	16.44
5	17.00	16.96	17.07	17.08	17.37	17.31
10	17.76	17.67	17.83	17.87	18.03	18.04
15	18.36	18.25	18.46	18.50	18.57	18.65
20	18.82	18.72	18.96	18.98	19.01	19.14
25	19.12	19.10	19.36	19.32	19.36	19.52
30	19.27	19.41	19.65	19.49	19.65	19.79

* Mixture of salts according to Solution A (Table 4.16).
a Experimental (Duedall, 1968), b Ekman (1908),
c Wang Millero (1973).

$\bar{V}_i^{sw}$ at low temperatures given in section (5.3). Additional measurements at 0°C for salinities around 35-40‰ are needed to clarify this.

As shown by Fig. 5.1 the values of $\bar{V}_s^{sw}$ given by the E data, using the Tait Gibson equation (eqs 4.61.7, 4.61.9, and 4.61.10) are in close agreement with the experimental data at S = 30‰ and 35‰, with the exception of the region 0-5° C at 35‰. However, the experimental values are less reliable in this temperature region. The deviations at 0°C increase with salinity, but the same trends are shown by the WM data with which the E data is in good overall agreement.

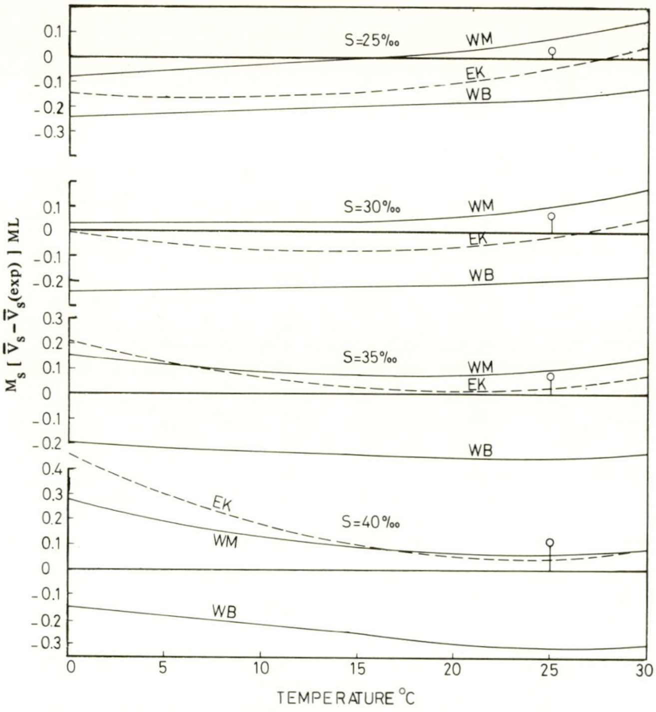

FIGURE 5.1

Differences in the calculated and experimental partial molal volumes of seasalt (for a solution containing 1 kg or 55.51 moles of water) : ⚲ From binary solution data eq 4.61.21
EK Ekman (as given by Li (1967))
WM Wang and Millero (1973)
WB Wilson and Bradley (1968)
Experimental $\bar{V}_s$ from Duedall (1968).

Values of $\bar{V}_s^{SW}$ and $\bar{V}_s^{I}$ for the experimental results are given in Table 5.8. At 25°C the comparative values from binary solution data indicate the small effect of mixing at this temperature. There are insufficient accurate data at the other temperatures for all the seawater components. However, an estimate for 0°C was made from binary solution data as given in Tables 5.2 and 5.4. The result is summarised in Table 5.9.

TABLE 5.8 The mean partial molal volume of seasalt* per g seawater, and per mole of ions, as a function of salinity and temperature at 1 bar.

S‰	20		25		30		35		40	
t°C	a_3 x10^3	b	a_3 x10^3	b	a_3 x10^3	b	a_3 x10^3	b	a_3 x10^3	b
0	4.694	7.334	5.887	7.359	7.087	7.383	8.296	7.408	9.513	7.433
5	4.945	7.726	6.211	7.764	7.489	7.801	8.779	7.839	10.081	7.876
10	5.151	8.049	6.479	8.099	7.823	8.150	9.183	8.200	10.559	8.250
15	5.313	8.301	6.691	8.365	8.090	8.428	9.509	8.491	10.949	8.555
20	5.430	8.484	6.848	8.560	8.291	8.636	9.757	8.713	11.249	8.789
25	5.502	8.597 (8.538)	6.949	8.686 (8.664)	8.424	8.775 (8.779)	9.927	8.864 (8.885)	11.459	8.954 (8.986)
30	5.529	8.640	6.993	8.742	8.490	8.844	10.019	8.946	11.581	9.048

* Experimental results (Duedall 1968), values independent of solution of salts chosen.

a ml/g seawater $(\bar{V}_s^A S/68.0811(10^3+0.0049S))$

b ml/mol ions $(\bar{V}_s^A/2.178796)$

Bracketed values at 25°C are from binary solution data.

The value of $\bar{V}_s$ is very close ($\pm$0.05mlmol^{-1}) to the values from Dorsey (1940, Table 108) and Ekman (1908). It is also in good agreement with the results from the Wang-Millero (1973) equation. This suggests that the experimental $\bar{V}_s$ is about 0.4mlmol^{-1} too low, that is, assuming the total mixing effect is negligible.

In the next chapter, the effect of pressure on $\bar{V}_s$ will be considered and the $\bar{V}_s$ contour covering the average oceanographic pressure-temperature areas will be given.

TABLE 5.9 Estimate of the mean partial molal volume of seasalt from binary solution data.

$ml mol^{-1}$(salts) at 1 bar,

S = 35‰ (I = 0.723 mol/kgwater)

0°C

Salt	$f_i = m_i/\Sigma m_i$	$\bar{v}_i$	$f_i \bar{v}_i$
NaCl	0.797397	16.50	13.157
KCl	0.017666	27.21	0.481
$NaHCO_3$	0.004539	22.77	0.103
KBr	0.001602	33.3	0.053
$MgCl_2$	0.103630	16.01	1.658
$CaCl_2$	0.020271	19.0	0.385
Na_2SO_4	0.054895	14.0	0.769
		$\bar{V}_s^{A'} = \Sigma f_i \bar{v}_i$	16.606

Solution A used as a basis for the electrolyte mixture (Table 4.16), but with NaCl replacing NaF and H_3BO_3 contributions to m_s (represents a change in f_{NaCl} of only 0.000957). Also, $SrCl_2$ (f_{SrCl_2} = 0.000173) replaced by $CaCl_2$.
The values of $\bar{v}_i$ are from Tables 5.2 or 5.4.

5.51 The apparent specific volume, asv, and the partial molal volume pmv, of water in seawater, 0-30°C.

The equations for calculating the asv and pmv of water in seawater have been given in Section 4.7. The asv, or Ψ_w, was shown to be equivalent to the specific volume of pure water under an applied pressure of p_e bars. Since,

$$p_e = 6.8966S - 0.0703St \tag{5.51.1}$$

the compressional effect of the salt on the water decreases with temperature, as shown by Fig. 4:9.

The pmv changes with concentration are only small due to the opposing gradients of ψ_w, which decreases as the salinity increases, and the term $(v_w^{(1)}(I\partial B^*/\partial I)/(B^*+p))$ which increases (see eq 4.72.3 and Fig. 4.10).

The changes with temperature for salinities 30,35 and 40‰ are shown in Table 5.10. The values show a minimum around 4°C as expected and then increase. The temperature gradient at 25°C is around 4.8×10^{-3} $ml deg^{-1} mol^{-1}$ (there is a small increase in this gradient with salinity, ca.2% from 30‰ to 40‰ in salinity). This value is to be compared with the corresponding partial molal expansibility, $\bar{E}$, (at 25°C, I = $0.72 mol kg^{-1}$) for the solutes NaCl and Na_2SO_4 of 75×10^{-3} and 143×10^{-3} $ml deg^{-1} mol^{-1}$, respectively. The mole ratio water: solute is, however, about 80:1 in such solutions.

TABLE 5.10 The partial molal volume of water in seawater as a function of temperature

1 bar pressure

	$\bar{V}_w^{(1)}$ $ml mol^{-1}$		
S‰	30	35	40
t°C			
0	18.0057	18.0015	17.9969
5	18.0043	18.0007	17.9966
10	18.0103	18.0071	18.0034
15	18.0222	18.0193	18.0160
20	18.0392	18.0366	18.0337
25	18.0610	18.0587	18.0561
30	18.0871	18.0850	18.0826

The mean change of the pmv of the water over 30° is about $0.08 ml mol^{-1}$ compared to a change of $-0.02 ml mol^{-1}$ over the salinity range 5-40‰.

5.6 The mean partial molal expansibility and mean apparent molal expansibility of seasalt.

The direct experimental results (Duedall 1968) for $\bar{V}_s^A$, summarised by the empirical equation (4.61.20) can be differentiated with respect to temperature to give $\bar{E}_s^A = (\partial\bar{V}_s^A/\partial t)_p$ = 16.36x10^{-2}-0.608x10^{-2}t+ 0.112x10^{-2}S (5.6.1)

However, in view of the somewhat limited experimental data (S=30‰, t=0 and 24.5°C; S=34.5‰, t=0,10,24.5°C; and S=40‰, t=24.5°C) and the discrepancy noted for the higher salinities at 0°C (Section 5.3), this equation is possibly not very accurate. In addition, a more general equation to cover changes in pressure (up to ca.1000bars) is required.

5.61 Tait-Gibson equation. The pme can be obtained by differentiating eq 4.61.10 (at constant pressure and salinity) to give

$$\bar{E}_s^{(p)} = M_t(\partial\psi_t/\partial t) + \left[A(1-c)^2 v_w^{(1)}/(B^*+p)\right] \times \left[G^*\left(\frac{\partial B^*/\partial t}{(B^*+p)} - \alpha_w^{(1)}\right) - \partial G^*/\partial t\right] \quad (5.6.2)$$

where A = 434.3M_t'x0.315 = 9.3138x10^3(solution A),

and since $B^* = 2670.8 + 6.8966S + (19.39-0.0703S)t - 0.223t^2$

$$G^* = \partial B^*/\partial S = 6.8966 - 0.0703t$$

$$M_t(\partial\psi_t/\partial t) = -0.1498$$

$$\partial G^*/\partial t = -0.0703, \quad \partial B^*/\partial t = 19.39-0.0703S-0.446t$$

$v_w^{(1)}$ is the specific volume of pure water at 1 bar (eq 5.72.5, Table 5.18) and $\alpha_w^{(1)}$ is the thermal expansivity of pure water, $((1/v)\partial v/\partial t)_p$, which is obtained from $v_w^{(1)}$ by differentiation.

The pme can also be obtained from the mean apparent molal expansibility, Φ_E, via

$$\bar{E}_s = \Phi_E + I\partial\Phi_E/\partial I$$
$$= \Phi_E + S(1-c)\,\partial\Phi_E/\partial S \quad (5.6.3)$$

The expression for Φ_E is simply obtained by differentiating eq 4.62.7 w.r.t. temperature,

$$\Phi_E = (\partial\Phi_v/\partial t)_p = M_t'\frac{(1-c)}{c}(\partial\psi_w/\partial t - \alpha_w v_w) + M_t\partial\psi_t/\partial t \quad (5.6.4)$$

The terms for any pressure, are given by:

$$\psi_w^{(p)} = v_w^{(1)}\left(1-0.315\log\frac{(B^*+p)}{(B+1)}\right)$$

$$\psi_t = 0.54214 - 2.1896\text{x}10^{-3}t$$

$$\partial\psi_w/\partial t = \alpha_w^{(1)}\psi_w^{(p)} + v_w^{(1)}0.1368D_1$$

from Section 4.62 (5.6.5)

with $D_1 = (\partial B/\partial t)/(B+1) - (\partial B^*/\partial t)/(B^*+p)$

and $\partial B/\partial t = (\partial B^*/\partial t)_{S=0}$, $B = B^*_{S=0}$

so that $$\Phi_E = 0.315 M'_t \frac{(1-c)}{c} v_w^{(1)} \{0.4343 D_1 - \alpha_w^{(1)} \log \frac{(B^*+p)}{(B+1)}\} + M_t \partial\psi_t/\partial t \qquad (5.6.6)$$

By differentiating this equation w.r.t. S, eq 5.6.2 for $\bar{E}_s$ is obtained via eq 5.6.3.

In general, Φ_E may be obtained from any data set via the equations

$$\Phi_E = M'_t(1-c)(\alpha_{sw} - \alpha_w)/cd_w + \alpha_{sw}\Phi_v \qquad (5.6.7)$$

or $$\Phi_E = M'_t(1-c)\ ((\alpha v)_{sw} - (\alpha v)_w)/c + M_t(\alpha v)_{sw} \qquad (5.6.8)$$

and $\bar{E}_s$ from (5.6.3), to give, for any pressure p

$$\bar{E}_s^{(p)} = (\alpha v)_{sw}^{(p)} M_t + A_s(\partial(\alpha v)_{sw}^{(p)}/\partial S)_{T,p} \qquad (5.6.9)$$

where $A_s = S(1-c)M_t + 10^3 M'_t(1-c)^2$

and α_{sw} represents the coefficient of thermal expansion of seawater, and

$$\alpha v = (\partial v/\partial t)_p$$

As an illustration of the use of (5.6.8) and (5.6.9) the equation of state developed by Wang and Millero (1973), and the direct experimental measurements of the thermal expansion will be used.

5.62 Φ_E and $\bar{E}_s$ from the Wang-Millero equation of state

The WM equation has been given previously (Section 4.61) and only a summary is given here, the coefficients are listed in Table 4.18. The basic equation is:

$$v_{sw}^{(p)} = v_{sw}^{(1)} - (p-1)v_w^{(1)}/R_{sw}$$

with $R_{sw} = B_{sw}(S,t) + A_{sw_1}(S,t)(p-1) + A_{sw_2}(S,t)(p-1)^2$

(22 Coefficients)

$$v_{sw}^{(1)} = 1.000027/(1+\sigma_t 10^{-3})$$

with $\sigma_t = f(S,t)$ (16 coefficients)

and similarly for water, with

$$v_w^{(1)} = f(t) \qquad \text{(8 coefficients)}$$

$R_w = f(t,p)$ (14 coefficients)

The terms in eqs (5.6.8) and (5.6.9) are obtained by differentiating w.r.t. temperature and then salinity, viz, with p' = (p - 1),

$$(\alpha v)^{(p)}_{sw} = (\alpha v)^{(1)}_{sw} - (\alpha v)^{(1)}_{w} p'/R_{sw} + (v^{(1)}_{w} p' \frac{\partial R_{sw}}{\partial t})/R^2_{sw} \quad (5.62.1)$$

$$\frac{\partial(\alpha v)^{(p)}_{sw}}{\partial S} = \frac{\partial(\alpha v)^{(1)}_{sw}}{\partial S} + (p'/R^2_{sw})\Big[\frac{\partial R_{sw}}{\partial S}\{(\alpha v)^{(1)}_{w} - (2v^{(1)}_{w}/R_{sw})(\frac{\partial R_{sw}}{\partial t})\} + v^{(1)}_{w} \frac{\partial(\partial R_{sw}/\partial t)}{\partial S}\Big] \quad (5.62.2)$$

with

$$\frac{\partial(\alpha v)^{(1)}_{sw}}{\partial S} = \frac{10^{-3}x1.000027}{(1 + 10^{-3}\sigma_t)^2}\Big[\frac{2x10^{-3}}{(1 + 10^{-3}\sigma_t)}\{\frac{\partial\sigma_t}{\partial t}\frac{\partial\sigma_t}{\partial S}\} - \frac{\partial(\partial\sigma_t/\partial S)}{\partial t}\Big]$$

Values of $\bar{E}_s$ and ϕ_E calculated from these equations are considered below.

5.63. Φ_E and $\bar{E}_s$ from thermal expansion measurements.

Bradshaw and Schleicher (1970) presented their experimental results in the form of a 27-term polynomial. Later, Bryden (1973) developed a 22-term polynomial (based on a least-squares analysis of the same experimental data). The relationship is

$$(\alpha v)^{(p)}_{sw} = a_o(p,S) + a_1(p,S)t + a_2(p,S)t^2 + a_3(p,S)t^3 \quad (5.63.1)$$

The various terms of Bryden's polynomial are listed in Table 5.11. Values of $\bar{E}_s$ and Φ_E can be more simply calculated from this relationship, and the differential w.r.t. S, than from the WM equation of state.

The salinity differential is given by

$$(\partial(\alpha v)_{sw}/\partial S)_{T,P} = g_o(p.S) + g_1(p)t + g_2(p)t^2 \quad (5.63.2)$$

where, e.g.

TABLE 5.11 Coefficients for the equation 5.63.1 (cm^3/g.deg)

$$(\alpha v)_{sw} = \partial v(T,p,S)/\partial T = a_o(p,S)+a_1(p,S)t + a_2(p,S)t^2 + a_3(p,S)t^3$$

with $10^6 a_n(p,S) = b_o+b_1p+b_2p^2+b_3p^3+(d_o+d_1p+d_2p^2)\ (S-35) + e(S-35)^2$

n	0	1	2	3
b_o	51.901	12.228	-0.13733	0.11886 x10^{-2}
b_1	0.27073	-0.011607	0.16712 x10^{-3}	-0.11078 x10^{-5}
b_2	-0.89942 x10^{-4}	0.42208 x10^{-5}	-0.44555 x10^{-7}	-
b_3	0.12001 x10^{-7}	-0.55154 x10^{-9}	-	-
d_o	2.8173	-0.11179	0.12329 x10^{-2}	-
d_1	-0.26390 x10^{-2}	0.10806 x10^{-3}	-0.12373 x10^{-5}	-
d_2	0.67564 x10^{-6}	-0.23520 x10^{-7}	-	-
e	-0.81218 x10^{-2}	-	-	-

Temperature, t (°C), salinity, S(‰) and absolute pressure, p (bars) The two-standard deviation uncertainty of this equation is ($3.31\text{x}10^{-6}cm^3$/ g.deg) $30‰ \leq S \leq 40‰$ (Bradshaw and Schleicher data (1970) as analysed by Bryden (1973))

$$10^6 g_o = 2.8173 - 0.26390\text{x}10^{-2}p + 0.67564\text{x}10^{-6}p^2 - 1.62436\text{x}10^{-2}(S-35)$$

$$10^6 g_1 = (-0.11179 + 0.10806\text{x}10^{-3}p - 0.2352\text{x}10^{-7}p^2)$$

$$10^6 g_2 = (0.12329\text{x}10^{-2} - 0.12373\text{x}10^{-5}p)$$

from Bryden's polynomial. $\bar{E}_s$ can be calculated by substituting eqs 5.63.1 and 5.63.2 into eq 5.6.9.

Values of Φ_E and $\bar{E}_s$ at 1 bar, calculated from the various data sets and equations, are compared in Tables 5.12 and 5.13. For Φ_E, the overall agreement is good.

TABLE 5.12 The mean apparent molal expansibility of seasalt* (Φ_E) $ml\,mol^{-1}deg^{-1}$ (S=34.702‰, 1 bar)

t °C	WM	BS1	BS2	EK
0	0.2258	0.2298	0.2296	0.2233
5	0.1859	0.1842	0.1841	0.1833
10	0.1519	0.1499	0.1496	0.1508
15	0.1232	0.1227	0.1225	0.1241
20	0.0994	0.1002	0.1000	0.1023
25	0.0805	0.0811	0.0808	0.0843

* solution A (Table 4 .16)
WM Wang and Millero (1973), BS1 Bradshaw and Schleicher (1969) with Bryden's polynomial (1973), BS2 Bradshaw and Schleicher (1969), Ekman (1908)

For $\bar{E}_s$ differences of 7% arise at the higher temperatures between the two polynomials representing the experimental thermal expansion, otherwise the differences in these two sets are 2-3%. The values from the TG equation are generally lower than the other sets. The TG results at 1 bar derive from Knudsen's density data and cover a wider salinity range than the BS set (viz. 2-41‰ versus 30-40‰) so that the salinity derivative is probably more reliable. The differences in the temperature derivatives are only about 2%, as will be shown in the next section, and as indicated by the Φ_E values. The latter can be simply represented as a linear function of salinity at 1 bar (Appendix 5).

The temperature derivative of the experimental $\bar{V}_s$ cannot be expected to be very reliable because of the limited temperature range covered. When further data become available for $\bar{v}_1^{sw}$, especially at the lower temperatures, a more reliable value of $(\partial\bar{V}_s(exp)/\partial t)$ can be obtained. This will be useful for deciding on the accuracy of the seawater data, especially the salinity derivatives.

TABLE 5.13 The mean partial molal expansibility of seasalt*
$ml\,mol^{-1} deg^{-1}$
(S=34.702‰, 1 bar)

t °C	$\bar{E}_s^A$					
	WM	BS1	BS2	EK	DW1	DW2
0	0.18785	0.18884	0.19265	0.18095	0.202	0.166
5	0.16026	0.15812	0.15934	0.14745	0.172	0.147
10	0.13431	0.13105	0.12995	0.12001	0.142	0.129
15	0.10980	0.10767	0.10852	0.09742	0.111	0.110
20	0.08657	0.08805	0.09148	0.07881	0.081	0.092
25	0.06452	0.07225	0.07721	0.06350	0.051	0.075
30	0.04350	0.06034	0.06404	0.05101	0.020	0.055

* solution A (Table 4.16)
WM Wang and Millero (1973), BS1 Bryden (1973), data of Bradshaw and Schleicher (1969), BS2 Bradshaw and Schleicher (1969),
EK Ekman (1908), DW1 Duedall (1968), eq 5.6.1.
DW2 data of Duedall and Weyl (1967) for 10° and 25°

This is the most direct approach for comparison, and avoids any assumptions regarding mixing effects. However, as for the pmv's it is also useful to compare the results based on binary solution data.

5.64 Redlich Equation

Millero (1974) has calculated the apparent equivalent expansibility, aee, of seasalt by differentiating the apparent equivalent volume, aev, (expressed as a function of temperature). As for the volume quantities he has analysed the results on the basis of the Masson and Redlich equations, viz.

$$\Phi_E^e = \Phi_E^{e^o} + s' I_v^{1/2} \quad \text{(Masson)} \qquad (5.64.1)$$

$$\Phi_E^e = \Phi_E^{e^o} + s I_v^{1/2} + b I_v \quad \text{(Redlich)} \qquad (5.64.2)$$

where I_v is the volume ionic strength. The superscript e indicates that the quantity is in the equivalent form. The coefficients of these equations have been compared with theoretical estimates, on the basis of

Young's rule. The results indicate that prediction can be made to within the experimental error of the single electrolyte data.

For interest, values of Φ_E (the mean apparent molal value) calculated from the EK data set have been fitted to the ionic strength; mol/kg pure water, rather than I_v. The differences in these ionic strengths are only small (See Table 6.1 and Fig 6.4) so that the coefficients for the molal equations of the Masson and Redlich form, using I and I_v, should not differ very much. Table 5.14 lists the coefficients for a range of temperatures.

TABLE 5.14 Coefficients for the Redlich and Redlich-extended forms of equation for the mean apparent molal expansibility of seasalt* (1 bar)

$$\Phi_E = a_o + a_1 I^{1/2} + a_2 I \qquad \text{Redlich}$$

$$\Phi_E = b_o + b_1 I^{1/2} + b_2 I + b_3 I^{3/2}$$

$ml\,mol^{-1} deg^{-1}$

t^oC	a_o / b_o	a_1 / $b_1 x10^3$	a_2 / $b_2 x10^3$	$b_3 x10^2$	std. dev.
0	0.2740	-15.31	-5.237		$2x10^{-4}$
	0.2710	4.206	-9.021	2.232	$7x10^{-6}$
5	0.2258	-12.19	-4.472		$1.4x10^{-4}$
	0.2235	3.439	-7.503	1.788	$6\ x10^{-6}$
10	0.1870	- 9.855	-3.864		$1.2x10^{-4}$
	0.1850	2.848	-6.328	1.453	$5\ x10^{-6}$
15	0.1553	- 8.083	-3.375		$9.6x10^{-5}$
	0.1537	2.385	-5.406	1.198	$4\ x10^{-6}$
20	0.1294	- 6.715	-2.979		$8x10^{-5}$
	0.1281	2.019	-4.673	0.9992	$3.8x10^{-6}$
25	0.1082	- 5.645	-2.654		$6.7x10^{-5}$
	0.1070	1.727	-4.084	0.8433	$3.3x10^{-6}$
30	0.0909	- 4.796	-2.387		$5.7x10^{-5}$
	0.0899	1.491	-3.607	0.7193	$2.9x10^{-6}$

* mixture of salts according to solution A (Table 4.16)
Equations based on the data of Ekman (1908).

In Table 5.15 the values at 25°C are compared with those for a few electrolytes in the binary solutions. The seasalt values fit the Redlich type equation very well (std.dev.10^{-4} ml$mol^{-1}$$deg^{-1}$). When sufficient accurate data (for a wide range of temperatures and for concentrations to 1 molkg^{-1} ionic strength) become available a more comprehensive analysis can be made.

TABLE 5.15 Coefficients for the Redlich form of equation for seasalt (solution A mixture of salts - see Table 4.16)
25°C 1 bar, Comparison with binary solutions

$$\Phi_E = a_o + a_1 I^{\frac{1}{2}} + a_2 I$$

ml$mol^{-1}$$deg^{-1}$

	a_o	$a_1 x10^2$	$a_2 x10^2$
Seasalt	0.1082	-0.5645	-2.654
NaCl	0.0930	-2.10	-
Na_2SO_4	0.2219	-9.33	-
KCl	0.0850	-2.20	-
LiCl	0.25	-0.65	-

5.65 Φ_E and $\bar{E}_s$ at pressures greater than 1 bar

The equation developed by Bryden (1973) from the data of Bradshaw and Schleicher (1970) is the most convenient to use and is considered quite adequate for Φ_E, on present evidence. The values of Φ_E for pressures above 1 bar have therefore been calculated from eq 5.6.8 using eq 5.63.1 for (αv), with the coefficients of Table 5.11.

The values of $\bar{E}_s^{(p)}$ were calculated from eq 5.6.9. The second term of this equation features the salinity differential of (αv). It is likely that the BS equation 5.63.2 will be unreliable as it is derived from measurements on only three salinities (30‰, 35‰ and 40‰). Therefore, values of $\bar{E}_s$ were also calculated from the WM equations (eqs 5.62.1, 5.62.2) for comparison (Table 5.16). The agreement between the two data sets is good at $0^{o}C$, generally better than 2-3%. However, large deviations occur as the temperature increases. At 1000 bars pressure, S = 35‰, the values derived from the <u>BS</u> data are 7-14% too low, and at 20° and $25^{o}C$, respectively, the deviations are 10% and 40% too high. These discrepancies are largely due to the differences in the salinity gradients (Table 5.17)

TABLE 5.16 The mean partial molal expansibility of seasalt* $\bar{E}_s^A$ $ml\,mol^{-1}deg^{-1}$ as a function of salinity, temperature and pressure

S‰		30			35			40		
°C	p(bars)	1	500	1000	1	500	1000	1	500	1000
0	<u>a</u>	0.1939	0.1264	0.0783	0.1885	0.1210	0.0729	0.1832	0.1157	0.0676
	<u>b</u>	0.1972	0.1247	0.0777	0.1873	0.1185	0.0737	0.1778	0.1126	0.0698
5	<u>a</u>	0.1631	0.1088	0.0707	0.1578	0.1035	0.0654	0.1525	0.0981	0.0601
	<u>b</u>	0.1665	0.1125	0.0763	0.1599	0.1084	0.0739	0.1536	0.1045	0.0715
10	<u>a</u>	0.1361	0.0931	0.0631	0.1307	0.0878	0.0578	0.1254	0.0825	0.0524
	<u>b</u>	0.1385	0.0964	0.0673	0.1341	0.0939	0.0661	0.1299	0.0915	0.0649
15	<u>a</u>	0.1127	0.0794	0.0555	0.1074	0.0741	0.0501	0.1020	0.0687	0.0448
	<u>b</u>	0.1129	0.0787	0.0542	0.1096	0.0770	0.0534	0.1065	0.0753	0.0528
20	<u>a</u>	0.0931	0.0675	0.0478	0.0877	0.0622	0.0425	0.0824	0.0568	0.0371
	<u>b</u>	0.0896	0.0610	0.0395	0.0864	0.0590	0.0384	0.0833	0.0571	0.0374
25	<u>a</u>	0.0773	0.0575	0.0401	0.0719	0.0522	0.0347	0.0666	0.0469	0.0294
	<u>b</u>	0.0685	0.0443	0.0253	0.0643	0.0411	0.0229	0.0602	0.0381	0.0206

* based on a salt mixture as for solution A (Table 4.16)

$\bar{E}_s^A/2.178796$ gives $ml\,mol^{-1}(ions)deg^{-1}$

$\bar{E}_s^A$ $(S/68.0811(10^3+0.0049S))$ gives $ml\ g^{-1}(seawater)\ deg^{-1}$

<u>a</u> Data of Bradshaw and Schleicher in polynomial of Bryden (1973)

<u>b</u> Wang and Millero (1973)

It is concluded that the <u>BS</u> equation, (5.63.2), is not as reliable as the <u>WM</u> and <u>TG</u> equations for calculating $\bar{E}_s$ except at low temperatures. This limitation of eq 5.63.2 showed up in the entropy calculations (Chapter 3); as might be expected, since the partial molal expansibility is the pressure derivative of entropy.

TABLE 5.17 Comparison of the Salinity Gradient of the Expansivity

$10^8 \left(\frac{\partial(\alpha v)}{\partial S} \right)_{P,T}$ ml/g.deg.(‰)

S - 35‰

t°C	p bars	250	500	1000
0	WM	216	163	87
	BS	220	167	85
5	WM	184	144	85
	BS	180	136	72
10	WM	150	118	71
	BS	143	109	58
15	WM	116	90	49
	BS	111	85	44
20	WM	81	59	24
	BS	85	64	31
25	WM	47	29	-1
	BS	62	46	17

WM Wang and Millero (1973) eq 5.62.2
BS Bradshaw and Schleicher (1970) eq 5.63.2

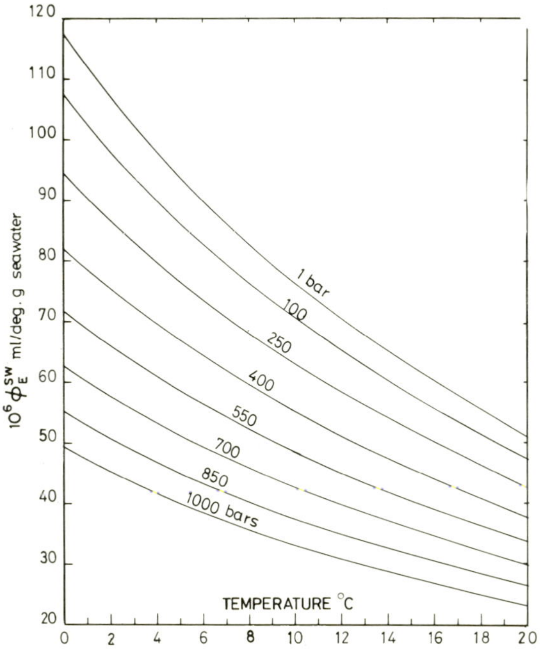

The mean apparent molal expansibility of seasalt, estimated from the data of Bradshaw and Schleicher (1970) using eqs 5.6.8, 5.63.1.

S = 34.702‰.

The same quantity in $ml deg^{-1} mol^{-1}$(ions) Φ_E^I, is given by $\Phi_E^{SW}(31.24712(10^3 + 0.0049S)/S)$.

FIGURE 5.2

Figure 5.2 shows the effect of temperature and pressure on Φ_E at a salinity of 34.702‰. The effect of pressure on $\bar{E}_s$ at the same salinity, and over the same temperature range is shown in Figure 5.3.

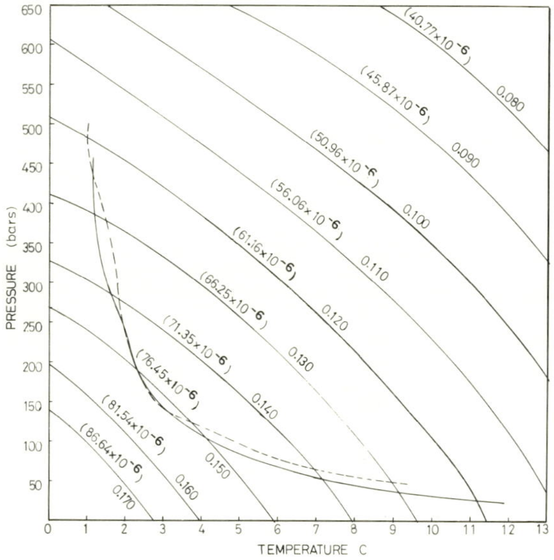

FIGURE 5.3

The mean partial molal expansibility of seasalt (2.178796x mldeg^{-1}mol^{-1}ions) i.e. $\Sigma m_i \bar{E}_i / \Sigma m_i$, solution A (Table 4.16) eqs 5.6.9, 5.63.1 and 5.63.2.

bracketed values in mldeg^{-1}g^{-1}(seawater)

Data of Bradshaw and Schleicher (1970)

S = 34.702‰.

P-T characteristics of two south-west Pacific stations are superimposed (salinity range 34.4-34.7‰)

Station G3/62/65 (24 February 1965) lat. 33 52S long. 153 27E, CSIRO (1969). ———

Station G3/93/63, (23rd July 1963) lat. 27 45S, long. 164 57W, CSIRO (1968). --------

Under average oceanographic conditions the value of $\bar{E}_s$ does not change very much, especially in the region 800-3,000 metres. The P-T characteristics of two southwest Pacific stations have been superimposed to illustrate the changes to be expected.

5.7 The thermal expansion of water and seawater.

5.71 The coefficient of thermal expansion of water.

The specific volume of pure water at 1-atm has been given by Kell (1967) in the form

$$v_w^{(1)} = (1+bt)/P_5(t) \qquad (5.71.1)$$

with the coefficients for the polynomial $P_5(t)$, and b listed in Table 5.18 (units are in cm^3/g). Differentiating with respect to temperature, gives

$$\frac{dv_w^{(1)}}{dt} = (\alpha v)_w^{(1)}$$

$$= (1/P_5(t))(b - (1+bt)P_4(t)/P_5(t)) \qquad (5.71.2)$$

The coefficients for P_4 are given in Table 5.18.

TABLE 5.18 Coefficients for the density* of water equation eq(5.72.5) and the coefficient of thermal expansion (eq 5.72.6)

	1 bar $P_5(t)$	$P_4(t)$
a_o	0.9998396	0.018224944
a_1	18.224944×10^{-3}	-1.584442×10^{-5}
a_2	-7.922210×10^{-6}	$-1.6634538 \times 10^{-7}$
a_3	-55.44846×10^{-9}	5.990248×10^{-10}
a_4	14.97562×10^{-11}	$-1.966476 \times 10^{-12}$
a_5	$-3.932952 \times 10^{-13}$	-
b	18.159725×10^{-3}	18.159725×10^{-3}

Kell (1967) * 1 ml = 1.000028 cm^3

More recently, Gildseth, Habenschuss and Spedding (1972) gave their precise density measurements of water in the form

$$d_w^{(1)} = 1 - \frac{(t-a_1)^2(t+a_2)}{a_4(t+a_3)} \qquad 0 \leqslant t \leqslant 40^{\circ}C \qquad (5.71.3)$$

with

a_1=3.9863, a_2=288.9414, a_3 = 68.12963

a_4 = 508929.2 and a_5 = 220.81177 (in g/ml)

Above 40°C an exponential term must be added to eq 3 (viz., 0.011445exp (-374.3/t)). The derived equation for the expansivity is,

$$(\alpha v)_w^{(1)} = ((1-d_w)/d_w^2)\ (2/(t-a_1) - a_5/(t+a_3)(t+a_2)) \qquad (5.71.4)$$

in ml/g.deg (note that d_w must be consistent with the coefficients i.e. d_w is in g/ml).

For the range 0-30°C, the density values calculated from the (1967) and (1972) equations are identical to 1 ppm (normalising for the different units, since 1 ml=1.000028cm^3). At higher temperatures the values from eq 3 (with the exponential term) are up to 10ppm higher (over the 70-80°C range). Table 5.19 lists values of $v_w^{(1)}$ and $(\alpha v)_w^{(1)}$ calculated from the two equations.

TABLE 5.19 Volume properties of ordinary water at 1-bar pressure

t°C		$v_w^{(1)}$	$10^6(\alpha v)_w^{(1)}$	$10^6\alpha_w$ deg^{-1}
0	a	1.000160	-68.15	-68.14
	b	1.000132	-67.94	
5	a	1.000036	15.99	15.98
	b	1.000008	15.93	
10	a	1.000300	87.93	87.90
	b	1.000272	87.90	
15	a	1.000900	150.87	150.73
	b	1.000872	150.90	
20	a	1.001797	206.98	206.61
	b	1.001769	207.01	
25	a	1.002961	257.81	257.05
	b	1.002933	257.80	
30	a	1.004369	304.47	303.14
	b	1.004340	304.40	

a: eq 5.71.1 and Table 5.18 Kell(1967)

b: eq 5.71.3 Gildseth, Habenschuss and Spedding (1972)

units: a: v in cm^3/g; b: v in ml/g

These values illustrate the differences arising from the different units. For specific volume, it is only necessary to add 0.000028 in order to convert the ml/g values to cm^3/g. For the expansivity, the difference is negligible, except at 0^oC. The values of $(\alpha v)_w^{(1)}$ from eq 5.71.2 are probably reliable to $0.1x10^{-6}$ ml/g.deg.

The effect of pressure on $(\alpha v)_w$ is most simply obtained from the Tait equation, using the B parameter derived from acoustic data (see Sections 6.31, 6.7 for details). This equation is especially convenient for developing the entropy equations for water (Chapter 3). As will be shown in Chapter 6, the specific volume of water can be adequately represented (to±20ppm or better) by the Tait equation. From the Tait equation (the form most commonly used, Harned and Owen 1958)

$$v_w^{(p)} = v_w^{(1)}(1 - 0.315\log\frac{(B+p)}{(B+1)}) \qquad (5.71.5)$$

so that, on differentiating with respect to temperature, at constant pressure,

$$(\alpha v)_w^{(p)} = (\alpha v)_w^{(1)}(1 - 0.315\log\frac{(B+p)}{(B+1)})$$
$$+ 0.315v_w^{(1)}\frac{dB}{dt}(P-1)/2.303(B+1)(B+p) \qquad (5.71.6)$$

with $B = 2671.8 + 19.454t - 0.27028t^2 + 9.798x10^{-4}t^3$

$$0 \le t \le 45^oC$$

Values of $(\alpha v)_w^{(p)}$, for a range of temperatures and pressures are listed in Appendix 5.

Values of the coefficient of expansion, $\alpha_w^{(p)}$, are listed in Table 5.20. The values of $\alpha_w^{(p)}$ tabulated by Fine and Millero (1973, Table VI) have been subtracted from the corresponding values of Table 5.20 to give the deviations shown in brackets. The differences in the two sets of α_w are generally less than 1 ppm, except at low temperatures. However, the deviations are obviously not random - they have the same trends as the deviations of the acoustic data from the measured values of $\alpha_{sw}^{(p)}$, the expansivity of seawater (S=35‰). This is illustrated in Fig.5.4, where the bracketed values of Table 5.20 (times -1 to convert to $\alpha_{FM}-\alpha_T$) at 0°C are plotted against the corresponding values for seawater (S=35‰).

TABLE 5.20 The expansivity ($x10^6 deg^{-1}$) of pure water

p bars (absolute) \ t°C	0	5	10	15	20	25	30
1	-68.14	15.78	87.90	150.73	206.61	257.05	303.14
51	-49.60	30.77	99.85	160.34	214.19	262.85	307.38
101	-31.85 (-0.6)	45.15 (-0.4)	111.46 (0.0)	169.64 (0.2)	221.53 (0.3)	268.49 (0.4)	311.55 (0.5)
151	-14.63	59.12	122.75	178.68	228.66	273.98	315.62
201	2.07 (-1.4)	72.67 (-0.8)	133.71 (-0.2)	187.46 (0.2)	235.59 (0.5)	279.32 (0.6)	319.57 (0.7)
251	18.28	85.84	144.37	196.01	242.34	284.52	323.42
301	34.02 (-2.1)	98.65 (-1.2)	154.73 (-0.5)	204.33 (0.1)	248.91 (0.4)	289.59 (0.5)	327.17 (0.6)
351	49.31	111.10	164.82	212.43	255.32	294.52	330.83
401	64.18 (-2.4)	123.21 (-1.5)	174.65 (-0.7)	220.32 (-0.1)	261.56 (0.2)	299.33 (0.3)	334.40 (0.5)
451	78.65	135.00	184.21	228.01	267.64	304.03	337.88
501	92.72 (-2.4)	146.49 (-1.5)	193.54 (-0.7)	235.51 (-0.3)	273.57 (0.0)	308.61 (0.1)	341.27 (0.2)
601	119.78 (-1.9)	168.59 (-1.2)	211.50 (-0.6)	249.96 (-0.3)	285.02 (-0.1)	317.44 (-0.2)	347.83 (-0.1)
701	145.48 (-0.9)	189.60 (-0.6)	228.59 (-0.3)	263.73 (-0.2)	295.93 (-0.2)	325.87 (-0.3)	354.08 (-0.4)
801	169.92 (0.6)	209.62 (0.5)	244.90 (0.4)	276.87 (0.2)	306.35 (0.0)	333.93 (-0.4)	360.06 (-0.7)
901	193.22 (2.8)	228.72 (2.0)	260.46 (1.4)	289.43 (0.8)	316.31 (0.2)	341.63 (-0.4)	365.78 (-0.8)
1001	215.45 (5.6)	246.96 (4.0)	275.35 (2.8)	301.45 (1.8)	325.85 (0.8)	349.01 (-0.1)	371.27 (-0.8)
1051	226.19	255.79	282.55	307.27	330.48	352.59	373.92

Values calculated from eq 5.71.6 - Tait equation
Bracketed values are deviations from the Tait values of the values calculated by Fine and Millero (1973, Table VI), i.e. $\alpha_{Tait} - \alpha_{Fine\text{-}Millero}$

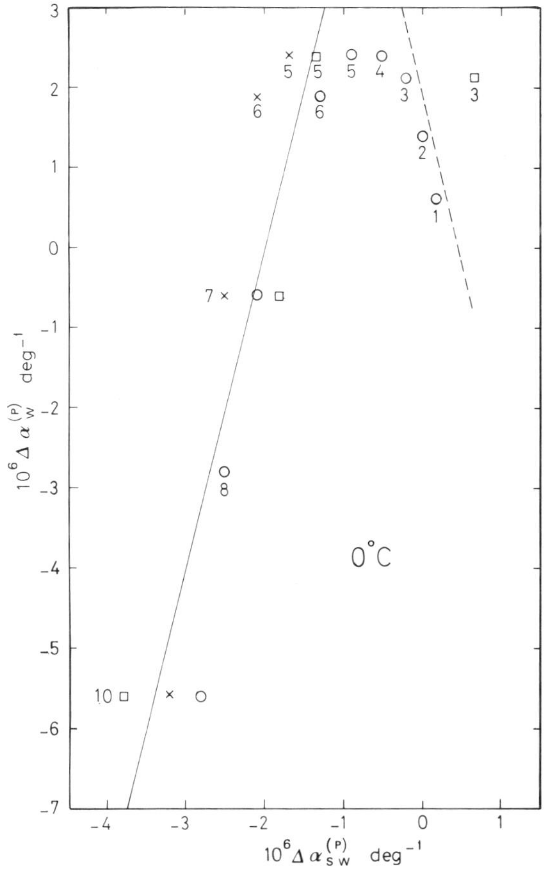

FIGURE 5.4

The expansivity differences (acoustic-Tait) for water, $\Delta\alpha_w^{(p)}$, versus the expansivity differences (acoustic - experimental) for seawater, $\Delta\alpha_{sw}^{(p)}$, at 0°C for pressure to 1000 bars (numbers are bars x 10^{-2})

$\Delta\alpha_w^{(p)}$ from Table 5.20

$\Delta\alpha_{sw}^{(p)}$ 0 Bradshaw and Schleicher (1970), as given by Fine, Wang and Millero (1974).

X Bradshaw and Schleicher (Table 5.22, herein)

□ Caldwell and Tucker (1970) (Table 5.22 - 2/5ths of actual values).

These latter values were taken from the compilation of Fine, Wang and Millero (1974, Table VII), and from Table 5.22 of the next section (5.72). From 500-1000 bars the values are well represented by

$$\Delta\alpha_{w}^{(p)} = 4(2+\Delta\alpha_{sw}^{(p)}) \qquad (5.71.7)$$

where $\Delta\alpha$ (in ppm) represents the acoustic value minus the corresponding measured value (or value calculated from the Tait equation, in the case of water). At lower pressures the trend is in the opposite direction (dashed line). Thus, the deviations appear to be due to a peculiarity of the acoustic data. As will be explained in detail in Chapter 6 (Sections 6.31 and 6.7) the Tait equation and the method of deriving its parameter B, smooths out and practically eliminates these peculiarities. The latter arise from a perturbation in the structure of water that is associated with small entropy changes, and hence with non-equilibrium adiabatic compression. The thermodynamic relationships used to derive quantities such as compressibilities (isothermal) and expansivity from sound velocity data do not strictly hold, as the conditions are not isentropic. As is well known, water is a unique liquid, having many properties not exhibited by any other liquid (Erlander 1968). For example, unlike other liquids, the pressure dependence of the viscosity of water, relative to its value at 1-atm, exhibits a minimum. Horne and Johnson (1966,1967) have shown that near 10°C the minimum is around 700 bars. The addition of an electrolyte (e.g. 1 m NaCl) has the effect of broadening or smearing out this minimum, and shifting it to a lower pressure (around 400-500 bars for 1 m NaCl). However, the characteristics of the water viscosity curve are preserved. If the water curve is shifted back by about 250 bars (i.e. it is assumed than an "effective" pressure of 250 bars is present at atmospheric pressure - as predicted by Tammann's Principle - Chapters 4 and 6) the water and NaCl curves are almost identical. Thus, the two curves are essentially reflecting the effect of the structural properties of water on its physical properties, just as the curves of $(\partial v/\partial t)_p$ versus temperature do (Section 5.73, Fig. 5.7). Seawater shows similar characteristics to NaCl solutions in this respect (Horne and Johnson 1966). The minimum is most pronounced at temperatures near $0^{\circ}C$ and flattens out and moves to lower pressures as the temperature rises to $20^{\circ}C$. The $\Delta\alpha$ values exhibit very similar characteristics in regard to their temperature and pressure dependences.

Another property of seawater related to the structure of its major component, water, is X_s, the difference in the pressure increments of the partial molal and apparent molal entropy of water-in-seawater (Chapter 3). This quantity is proportional to the "structure" parameter A of the equation $\phi_v^o = Ar^3 - B'z^2/r$, which was discussed in Chapter 4, but gives the pressure characteristics as well as those of temperature.

Values of X_s at 500 and 1000 bars for the temperatures 0-30°C have been plotted against the corresponding values of $\Delta\alpha_w^{(p)}$ Fig. 5.5. At 1000 bars the values of $\Delta\alpha_w^{(p)}$ are given by

$$10^6\Delta\alpha_w^{(p)} = 11.62 - 2.79\text{x}10^2 X_s \qquad (5.71.8)$$

to within 0.1 ppm.

The values at 500 bars change in the opposite direction with temperature, and remain positive or otherwise very close to zero. This is true for pressures up to about 700 bars, as can be seen from Table 5.20 which lists ($-\Delta\alpha_w^{(p)}$). In other words, the "acoustic" values are generally higher. They become lower at the higher pressures. It may be significant that the relative-viscosity versus pressure minimum, discussed above, is around 700 bars at 10-15°C.

According to Erlander (1968) the destruction of cluster aggregates (the dominant form of water at low temperatures, ca. 0-20°C) decreases the average distance between molecules, because the empty pockets in the aggregates are destroyed. One consequence is that the volume decreases.

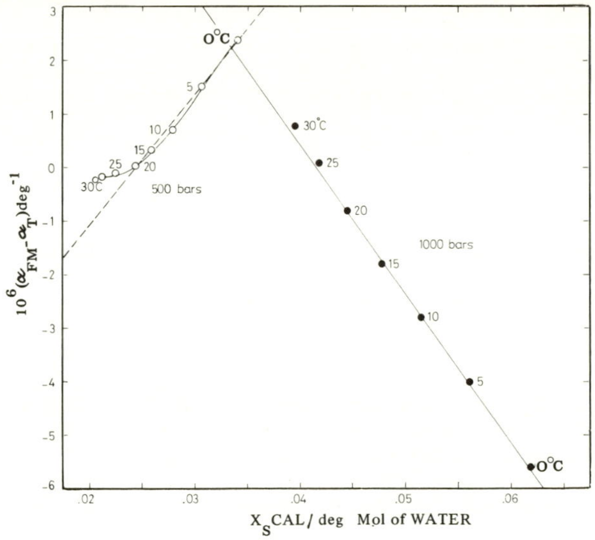

FIGURE 5.5

Differences in the "acoustic" expansivity (Fine and Millero 1973) and the "Tait" expansivity of water (eq 5.71.6), versus a "water-structure" function X_s. X_s is the difference between the partial molal and apparent molal entropy of water-in-seawater (Chapter 3). Values of $-(\alpha_{FM}-\alpha_T)$ are listed in Table 5.20 . The temperatures are as shown.

The acoustic measurements may be associated with destruction of some aggregates. The effect would be to increase $(\frac{1}{v_w}(\partial v_w/\partial t))$ or α_w. This would explain the slightly higher values of the acoustic α_w's. At higher pressure, as indicated by the changes of $(\partial v/\partial t)$, for water or seawater, with temperature (Section 5.73, Fig. 5.7), the characteristic structure of the higher temperature region (55-100°C) begins to prevail. That is, the $(\partial v/\partial t)$ versus temperature curve becomes more linear.

The destruction of clusters (the dominant form of water at the higher temperatures) increases the average distance between molecules, and the volume increases. Therefore, at a given temperature and pressure the α_w changes due to this cause are negative, i.e. α_w decreases. This explains the change in the sign of $\Delta\alpha_w$ at pressures above 700 bars.

The reduction in viscosity of water upon application of pressure at low temperatures is due to the decrease in the number of cluster aggregates, that is, the elimination of empty pockets (Erlander 1968). Thus, it would be expected that the relative-viscosity versus pressure minimum would occur around the pressure coinciding with the linearising of the $\partial v/\partial t$ versus temperature curve.

5.72 <u>The coefficient of thermal expansion of seawater</u>

The coefficient α_{sw} can be obtained directly from seawater density data fitted to a suitable function of t,

$$\alpha = -(1/d)(\partial d/\partial T)_p \qquad (5.72.1)$$

Millero and Lepple (1973) have fitted their density data for artificial seawater, and density data of Knudsen and Cox et al. (1970) to equations of the form

$$d = A + Bt + Ct^2 + Dt^3 + Et^4 \qquad (5.72.2)$$

from which values of α_{sw} were calculated and compared. From the results they concluded that the temperature dependence of the densities of Knudsen is very precise and superior to that of Cox *et al*. Millero and Lepple (1973) also developed a relationship for the expansibility of seawater as a function of the volume chlorinity, (Cl_v or Cl‰ xd) which is equal to the chlorosity at 20°C.

$$\alpha_{sw}' = \alpha_w + A_E^{e'} Cl_v + B_E^{e'} Cl_v^{3/2} \qquad (5.72.3)$$

where $A_E^{e'}$ and $B_E^{e'}$ are analagous to the coefficients of eq 5.11.2 but are in terms of equivalents. As for the volume relationships, an equation based on the Debye-Hückel limiting law was also developed, viz.

$$\alpha_{sw} = \alpha_w + A_E^e Cl_v + B_E^e Cl_v^{3/2} + C_E^e Cl_v^2 \qquad (5.72.4)$$

where $A_E^e = 10^{-3}(\phi_E^o - \alpha_w \phi_v^o)k$, $B_E^e = 10^{-3}(S_E^e - \alpha_w s_v^e)k(k')^{\frac{1}{2}}$

and $C_E^e = 10^{-3}(b_E^e - \alpha_w b_v^e)k(k')$.

the small superscript e is used to emphasise the different concentration scale to the one used throughout this book. k and k' are concentration weighting terms, viz.

$k = N_T/Cl_v = 0.0312867$ (N_T being the total normality - equivalents/liter of solution).

The reader is referred to the original papers for further details.

It has been found that a rational function rather than a polynomial is a better choice for summarising the density data of water as a function of temperature (Kell 1967). As indicated in the previous section, the values of $\alpha_w^{(1)}$, are based on the rational function

$$d_w^{(1)} = P_5(t)/(1+bt) \qquad (5.72.5)$$

where $P_5(t) = a^o + a_1 t + a_2 t^2 + \ldots \quad a_5 t^5$

with the coefficients as listed in Table 5.18. Differentiation with respect to temperature gives

$$\alpha_w^{(1)} = b/(1+bt) - P_4(t)/P_5(t) \qquad (5.72.6)$$

For seawater, the rearranged form of eq 5.1.5a would be compatible with this form of function, viz.

$$\alpha_{sw}^{(1)} = (\alpha_w + A_s \Phi_E d_w)/(1+A_s \Phi_v d_w) \qquad (5.72.7)$$

where $A_s = c/(1-c)M_t'$, $c = 10^{-3}S$ (‰)

or $$\alpha_{sw} = \left(b - \frac{P_4(t)}{P_5(t)}(1+bt) + A_s \Phi_E P_5(t)\right) \times \left(1 + bt + A_s \Phi_v P_5(t)\right)^{-1} \qquad (5.72.8)$$

The amv, Φ_v, as a function of t (eq 4.61.2), is largely derived from Knudsen's density data (Dorsey, Table 108,1940) which has a precise temperature dependence, so that eq 5.72.7 might be expected to give a good representation of α_{sw} at 1 bar. Values are given in Table 5.21 along with values given previously by Millero and Lepple (1973) and other values calculated from various data sets as outlined below. The results suggest that the values derived from the polynomial eq 5.72.2 are too high, except at 0^oC.

TABLE 5.21 The Coefficient of thermal expansion of seawater

$\alpha_{sw} 10^6 (deg^{-1})$

(1 bar, S = 35‰)

t^oC	a A	a B	b	c	d	e	f	g
0	51.8	52.1	51.9	49.5	51.9	53.6	51.5	50.7
5	117.2	116.8	117.3	115.9	114.2		113.9	113.0
10	172.3	171.6	172.1	172.7	167.3	166.2	167.2	167.0
15	220.4	219.4	219.8	221.8	213.5		214.2	215.1
20	263.5	262.8	263.0	264.0	256.0	257.4	256.9	258.9
25	303.0	302.8	303.4	299.7	296.9	297.2	296.8	299.2
30	340.2	339.7			336.1		335.0	337.1
35	375.9	375.3			369.3		372.4	

a,b,c Millero and Lepple (1973) (A Copenhagen seawater, B artificial seawater, b Knudsen, c Cox et al.)
d Dorsey (1940, Table 108, Knudsen *et al* data) eq 5.72.7
e Bradshaw and Schleicher (1970), direct measurement
f Wang and Millero (1973) eq differentiated w.r.t. temperature
g Ekman (1908) eq 5.72.11

During 1969, two pairs of workers measured the thermal expansion of seawater, using different methods. Caldwell and Tucker (1970) obtained α_{sw} by observing the onset of convection in a Rayleigh Benard device, while Bradshaw and Schleicher (1970) measured the expansion via change of height of mercury in a dilatometer. The differences in the two data sets were generally less than 2%, although the accuracy claimed by Caldwell and Tucker was only 3%. Bradshaw and Schleicher presented their results in the form

$$\partial v(T,P,S)/\partial t = a_o(P,S) + a_1(P,S)t + a_2(P,S)t^2 + a_3(P,S)t^3 \quad (5.72.9)$$

However, these polynomials did not give the best fit to the data and Bryden (1973) later developed a new polynomial, based on a least squares analysis. The two sets of coefficients were used previously in deriving values of $\bar{E}_s$ (Table 5.13). Bryden's set is listed in Table 5.11.

The value of α_{sw} can be derived from the Tait-Gibson equation by differentiating with respect to temperature. Since, from Section 4.62

$$v_{sw}^{(p)} = (Q\psi_w^{(p)} + \psi_t)/(1+Q) \qquad (5.72.10)$$

$$\alpha_{sw}^{(p)} = (Q\partial\psi_w^{(p)}/\partial t + \partial\psi_t/\partial t)/(Q\psi_w^{(p)} + \psi_t) \qquad (5.72.11)$$

where $\psi_w^{(p)} = v_w^{(1)}\ (1-0.315\log(B^*+p)/(B+1))$ (5.72.12)

and $\psi_t = 0.54214 - 2.1896\text{x}10^{-3}t$ (5.72.13)

$$Q = (M_t'/M_t)(1-c)/c = 0.99512(1-c)/c, \text{ with } c=10^{-3}S$$

Most of the quantities derived for the TG equation were based on <u>$Q = (1-c)/c$</u> (Li 1967) so that this approximation is used here.

The quantity ψ_t is assumed to change negligibly with pressure.

$$\partial\psi_w^{(p)}/\partial t = \alpha_w^{(1)}\ \psi_w^{(p)} + v_w^{(1)}0.1368D_1 \qquad (5.72.14)$$

with $D_1 = ((\partial B/\partial t)/(B+1) - (\partial B^*/\partial t)/(B^*+p))$

and $B^* = 2670.8 + 6.8966S + 19.39t - 0.0703St - 0.233t^2$ (4.61.7)*

$B = B^*(S=0)$

Values of $v_w^{(1)}$ (or $1/d_w^{(1)}$) and $\alpha_w^{(1)}$ are given by eqs 5.72.5 and 5.72.6 and Table 5.18; $v_w^{(1)}$ should be converted to ml g^{-1}, to conform to the other terms. Alternatively, eqs 5.71.3 and 5.71.4 may be used directly.

The WM equation of state outlined in the previous section (5.62) was also used to estimate α_{sw}. This equation is even more complex than that of Bryden's (Table 5.11). This is in contrast to the TG equation with its relatively few coefficients. The calculated values are compared in Table 5.22, over the P-T-S range given by Caldwell and Tucker (1970). This covers the area of general oceanographic interest.

The agreement between the experimental and derived values is good. The values calculated from the TG equation give better agreement with the experimental results (at high pressures, low temperatures) than the corresponding values for the WM equation of state. Values given previously by Caldwell and Tucker (Crease (1962) and Oceanographic Tables) are also

TABLE 5.22 The Thermal Expansivity of seawater as a function of pressure and temperature, S = 34.702‰
Comparison of directly measured and calculated values.

P(bars)	336.9	521.5	551.8	621.0	762.8	825.7	826.3	1058.6	1179.8
t^oC	1.252	0.949	0.987	0.877	0.837	0.750	0.754	0.656	0.624
				$\alpha_{sw} x 10^6$ of eq^{-1}					
O.T.	149.7	185.5	191.7	204.1	229.3	238.9	239.0	274.6	291.1
WM	148.7	184.1	190.2	202.2	226.6	236.1	236.3	269.2	284.3
Cr	147.7	182.9	188.9	200.9	225.0	234.5	234.6	266.9	281.3
BS	149.9	185.8	192.0	204.3	229.0	238.8	238.9	272.4	287.5
CT	147.1	187.6	193.3	203.0	231.6	239.4	240.9	278.8	299.0
EK	147.3	183.4	189.8	202.5	228.8	239.5	239.6	278.3	297.2

P(bars)	401.6	673.7	1045.2	1.0	385.8	700.6	1032.7
t^oC	2.617	2.393	2.246	17.496	17.415	17.385	17.342
				$\alpha_{sw} x 10^6$ deg^{-1}			
O.T.	175.9	224.7	280.4	235.9	280.1	312.6	345.5
WM	174.9	222.8	276.5	235.5	279.5	310.9	340.3
Cr	173.8	221.4	273.8	233.9	278.0	309.0	337.8
BS	175.0	223.3	288.6	235.8	279.5	310.6	337.6
CT	177.7	223.8	282.5	239.2	284.3	319.5	355.5
EK	173.7	223.6	284.0	237.0	281.3	312.1	340.5

O.T. Oceanographic tables
WM Wang and Millero (1973)
Cr Crease (1962)
BS Bradshaw and Schleicher (1970) - measured
CT Caldwell and Tucker (1970) - measured
EK Ekman (1908) Tait-Gibson equation (Li 1967).

included. The data on which the TG equation is based (Ekman's bulk compression data and Knudsen et.al's 1-atm data, as listed by Dorsey 1940 - Table 108) are essentially the same as those used for the oceanographic tables. The agreement in α_{sw} is therefore to be expected. However, it is interesting that the TG equation shifts the values closer to the WM ones, except at high pressures with low temperatures. In this P-T area the WM values are apparently less accurate. The deviations of the different results at 1^oC, over the pressure range 300-1200 bars, are summarised in Fig. 5.6.

The conclusion is that the TG form of equation (viz. eq 5.72.11)

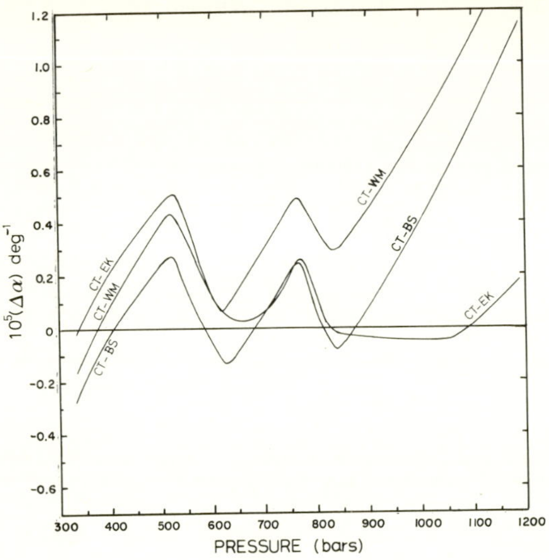

The difference in the measured expansibilities of seawater and the values calculated from the data of Ekman (as given by Li (1967)), coded EK, and acoustic data analysed by Wang and Millero (1973), coded WM,

CT Caldwell and Tucker (1970)

BS Bradshaw and Schleicher (1970)

Temperature = 1°C

FIGURE 5.6

for α_{sw} quite adequately represents the experimental results (Table 23). The data of Knudsen et al and Ekman, as given by Dorsey (1940, Table 108) and Li (1967) were used to calculate the parameters of the TG equation. The above text was written during early 1974. Publications since then do not, however, change the overall picture. Further publications on the "acoustic" equation of state (Fine et al, 1974) and direct measurements of the specific volume of seawater (relative to the specific volume of water from acoustic data) as given by Emmet and Millero (1974) do not change the above conclusions.

5.73 The Changes of expansivity with temperature - the relationship with water structure.

Finally, it is of interest to look at the transitional changes that occur in the thermal expansivity of water and seawater as the temperature changes. Such changes are related to the peculiar structural properties of water.

TABLE 5.23 The Thermal Expansivity of seawater as a function of salinity, temperature and pressure(absolute)

$\alpha_{sw} \times 10^6 \text{ deg}^{-1}$

t^oC	P(bars)	1	100	200	400	500	600	800	1000
		S = 30‰							
0	WM	36.6	64.9	91.7	140.5	162.6	183.2	220.2	251.9
	EK	36.2	63.9	90.4	139.2	161.7	183.2	223.1	259.6
10	WM	156.9	175.2	192.8	225.3	240.2	254.4	280.4	303.7
	EK	157.6	176.2	193.9	226.8	242.1	256.6	283.8	308.7
20	WM	250.5	261.5	272.2	292.3	301.9	311.2	328.8	345.5
	EK	252.8	264.2	275.0	295.1	304.5	313.4	330.1	345.5
25	WM	292.1	300.2	308.1	323.4	330.9	338.2	352.6	366.8
	EK	294.5	302.6	310.4	324.9	331.6	338.1	350.1	361.2
		S = 35‰							
0	WM	51.5	78.5	104.2	150.9	172.0	191.7	227.2	257.6
	EK	50.7	77.3	102.6	149.5	171.2	191.8	230.3	265.5
10	WM	167.2	184.9	201.8	233.0	247.4	261.0	286.1	308.5
	EK	167.0	184.8	201.7	233.2	247.8	261.7	287.7	311.6
20	WM	256.9	267.4	277.7	297.0	306.1	315.0	332.0	348.0
	EK	258.9	269.5	279.8	298.8	307.6	316.1	331.9	346.4
25	WM	296.8	304.5	312.0	326.6	333.7	340.7	354.4	368.0
	EK	299.2	306.8	314.1	327.5	333.8	339.8	351.1	361.4
		S = 40‰							
0	WM	65.8	91.6	116.1	160.8	181.0	199.9	233.8	263.0
	EK	64.4	89.9	114.2	159.2	180.0	199.9	236.9	270.8
10	WM	177.3	194.3	210.5	240.6	254.4	267.5	291.7	313.3
	EK	175.9	192.8	209.0	239.1	253.0	266.4	291.3	314.1
20	WM	263.2	273.2	283.0	301.5	310.3	318.8	335.0	350.4
	EK	264.4	274.5	284.1	302.1	310.4	318.4	333.3	347.1
25	WM	301.2	308.5	315.7	329.5	336.3	342.9	356.1	369.1
	EK	303.6	310.6	317.3	329.8	335.6	341.2	351.7	361.3

WM Wang and Millero (1973) equation, differential wrt t eq 5.6.21

EK Ekman (1908)
eqs 5.72.11, 5.72.12, 5.72.13.

Values of $\alpha v, (\partial v/\partial t)$, for water at 1 bar and 500 bars have been plotted as a function of temperature in Fig. 5.7. At temperatures around 50-100°C the curves are linear. Erlander (1968) interprets this linearity as due to the formation or destruction of water clusters and a change in the thermal motion of water molecules and its aggregates.

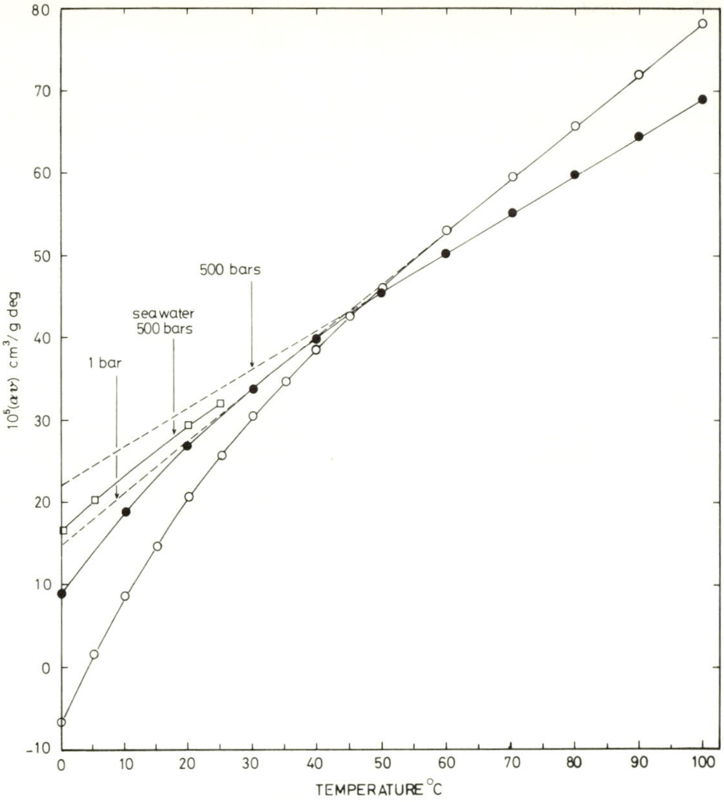

FIGURE 5.7

The Change in $\partial v/\partial t$ with temperature

○ water at 1 bar

● water at 500 bars

□ seawater (S=35‰) at 500 bars

----- extrapolated linear sections (ca. 50-100°C) of curves.

At lower temperatures the water clusters aggregate, and empty spaces form at the junctions. The empty space is estimated to be 10% of the aggregate's volume near 0°C, 1 bar pressure.

At 500 bars the non-linearity persists at the lower temperatures, although the curve is beginning to flatten out. Seawater, at the same pressure, exhibits an even flatter curve. This might be expected since the compressional effect of the solute raises the "effective" pressure to around 750-700 bars for the temperature range 0-20°C.

The $(\partial v/\partial t)$ versus temperature curve signifies the formation of expanded ice-like structures at low temperatures and these expanded water clusters begin to gradually disappear as the temperature is raised from 20°C to about 50 or 60°C.

From the viscosity-pressure isotherm studies of Horne and Johnson (1966, 1967), it was shown that the break-up of the expanded water clusters is enhanced by the application of pressure. This is illustrated in Figure 5.7 where the deviation from linearity is much less at the higher pressures.

The interesting thermal properties of water and seawater will be referred to again in Chapter 6. There, the important consequences of "water structure" will be discussed in relation to the interpretation of data for the calculation of compressibility. The more recent models of liquid water will be mentioned; in particular, the "LHSN cluster" model (Hagler, Scheraga and Nemethy's model as modified by Lentz, Hagler and Scheraga 1974). This model differs from the "classical" concept of larger clusters breaking down into smaller ones as the temperature increases. The LHSN cluster is only small (7 moles or less) and the temperature effect is dominated by structural rather than size changes. There is a conversion from rigid to less rigid structures as the temperature rises. Nevertheless, in the broad view, the temperature patterns of the two "cluster" models follow the same trends. In addition, the LHSN model provides useful structural parameters that are functions of temperature (see Section 6.74).

Hepler (1969) has used the equations

$$(\partial C_p/\partial p)_T = -T(\partial^2 V/\partial T^2)_p$$

and

$$(\partial^2V/\partial T^2)_p = \frac{M}{d_w}(\alpha^2 + (\partial\alpha/\partial T)_p)$$

where M is the molecular weight, to relate thermal expansion to structure in water (up to 1000 bars and over the temperature range 0-140°C). He extended the analysis to aqueous solutions, using the partial molal quantities for the components. This enabled some interesting insights to be gained on the structure effects of different solutes. His approach might usefully be applied to seawater.

REFERENCES Chapter 5.

Bradshaw A. and K.E. Schleicher 1970.

Direct measurement of thermal expansion of seawater under pressure.
Deep-Sea Res. 17: 691-706.

Bryden H.L. 1973.

Polynomials for thermal expansion, adiabatic temperature gradient and potential temperature of seawater.
Deep-Sea Res. 20: 401-408.

Caldwell D.R. and B.E. Tucker 1970.

Determination of thermal expansion of seawater by observing onset of convection.
Deep-Sea Res. 17: 707-719.

Cox R.A., M.J. McCartney, and F. Culkin 1970.

The specific gravity (salinity)/temperature relationship in natural seawater.
Deep-Sea Res. 17: 679-689.

Crease J. 1962.

The specific volume of seawater under pressure as determined by recent measurements of sound velocity.
Deep-Sea Res. 9: 209-213.

CSIRO AUST. 1968.

Oceanographical observations in the Pacific Ocean in 1963. H.M.A.S. Gascoyne, Cruise G3/63.
CSIRO Aust. Oceanogr. Cruise Rep. 26.

CSIRO AUST. 1969.

Oceanographical observations in the Pacific Ocean in 1965. H.M.A.S. Gascoyne, Cruise G3/65.
CSIRO Aust. Oceanogr. Cruise Rep. 44.

Chapter 5.

Dorsey N.E. 1940.

"Properties of ordinary water-substance",
A.C.S. monograph series 81,
Reinhold, New York. 673pp.

Duedall I.W. 1968.

Partial molal volumes of 16 salts in seawater.
Environ.Sci.Technol. 2: 707.

Dunn L.A. 1968.

Apparent molar volumes of electrolytes part 2
- some 1-1 electrolytes in aqueous solution at 25°C.
Trans.Faraday Soc. 64: 1898-1903.

Ekman V.W. 1908.

Die Zusammendrueckbarkeit des Meerwassers.
Publ.Circonst.perm.Explor.Mer. 43: 1-47.

Emmet F.T. and F.J. Millero 1974.

High pressure density measurements of seawater -
preliminary results.
J.Geophys.Res. 79(24): 3463-3472.

Erlander Stig R. 1968.

The structure of water and its relationship to
hydrocarbon-water interactions.
J.Macromol.Sci.-Chem. A2(3): 595-621.

Fine R.A. and F.J. Millero 1973.

The compressibility of water as a function of
temperature and pressure.
J.Chem.Phys. 59(10): 5529-36.

Fine R.A., Dong-Ping Wang, and F.J. Millero 1974

The equation of state of seawater.
J.Mar.Res. 32(3): 433-456.

Gildseth W., A. Habenschuss, and F.H. Spedding 1972.

Precision measurements of densities and
thermal dilation of water between 5° and 80°C.
J.Chem.Eng.Data. 17(4): 402-409.

Harned H.S. and B.B. Owen 1958.

"The Physical Chemistry of Electrolyte Solutions",
3rd ed, A.C.S. Monograph, Reinhold, New York, 803pp.

Hepler L.G. 1969.

Thermal expansion and structure in water and
aqueous solutions.
Can.J.Chem. 47: 4613-4617.

Horne R.A. and D.S. Johnson 1966.

The viscosity of compressed seawater.
J.Geophys.Res. 71(22): 5275-77.

Chapter 5.

Horne R.A. and D.S. Johnson, 1967.

The effect of electrolyte addition on the viscosity of water under pressure.
J.Phys.Chem. 71(4): 1147-1149.

Kell G.S. 1967.

Precise representation of volume properties of water at one atmosphere.
J.Chem.Eng.Data. 12(1): 66-69.

Kester D.R., I.W. Duedall, D.N. Connors, and R.M. Pytkowicz 1967.

Preparation of artificial seawater.
Limnol.Oceanogr. 12: 176-179.

Knudsen M. (Editor) 1901.

Hydrographic Tables According to the Measurings of Carl Forch, P. Jacobson, M. Knudsen and S.P.L. Sorensen. Williams Norgate, Copenhagen, 63 pp.

Lentz B.R., A.T. Hagler, and H.A. Scheraga 1974.

Structure of liquid water. II. Improved statistical thermodynamic treatment and implications of a cluster model.
J.Phys.Chem. 78(15): 1531-1550.

Leyendekkers J.V. 1974.

Prediction of the partial molal volumes of electrolytes in seawater.
Mar.Chem. 2(2): 89-110.

Li Yuan-Hui 1967.

Equation of state of water and sea water.
J.Geophys.Res. 72(10): 2665-2678.

Millero F.J. 1968.

Apparent molal expansibilities of some divalent chlorides in aqueous solutions at 25°C.
J.Phys.Chem. 72(13): 4589-4593.

Millero F.J. 1969.

The partial molal volumes of ions in seawater.
Limnol.Oceanogr. 14: 376-385.

Millero F.J. 1974.

Seawater as a multicomponent electrolyte solution, in The Sea - ideas and observations, vol. 5, E.Goldberg, editor, Chapter 1, pp1-55, Interscience, New York.

Millero F.J. and F.K. Lepple 1973.

The density and expansibility of artificial seawater solutions from 0 to 40°C and 0 to 21‰ chlorinity.
Mar. Chem. 1(1972-73): 89-104.

Chapter 5.

Millero F.J. and W.L. Masterton 1974.

Volume change for the formation of magnesium sulfate ion pairs at various temperatures.
J.Phys.Chem. 78(13): 1287-1294.

Wang Dong-Ping and F.J. Millero 1973.

Precise representation of the P-V-T properties of water and seawater determined from sound speeds.
J.Geophys.Res. 78(3): 7122-7128.

APPENDIX 5.

TABLE 5A.1

(a) Apparent molal and partial molal expansibilities of aqueous solutions of salts.

25°C 1-atm

Coefficients for the equations

$$\phi_E = \phi_E^o + S_E c^{\frac{1}{2}} \qquad cm^3 deg^{-1} mol^{-1}$$

$$\bar{E} = \phi_E^o + 1.5 S_E c^{\frac{1}{2}}$$

Salt	$\phi_E^o x10^2$	$-S_E x10^2$
HCl	3.4	0.02
KCl	8.50	2.20
LiCl	2.5	0.65
LiOH	4.5	1.8
NaCl	9.30	2.10
NaOH	11.7	3.5
Na_2SO_4	22.19	9.33

Parsons (1959), c in mol/litre

(b) Values of ϕ_E as a function of molality, m, at 25°C, 1-atm

ϕ_E in $cm^3 deg^{-1} mol^{-1}$

$BaCl_2$		$LaCl_3$		KCl	
m	$10^2 \phi_E$	m	$10^2 \phi_E$	m	$10^2 \phi_E$
0.01005	10.7	0.06464	7.97	0.06022	7.37
0.06028	13.74	0.09045	8.31	0.10053	7.13
0.10054	13.57	0.16154	8.17	1.0308	6.18
0.25239	13.34	0.36453	7.79	3.2731	4.54
0.50852	12.59	0.70812	7.04		
1.03382	11.13	1.04320	6.24		

Parsons (1959)

TABLE 5A.1 (c) The partial molal expansibility of ions in water, ($\text{ml mol}^{-1}\text{deg}^{-1}$) 25°C 1-atm

Cation	$\bar{E}^o_{ion}$	Anion	$\bar{E}^o_{ion}$
Ba^{2+}	-0.012	Br^-	0.050
Bu_4N^+	0.177	Cl^-	0.046
Ca^{2+}	-0.042	F^-	0.035
Cs^+	0.012	I^-	0.069
Et_4N^+	0.054	NO_3^-	0.100
H^+	-0.012	OH^-	0.060
K^+	0.023	SO_4^{2-}	0.107
Li^+	-0.014		
Me_4N^+	0.033		
Mg^{2+}	-0.105		
Na^+	0.032		
NH_4^+	-0.028		
Pr_4N^+	0.095		
Rb^+	0.016		
Sr^{2+}	-0.010		

From the compilation of Millero (1968)

TABLE 5A.2 The temperature differential of the specific volume of pure water.

$$(\partial v_w/\partial t)_p = (\alpha v)_w 10^6 cm^3/g\ deg.$$

p bars (absolute)	t°C 0	5	10	15	20	25	30
1		Table 5.19					
51	-49.49	30.69	99.64	160.11	214.09	263.03	308.03
101	-31.69	44.94	110.97	169.00	220.92	268.08	311.53
201	2.05	71.98	132.50	185.92	233.90	277.68	318.17
301	33.53	97.26	152.65	201.76	246.07	286.66	324.37
401	62.97	120.94	171.55	216.63	257.48	295.09	330.18
501	90.56	143.16	189.30	230.60	268.22	303.00	335.62
601	116.48	164.06	206.02	243.76	278.32	310.45	340.74
701	140.87	183.75	221.77	256.18	287.85	317.48	345.54
801	163.86	202.34	236.66	267.90	296.86	324.10	350.07
901	185.58	219.91	250.73	279.00	305.38	330.37	354.35
1001	206.11	236.54	264.07	289.52	313.45	336.29	358.38

From eq 5.71.6

TABLE 5A.3 The mean apparent molal expansibility of seasalt

(Solution A, Table 4.16)

at 1 bar pressure

Coefficients for the equations

$$\Phi_E^A = d_o + d_1 S \qquad \text{ml deg}^{-1}\text{mol}^{-1}\text{(salts)}$$

$$\Phi_E^A = a_o + a_1 S + a_2 S^2$$

$$\bar{E}_s^A = \Phi_E^A + S(1-10^{-3}S)(\partial\Phi_E^A/\partial S)$$

For higher pressures: use eqs 5.6.2 and 5.6.4

Temp °C	d_o	$-10^4 d_1$	$\pm 10^4$ st. dev.	a_o	$-10^4 a_1$	$10^7 a_2$	$\pm 10^5$ st. dev.
0	0.27040	13.5765	4	0.271419	14.9721	33.2276	1.0
5	0.22302	11.4395	3	0.223812	12.5245	25.8326	0.7
10	0.18468	9.7663	2	0.185307	10.6223	20.3820	0.5
15	0.15343	8.4387	2	0.153930	9.1230	16.2921	0.4
20	0.12784	7.3736	2	0.128250	7.9269	13.1735	0.3
25	0.10689	6.5111	1	0.107217	6.9630	10.7607	0.2
30	0.08978	5.8074	1	0.090051	6.1799	8.8688	0.2

Data of Ekman (1908), eq 5.6.4 with p = 1.

TABLE 5A.4 The partial molal expansibilities of alkali halides at 4°C - calculated from the temperature of maximum density of aqueous solutions.

Component, i of the binary aqueous solution	ϕ_v^o (0°C) a ml mol^{-1}	$10^3\phi_E^o/\phi_v^o$(4°C) b deg^{-1}	$10^2\phi_E^o$ (4°C) mlmol^{-1}deg^{-1}
CsCl	35.23 (36)	5.24	18.9
KBr	29.38 (30)	6.61	19.8
KCl	23.00 (23.7)	7.71	18.2
KI	40.05 (41)	5.98	23.8
LiBr	21.94 (22.2)	5.02	11.1
LiCl	15.30 (15.6)	5.94	9.2
LiI	32.50 (32.9)	3.75	12.3
NaBr	18.86 (19.8)	11.49	22.8
NaCl	12.36 (13)	15.45	20.1
NaI	29.34 (30.5)	8.84	27.0
RbBr	34.61 (35.2)	5.90	20.8
RbCl	28.11 (28.7)	6.36	18.3
RbI	45.15 (46)	5.39	24.8

a The bracketed values are estimates for 4°C, using the 0° and 25°C data (Parsons 1959).

b This is equivalent to $(1/\bar{v}_i^o)(\partial\bar{v}_i^o/\partial T)_{4°C}$ or $\bar{\alpha}^*$, the thermal coefficient of partial molal expansion at infinite dilution and 4°C (more precisely, 3.98°C). The equation used to calculate $\bar{\alpha}^*$ is:

TABLE 5A.4 (continued)

$$\bar{\alpha}^* = -2\bar{B}V_1^*/\bar{v}_i^{o*}(\Delta\theta/x)_L \qquad (5A.4.1)$$

where $\bar{B}$ is the coefficient in the parabolic relation to temperature of the molar volume of water, near 3.98°C.

V_1^* and $\bar{v}_i^{o*}$ are the molar volume of water and the limiting partial molar volume of the component i, respectively, at 3.98°C. $(\Delta\theta/x)_L$ is known as the Despretz constant. $\Delta\theta$ is the shift in the temperature of maximum density, relative to that of water, x is the mole fraction of solute. For further details of these interesting relationships (including data for the tetralkyl ammonium salts) the reader is referred to J. R. Kuppers (1974).

> Partial molal expansibilities from the temperature of maximum density of aqueous solutions.
>
> J.Phys.Chem.78: (10) 1041-2.

from whom the data in column b were obtained.

Also of interest, in this regard, is the paper by K. R. Foster and E. D. Finch (1974).

> An acoustical technique for measuring the temperature of maximum density of dilute aqueous solutions.
>
> J.Phys.Chem. 78(22): 2305-6.

REFERENCES Appendix 5.

Ekman V.V. 1908.

Die Zusammendrueckbarkeit des Meerwassers.
Pub.Circonst.perm.Explor.Mer. 43: 1-47.

Millero F.J. 1968.

Apparent molal expansibilities of some divalent chlorides in aqueous solutions at 25°C.
J.Phys.Chem. 72(13): 4589-4593.

Parsons R. 1959.

"Handbook of electrochemical constants", Butterworths, London, 113pp.

CHAPTER 6

COMPRESSIBILITY

The Pressure Derivative of the Partial Molal Volume

6.1 BASIC EQUATIONS.

The change of the pmv of a component i with pressure, at constant temperature, is called the partial molal compressibility $\bar{K}_i$, the formal definition being

$$-\bar{K}_i \equiv (\partial\bar{v}_i/\partial p)_T \qquad (6.1.1)$$

The apparent molal compressibility of the solute is defined by (for a binary solution)

$$-\phi_k \equiv (\partial\phi_v/\partial p)_T \qquad (6.1.2)$$

and since

$$\phi_v = (V - n_1V_1^o)/n_2 \qquad (4.2.1)*$$

$$-\phi_k = \left[(\partial V/\partial p)_T - n_1(\partial V_1^o/\partial p)_T\right]/n_2 \qquad (6.1.3)$$

with n representing the number of moles of the subscripted component.

The coefficients of isothermal compressibility are defined by

$$\beta \equiv -(1/V)(\partial V/\partial p)_T$$
$$= (1/d)(\partial d/\partial p)_T \qquad (6.1.4)$$

for the solution.

For the pure solvent, water (represented by the subscript w: Note that the same comments, regarding the subscript of the solvent, apply as those given in Section 4.2)

$$\beta_w = -(1/V_1^o)(\partial V_1^o/\partial p)_T$$
$$= (1/d_w)(\partial d_w/\partial p)_T \qquad (6.1.5)$$

where V is the volume of the solution, d the density.
The subscripts 1 and 2 indicate the solvent and solute, as usual for the binary solution, V_1^o is the volume of 1 mole of the pure solvent, water, n the number of moles.
Using eqs 6.1.4 and 6.1.5, equation 6.1.3 becomes

$$-\phi_k = (-V\beta + n_1V_1^o\beta_w)/n_2 \qquad (6.1.6)$$

For 1000 g water and m moles of salt

$$-\phi_k = \left[-V\beta + 55.51V_1^o\beta_w\right]/\underline{m} \qquad (6.1.7)$$

From eqs 4.2.8, 6.1.2, 6.1.4 and 6.1.5:

$$\phi_k = (10^3/\underline{m}dd_w)(d_w\beta - d\beta_w) + \beta M_2/d \qquad (6.1.8)$$
$$= (10^3/\underline{m}d_w)(\beta-\beta_w) + \beta\phi_v$$

From eqs 4.2.4, 4.2.6, 4.2.7 respectively:

$$\bar{K}_2 = \phi_k + I\partial\phi_k/\partial I \qquad (6.1.9)$$
$$\bar{K}_1 = \beta_wV_1^o - \left[I^2/w(55.51)\right]\partial\phi_k/\partial I \qquad (6.1.10)$$

(where $\beta_wV_1^o = \bar{K}_1^o$, $w = \frac{1}{2}z_+z_-\nu$)

$$\bar{K}_2 - \phi_k = (\beta_wV_1^o - \bar{K}_1)55.51w/I \qquad (6.1.11)$$

Molar (volume) scale. The corresponding equations on the molar scale are

$$\phi_k = (10^3/\underline{c}d_w)(d_w\beta - d\beta_w) + \beta_w M_2/d_w$$
$$= (10^3/\underline{c})(\beta-\beta_w) + \beta_w\phi_v \qquad (6.1.12)$$

Differentiating the Masson eq (4.2,10), gives

$$(\partial\phi_v/\partial p)_T = (\partial\phi_v^o/\partial p)_T + ((\partial s_v/\partial p)_T + s_v\beta/2)\,\underline{c}^{1/2} \qquad (6.1.13)$$

since $\beta = (1/\underline{c})(\partial\underline{c}/\partial p)_T$,

Thus $$\phi_k = \phi_k^o - ((\partial s_v/\partial p)_T + s_v\beta/2))\,\underline{c}^{1/2} \qquad (6.1.14)$$

or $$\phi_k = \phi_k^o + s_k\underline{c}^{1/2}$$

where $$s_k = -((\partial s_v/\partial p)_T + \tfrac{1}{2}s_v\beta) \qquad (6.1.15)$$

For many electrolytes, β in eq 6.1.14 may be replaced by β_w, at moderate concentrations (ca. 2.5N for 1:1 electrolytes). Using this approximation, β may be expressed in terms of the Masson parameters, in the same form as for the thermal expansion coefficient (eq 5.11.2)

$$\beta = \beta_w + (\phi_k^o - \beta_w\phi_v^o)10^{-3}\underline{c} + (s_k - \beta_w s_v)10^{-3}\underline{c}^{3/2} \qquad (6.1.16)$$

Lepple and Millero (1971) and Millero (1973, 1974) have derived the coefficients of eq 6.1.16 for seasalt in terms of equivalents, using binary solution data. The resulting values of $\beta - \beta_w$ at 25°C are within $\pm 0.09\times10^{-6}\,\text{bar}^{-1}$ of the experimental results, which suggests that the mixing effect is either negligible, or has been adequately accounted for by the method used (based on Young's rule). Extension to other temperatures (0-40°C) confirms the agreement as general (Millero 1973). More recently, Millero, Ward, et al (1974) have measured the isothermal compressibilities of aqueous NaCl, $MgCl_2$, Na_2SO_4 and $MgSO_4$ solutions from 0-45°C, at 1-atm. These data were fitted to eq 6.1.16 ; the coefficients are listed in a table (No.II) which appears in the microfilm edition of the journal in which the paper of Millero, Ward, et al (1974) was published.

6.2 Binary Solutions. The available data have generally been expressed in the form (Parsons 1959, Harned & Owen 1958):

$$\phi_{k_i} = \phi^o_{k_i} + s_k \underline{c}_v^{1/2} \qquad (6.2.1)$$

or

$$\phi_{k_i} = \phi^o_{k_i} + s'_k I_v^{1/2} \qquad (6.2.2)$$

where I_v is the molar ionic strength, $\frac{1}{2}(z_+z_-\nu)_i\underline{c}$, so that $s'_k = s_k/(\frac{1}{2}(z_+z_-\nu)_i)^{1/2}$

For the concentration range of seawater the differences in I_v and I (mol kg^{-1}water) are not very large for most electrolytes (Table 6.1). However, accurate conversion can be made via the relationship given previously,

$$c = d_w\underline{m}/(1 + 10^{-3}\underline{m}\phi_v d_w) \qquad (4.2.20)^*$$

TABLE 6.1 Effect of using I (weight ionic strength) instead of I_v (volume ionic strength) for the equations

$$\phi_{k_i} = \phi^o_{k_i} + s'_k I^{1/2}$$

$$\bar{K}_i = \phi^o_{k_i} + (3/2)s'_k I^{1/2}$$

(S = 35°/oo, I = 0.72321, 1-atm, 25^oC)

Salt	I_v	$-\phi_{k_i}$ ($ml\,mol^{-1}bar^{-1}$)		$-\bar{K}_i$ ($ml\,mol^{-1}bar^{-1}$)	
		with I_v	with I	with I_v	with I
NaCl	0.71175	41.90	41.82	37.05	36.93
Na_2SO_4	0.71810	124.16	124.06	109.84	109.68
$NaHCO_3$	0.70852	32.56	32.47	28.14	28.01
NaF	0.72153	62.66	62.65	56.84	56.82
KCl	0.70659	34.78	34.65	29.57	29.38
KBr	0.70317	25.74	25.61	20.96	20.76
$MgCl_2$	0.71811	87.19	87.14	79.99	79.91
$CaCl_2$	0.71744	71.60	71.54	64.40	64.31
$SrCl_2$	0.71718	72.20	72.14	65.01	64.91

ϕ^o_k and s'_k from Table 6A.1

Values of ϕ_k^o (or $\bar{K}^o$) and s_k are listed in Table 6A.2 for various electrolytes. The values for seawater salts and ions, (Millero 1973) for temperatures 0-40°C, are also listed (Table 6A.1). The ionic values will vary according to the convention chosen. In this case they are based on $\phi_k^o(H+) = s_k'(H+) = 0$, at all temperatures. The ϕ_k^o and most s_k' values are the same as for the binary solutions, but s_k' for sulfate is the weighted average derived from data on Na_2SO_4 and $MgSO_4$ solutions.

The temperature range of most interest in oceanography is 0-10°C, yet most of the available data are for 25°C. However, a guide to the values of many ions can be made using the data for seawater ions (for the range 0-40°C) and the ionic entropy (at 25°C). These values will, of course, be conventional ones based on $\phi_k^o(H+) = 0$ at all temperatures. Figures 6.1 and 6.2 illustrate the trends to be expected.

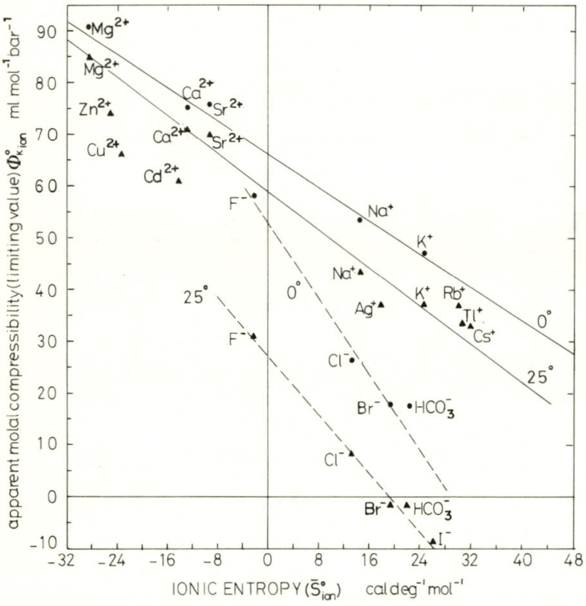

The limiting apparent molal compressibilities of ions as a function of the conventional ionic entropy. ϕ_k^o (ion) based on ϕ_k^o(H+)=0 at all temperatures, Millero (1973).

See Tables 6A.1 and 6A.1a

FIGURE 6.1

The ϕ_k^o of anions change more with temperature than the cations. The ϕ_k^o of the divalent cations only change by about 0.2mlbar^{-1}mol^{-1} per degree C, whereas the ϕ_k^o of oxyanions show changes ten times as large (assuming the sulfate ion is representative, i.e. 2.2mlbar^{-1}mol^{-1} per degree C). Since water becomes more structured at the lower temperatures, these results for ϕ_k^o are compatible with what might be expected. The ϕ_{k_i} of salts with oxyanions also exhibit greater sensitivity to changes in solute concentration (and to changes in pressure, as will be shown later).

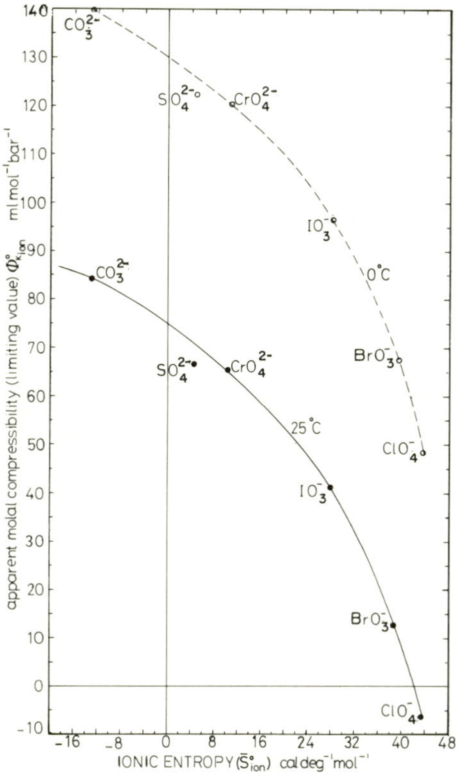

The limiting apparent molal compressibilities ϕ_k^o of oxyanions as a function of the conventional ionic entropies. Based on the data given by Millero (1973) and Parsons (1959). Ionic values based on ϕ_k^o(H+)=0 at all temperatures.

FIGURE 6.2

For the seawater concentration range (ionic strength 0.1-1 mol/kg water) the partial molal compressibility of a salt, i, is given by

$$\bar{K}_i \quad = \quad \phi^o_{k_i} \quad + \quad (3/2)s'_k I^{\frac{1}{2}} \qquad (6.2.3)$$

with $I = 0.01994S/(1-10^{-3}S)$, with S representing the salinity (‰) as usual.

Values of ϕ_{k_i} and $\bar{K}_i$ for the major salts of seawater are shown in Figure 6.3. These values have been computed from the ionic values given by Millero, so that they should be close to the actual values in the

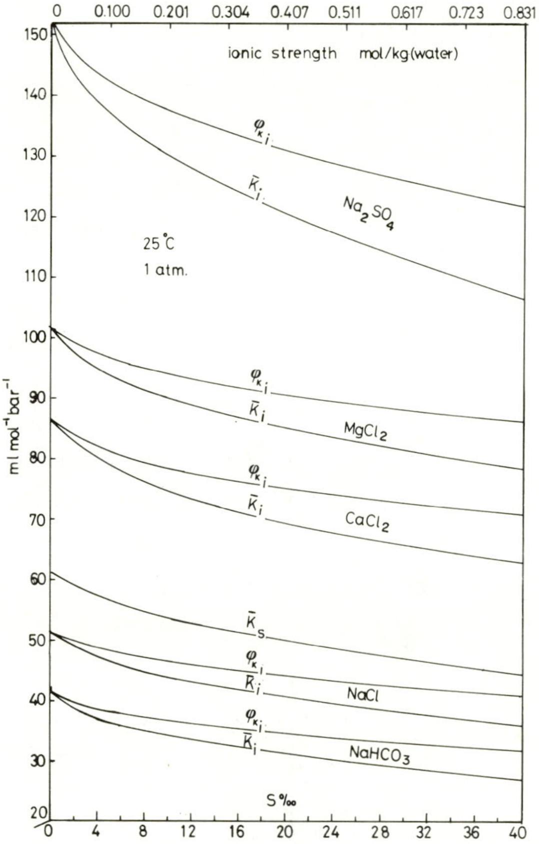

FIGURE 6.3

The apparent molal compressibility, ϕ_K, and partial molal compressibility, $\bar{K}$, of seawater components, data of Millero (1973), 25°C, 1 bar.

seawater medium. With the exception of Na_2SO_4 they are, in effect, the values in the binary solutions at the same ionic strength as the corresponding seawater solution. The weighting of the sulfate ion gives a larger value for s_k' of sodium sulfate than that of the single electrolyte solution (33.8 in seawater versus 23.8 $mol^{-\frac{1}{2}} l^{\frac{1}{2}}$ in the binary solution). The effect of this is to give, for a salinity of 35‰, a difference in ϕ_{k_i} of about 8%.

For 25°C, the volume ionic strength, equivalent to a particular salinity was computed from

$$I_v = Id_w/(1+10^{-3}(I/w_i)\phi_v d_w) \qquad (6.2.4)$$

where $w_i = \frac{1}{2}(z_+ z_- \nu)_i$

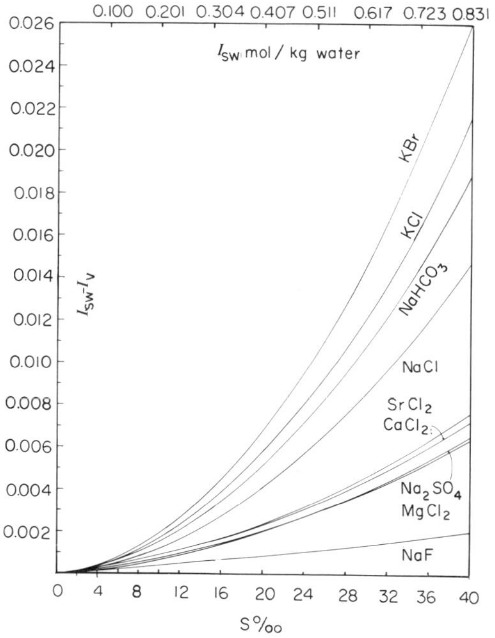

FIGURE 6.4

Differences in the weight ionic strength, I, and volume ionic strength, I_v, for a range of salinities ($I = 0.01994S/(1-10^{-3}S)$).
25°C, 1 bar.

The differences in the weight and volume ionic strengths are shown in Figure 6.4. In general, the differences are certainly not negligible at the high salinities. However, the effect on ϕ_{k_i} of neglecting these differences is not very great (Table 6.1).

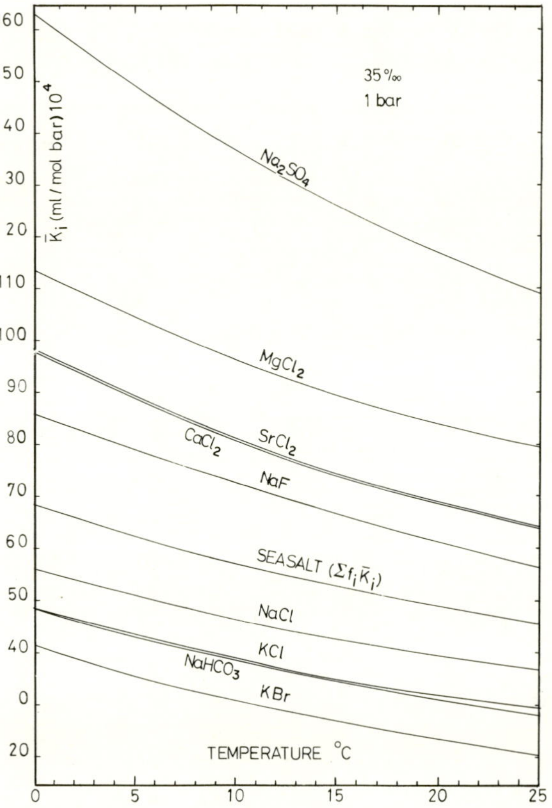

FIGURE 6.5

The partial molal compressibility of seawater components as a function of temperature (calculated from ϕ_k data of Millero (1973)) S = 35‰ and at 1 bar pressure.

The effect of temperature, at an ionic strength equivalent to S = 35‰, are shown in Figure 6.5. The weighted sum of the seawater salts ($\Sigma f_i \bar{K}_i$) is shown for comparison (values of f_i are for solution A, Table 4.16). Since the ionic values are only conventional, no attempt has been made to show their relative contributions.

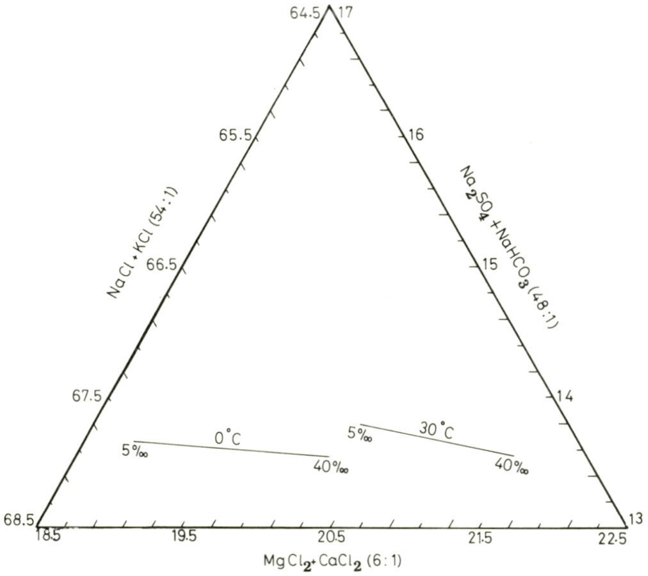

FIGURE 6.6

The relative percent contributions of the components to the partial molal compressibility of seasalt, (viz. $m_i\bar{K}_i/\Sigma m_i\bar{K}_i$) at 1 bar, for the temperatures 0° and 30°C, and salinities 5-40‰.
(i:j) after component names represents, in the same order, the ratios of their individual contributions.

The percent contribution of the various salts are illustrated in Figure 6.6. The largest changes are shown by magnesium chloride whose contribution increases with concentration and with temperature. Table 6.2 shows values at 25°C.

TABLE 6.2 Relative contributions of the salts to the total solute effect on compressibility ($\Sigma m_i \bar{K}_i$) (1 bar, 25°C)

$m_i \bar{K}_i / (\Sigma m_i \bar{K}_i)$

S°/oo	NaCl	Na_2SO_4	$NaHCO_3$	NaF	KCl	KBr	$MgCl_2$	$CaCl_2$	$SrCl_2$
5	0.6510	0.1330	0.0029	0.0002	0.0123	0.0008	0.1717	0.0277	0.0002
10	0.6495	0.1327	0.0029	0.0002	0.0121	0.0008	0.1737	0.0279	0.0002
15	0.6481	0.1325	0.0029	0.0002	0.0119	0.0008	0.1754	0.0280	0.0002
20	0.6469	0.1323	0.0028	0.0002	0.0118	0.0008	0.1769	0.0280	0.0002
25	0.6457	0.1321	0.0028	0.0002	0.0116	0.0008	0.1784	0.0281	0.0002
30	0.6446	0.1319	0.0028	0.0002	0.0115	0.0007	0.1798	0.0282	0.0002
35	0.6435	0.1317	0.0028	0.0002	0.0114	0.0007	0.1812	0.0283	0.0002
40	0.6424	0.1315	0.0028	0.0002	0.0112	0.0007	0.1826	0.0284	0.0002

The effect of temperature and salinity on $\bar{K}_i$ for sodium chloride and sodium sulfate at 1-atm are shown in Figure 6.7 and Figure 6.8.

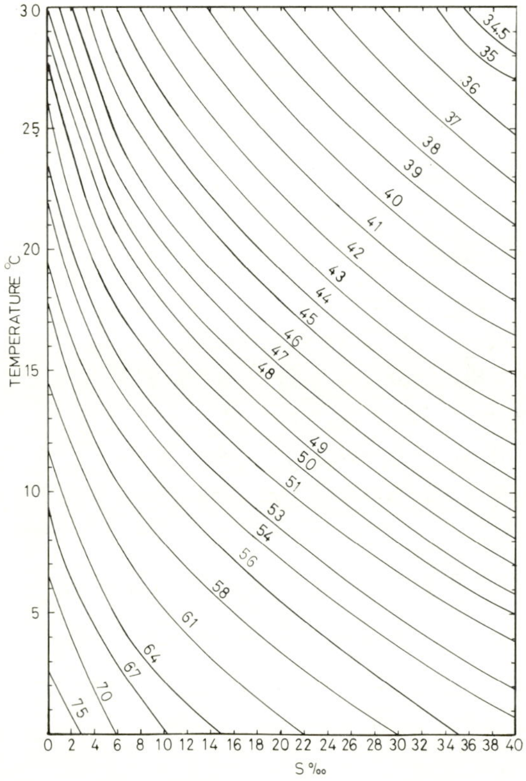

The partial molal compressibility of sodium chloride, $\bar{K}_{NaCl}$, at 1 bar, in the binary solution at the same ionic strength as the seawater $I = 0.01994S/(1-10^{-3}S)$. Calculated from the data given by Millero (1973), units 10^4 $ml mol^{-1} bar^{-1}$ (To avoid overcrowding, consecutively valued lines are not always shown).

FIGURE 6.7

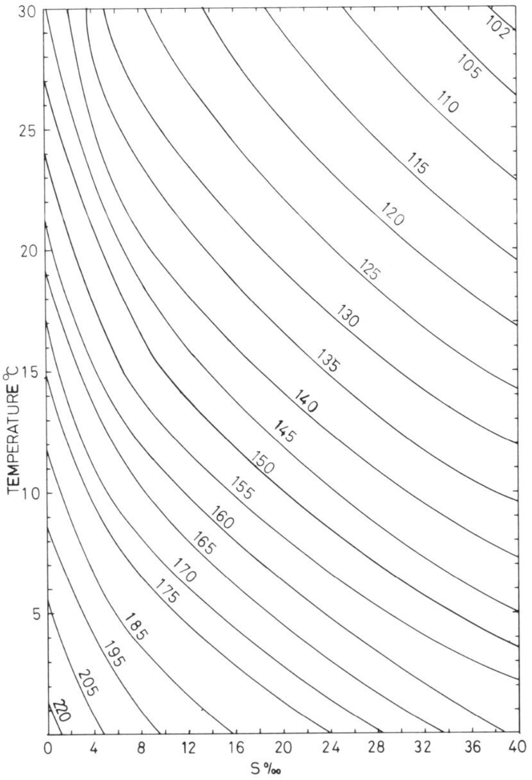

FIGURE 6.8

The partial molal compressibility of sodium sulfate, $\bar{K}_{Na_2SO_4}$, at 1 bar in the binary solution at the same ionic strength as the seawater, $I = 0.01994S/(1-10^{-3}S)$. Calculated from the data given by Millero (1973), which has been weighted to allow for ionic interactions in seawater. units $10^4 \mathrm{ml\,mol^{-1}\,bar^{-1}}$.

6.21. The limiting law equation for ϕ_k.

Although the concentration dependence of the ϕ_k of electrolytes follows the form of eqs 6.2.1 or 6.2.2 (in the seawater ionic strength range), the form compatible with the requirements of the Debye-Hückel limiting law is given by:

$$\phi_{k_i} = \phi^o_{k_i} + s^L_k \underline{c}^{\frac{1}{2}} + b_k \underline{c} \quad (6.21.1)$$

or

$$\phi_{k_i} = \phi^o_{k_i} + (s^L_k/w_i^{\frac{1}{2}}) I_v^{\frac{1}{2}} + (b_k/w_i) I_v \quad (6.21.2)$$

(cf the Redlich equation for ϕ_{v_i}, eq 4.2.18).

The limiting law slope is related to the slope s^L_v by

$$s^L_k = -((\partial s^L_v/\partial p)_T + \tfrac{1}{2}\beta_w s^L_v) \quad (6.21.3)$$

(cf eq 4.2.19). β_w is the isothermal compressibility of water.

Values of the limiting slope for eq 6.21.2 are listed in Table 6A.3.

Millero et al (1974) fitted their compressibility results (see Section 6.11) to equations of the form of both 6.2.2 and 6.21.2. Only data for concentrations $\leq$ 1M were used for the latter equation.

While more accurate values of ϕ^o_k are believed to be obtained from using eq 6.21.2 to extrapolate to infinite dilution, eq 6.2.2 can be used to estimate values of ϕ_k for the seawater concentration range ($0.1 \leq I \leq 1m$). The coefficients $\phi^o_{k_i}$ and s'_k are listed in Appendix 6, Table 6A.1. The ϕ^o_k values from eq 6.21.2 were as much as $39 \times 10^{-4} cm^3 mol^{-1} bar^{-1}$ lower than those from eq 6.2.2 (this large difference is for Na_2SO_4 at 0°C). Agreement with "acoustic" values was good, however (See Section 6.62).

Millero et al (1974) found that the values of b_k, for the salts studied, are related to the corresponding value for b_v at a given temperature (b_v is the Redlich deviation parameter, see eq 4.2.18). The relationship is:

$$b_k = a_1 + a_2 b_v \quad (6.21.4)$$

with the coefficients as listed in Table 6.2a. (b_k values are listed also).

TABLE 6.2a Coefficients for the equation

$$b_k = a_1 + a_2 b_v \quad \text{(eq 6.21.4)}$$

(The values of b_k and b_v are given by eqs 6.21.2 and 4.2.18 respectively)

	$10^4 a_1$	$10^4 a_2$	$t^\circ C$: 0	10	15	$10^4 b_k$ 20	25	30
NaCl	3.14	8.098	13.9	(8.8)	6.7	(5)	(3.3)	2.0
$MgCl_2$	2.72	1.242	5.9	(3.5)	2.4	(1.5)	(0.7)	0.3
Na_2SO_4	-1.80	1.570	15.5	(11.7)	9.4	(7)	(4.5)	1.6
$MgSO_4$	0.53	0.343	7.6		3.9			0.1

( ) interpolated values

From the data of Millero, Ward, Lepple and Hoff (1974).

An estimate of the b_k values for the total solute in seawater ("seasalt") was made by summing the weighted values (ionic strength fraction) of the components of a simplified recipe for seawater. The recipe is made up of only the four components, water, sodium chloride and sulfate and magnesium chloride. (Table 6.2b), viz.

$$(b_k)_{sw} = \Sigma \frac{I_i}{I} b_{k_i} \tag{6.21.5}$$

These seasalt values were found to be almost linearly related to the corresponding values (for a given salinity) of the effective pressure, p_e, exerted by the seasalt on the solution, with

$$p_e = 6.8966S - 0.0703St \tag{6.21.6}$$

$0 \leq t \leq 25$ from eq 4.61.7, since $p_e = B^* - B^*_{(S=0)}$, with the salinity S, in parts per thousand, ‰, and the temperature, t, in °C.

For a salinity of 35‰, the $(b_k)_{sw}$ values are related to the corresponding p_e values by

$$0.723(10^4(b_k)_{sw}) = -17.54 + 0.108p_e \tag{6.21.7}$$

TABLE 6.2b Relationship between b_k for "seawater" and the effective pressure of the solute on the solvent, water. 1-bar

		$10^4 b_{k_i} I_i$					
Salt i	I_i mol/kg	t 0°C	10	15	20	25	30°C
NaCl	0.437	6.07	3.85	2.93	2.19	1.44	0.87
$MgCl_2$	0.198	1.17	0.70	0.48	0.30	0.14	0.06
Na_2SO_4	0.088	1.36	1.03	0.83	0.62	0.40	0.14
	0.723						
	$\Sigma_i 10^4 b_{k_i} I_i$	8.60	5.58	4.24	3.11	1.98	1.07
	ditto, from eq 6.21.7	8.49	5.87	4.49	3.20	1.90	0.60
p_e bars eq 6.21.6		241	217	204	192	180	168

values of b_{k_i} from Table 6.2a (ϕ_k is given in $cm^3 mol^{-1} bar^{-1}$, and I_v in mol/l for eq 6.21.2)

to within about 5% (except at 30°C). The values calculated from eq 6.21.7 are compared with the summed values of the three salts in Table 6.2b.

Millero (1971) and Millero, Ward et.al (1974) have interpreted b_v and b_k in terms of ion-ion interactions. From eq 6.21.4 it appears that the ion-ion interactions responsible for the magnitude and temperature dependence of b_k are related to the behavior of b_v. From eq 6.21.7 b_k is related (in terms of the temperature dependence) to the "compressional" effect that the solute exerts on the solvent, water. This effect is ion-specific and, as will be shown in Section 6.31 the anion appears to contribute predominantly to this effect.

6.22 ϕ_k for components of systems exhibiting ion-pairing.

The calculation of the apparent molal compressibilities of ion pairs, and changes in ϕ_k^o due to ion pair formation (i.e. changes in $\bar{K}^o$) are discussed by Millero, Ward, et. al (1974). The methods outlined were applied to the aqueous $MgSO_4$ system, over the temperature range 0-45°C.

The effect of ion-pairing on the additivity principle, for apparent molal volumes, has been outlined in Section 4.21. The corresponding equations for ϕ_k may be obtained by differentiating eqs 4.21.2 to 4.21.5 w.r.t. pressure, at constant temperature. The aqueous $MgSO_4$ system may be considered as an example.

Since $\quad -\phi_{k_i} = (\partial\phi_{v_i}/\partial p)_T \quad$ and

$$\phi_v(MgSO_4)_{exp} = (1-\alpha)\phi_v(IP) + \alpha\phi_v(+-) \qquad (4.21.4)^*$$

then, differentiation gives, for the observed compressibility

$$\phi_k(MgSO_4)_{exp} = (1-\alpha)\phi_k(IP) + \alpha\phi_k(+-) + \Delta\phi_v(\partial\alpha/\partial p)_T \qquad (6.22.1)$$

with IP representing the ion pair, and (+-) the "free ions", Mg^{2+} and SO_4^{2-}, α is the fraction of free ions (or degree of dissociation), and

$$\Delta\phi_v = \phi_v(IP) - \phi_v(+-) \qquad (6.22.2)$$

Thus,

$$\phi_k(IP) = (\phi_k(MgSO_4)_{exp} - \alpha\phi_k(+-))/(1-\alpha) - \Delta_k \qquad (6.22.3)$$

with $$\Delta_k = \frac{\Delta\phi_v}{(1-\alpha)}(\partial\alpha/\partial p)_T \qquad (6.22.3\underline{a})$$

By definition, $$\Delta\phi_k = \phi_k(IP) - \phi_k(+-) \qquad (6.22.4)$$

thus, from eqs 6.22.3 and 6.22.4

$$\Delta\phi_k = (\phi_k(IP) - \phi_k(exp))/\alpha + \frac{\Delta\phi_v(\partial\alpha/\partial p)_T}{\alpha} \qquad (6.22.4\underline{a})$$

or,

$$\Delta\phi_k = (\phi_k(\exp) - \phi_k(+-))/(1-\alpha) - \frac{\Delta\phi_v}{(1-\alpha)}(\partial\alpha/\partial p)_T \quad (6.22.4\underline{b})$$

For 2:2 charge systems such as aqueous $MgSO_4$,

$$\alpha = 1 - K_A\gamma_\pm^2 m$$
$$= 1 - K_A\gamma_{MgSO_4}\underline{m} \quad (6.22.5)$$

where $\underline{m}$ is the molality of the $MgSO_4$ solution, K_A is the association constant and $\gamma_\pm$ the mean molal (stoichiometric) activity coefficient (which equals $\gamma_{MgSO_4}^{1/2}$, with γ representing the activity coefficient of the electrolyte). Differentiating w.r.t. pressure at constant temperature and constant $\underline{m}$, gives

$$(\partial\alpha/\partial p) = -K_A\underline{m}(\partial\gamma_{MgSO_4}/\partial p) - \underline{m}\gamma_{MgSO_4}(\partial K_A/\partial p)$$
$$= -K_A\gamma_{MgSO_4}\underline{m}(\partial\ln\gamma_{MgSO_4}/\partial p + \partial\ln K_A/\partial p)$$
$$= -(1-\alpha)\left(\frac{\partial\ln\gamma_{MgSO_4}}{\partial p} + \frac{\partial\ln K_A}{\partial p}\right) \quad (6.22.6)$$

From eq 4.1.1

$$\frac{\partial\ln\gamma_{MgSO_4}}{\partial p} = (\bar{v}_{MgSO_4} - \bar{v}^o_{MgSO_4})/RT \quad (4.1.1)*$$

with $\bar{v}$ and $\bar{v}^o$ representing the partial molal volumes in the solution and at infinite dilution, respectively. Also, (Harned and Owen 1958)

$$\partial\ln K_A/\partial p = -\Delta\bar{v}^o/RT \quad (6.22.7)$$

with

$$\Delta\bar{v}^o = \bar{v}^o(IP) - \bar{v}^o(+-)$$
$$= \phi_v^o(IP) - \phi_v^o(+-) \quad (6.22.8)$$

Thus, eq 6.22.6 becomes

$$(\partial\alpha/\partial p) = -\frac{(1-\alpha)}{RT}\left((\bar{v}-\bar{v}^o)_{MgSO_4} - \Delta\phi_v^o\right) \quad (6.22.6\underline{a})$$

For 25°C, from the data of Lee (1966),

$$(\bar{v}-\bar{v}^o)_{MgSO_4} = 26.58I^{\frac{1}{2}} - 36.86I + 30.75I^{3/2} - 9.15I^2 \qquad (6.22.9)$$

and $$\Delta\phi_v^o = 7.2 \text{ mlmol}^{-1} \qquad (6.22.10)$$

Also,

$$\phi_v(+-) = -7.157 + 6.426I^{\frac{1}{2}} - 0.0884I \qquad (6.22.11)$$

which is obtained from $\phi_v(MgCl_2) + \phi_v(Na_2SO_4) - 2\phi_v(NaCl)$, on the assumption that this addition gives a good estimate of the values of the free ions (Mg^{2+} and SO_4^{2-}), i.e. that ion-pairing, in the three electrolyte systems used, is negligible (for the seawater concentration range). Lee (1966) found that

$$\phi_v(IP) = 8.6I^{\frac{1}{2}} \qquad (6.22.12)$$

so that $$\Delta\phi_v = \phi_v(IP) - \phi_v(+-)$$
$$= 7.16 + 2.17I^{\frac{1}{2}} + 0.09I \qquad (6.22.13)$$

Lee (1966) also found that α had a value of 0.6 over a wide concentration range ($\underline{ca}$ 0.4 $\leq I \leq 3.0$).

With this value of α, and using eqs 6.22.6a, .9, .10 and .13, and the data for the ϕ_k's given by Millero, Ward et al (1974), values of $\phi_k(IP)$ can be estimated from:

$$0.4\phi_k(IP) = \phi_k(\exp) - 0.6\phi_k(+-) - \Delta\phi_v(\partial\alpha/\partial p)_T \qquad (6.22.3\underline{a})$$

The data for calculating $\phi_k(\exp)$ are listed in Table 6.2c. The corresponding value for the "free" ions, $\phi_k(+-)$ is obtained in a similar manner to $\phi_v(+-)$, that is, from

$$\phi_k(+-) = \phi_k(MgCl_2) + \phi_k(Na_2SO_4) - 2\phi_k(NaCl) \qquad (6.22.14)$$

At 25°C, from Table 6.2c and eq 6.22.14,

$$10^4\phi_k(+-) = -143.2 + 18.5I_v^{\frac{1}{2}} \qquad (6.22.15)$$

and $$10^4\phi_k(\exp) = -135.7 + 17.9I_v^{\frac{1}{2}} \qquad (6.22.16)$$

thus, $10^4(\phi_k(\text{exp}) - 0.6\phi_k(+-))$

$$= -49.8 + 6.8I^{\frac{1}{2}} \qquad (6.22.17)$$

TABLE 6.2c Coefficients for the equation

$$\phi_k = \phi_k^o + s_k I_v^{\frac{1}{2}}$$

		0	15	25	30	45	Av dev
$MgCl_2$	$-10^4\phi_k^o$	147.8	113.9	100.4	96.2	88.6	
		(129.4)	(103.6)	(95.3)	(93.7)	(89.9)	0.6
	$10^4 s_k$	26.6	19.0	16.0	14.7	12.5	
$MgSO_4$		183.4	151.3	135.7	129.6	118.2	
		(169.6)	(145.8)	(134.6)	(130.5)	(125.6)	0.5
		26.1	21.0	17.9	16.6	14.3	
NaCl		81.4	60.1	51.3	48.5	42.4	
		(75.6)	(55.6)	(45.9)	(42.2)	(39.0)	0.2
		21.0	14.9	12.2	11.3	9.5	
Na_2SO_4		230.8	168.4	145.4	139.2	122.1	
		(191.1)	(153.5)	(130.3)	(121.5)	(114.9)	0.6
		54.2	33.3	26.9	25.8	19.4	

(Column heading: t^oC)

from Millero, Ward *et.al.* (1974)
ϕ_k^o values at 25°C interpolated numerically by Millero *et al.*, s_k values at this temperature estimated from graphical interpolation by us. Bracketed values are based on data below 1*m* and using the Debye Hückel limiting law equation (eq 6.21.2)

Substituting this equation into 6.22.3a, and using eqs 6.22.6*a* and 6.22.13 to calculate the last term, gives the values of $\phi_k(\text{IP})$ listed in Table 6.2*d*. These values can be represented by

$$10^4\ \phi_k(\text{IP}) = -140.3 + 49.2I^{\frac{1}{2}} \qquad (6.22.18)$$

Values calculated from this equation are compared with the original estimates in Table 6.2d.

Equation 6.22.18 predicts $10^4\phi_k^o(IP)$ (i.e. $10^4\bar{K}^o(IP)$) as -140.3 $cm^3mol^{-1}bar^{-1}$. This value, however, is derived from a "high concentration" HC($I \geq 0.4$mol/kg) extrapolation and is probably too high (numerically). Combination with the HC extrapolation for $\phi_k(+-)$ from eq 6.22.15 gives

$$10^4\Delta\phi_k^o = 10^4(\phi_k^o(IP) - \phi_k^o(+-))$$

$$= -140.3 + 143.2$$

$$= 2.9\ cm^3mol^{-1}bar^{-1} \qquad (6.22.19)$$

Millero, Ward et al (1974) estimated a value of around 10 $cm^3mol^{-1}bar^{-1}$. They used similar equations to the above, but different data for α and neglected Δ_k.

TABLE 6.2d Estimates of ϕ_k for the $Mg\text{-}SO_4^o$ ion pair and elements of equation 6.22.3

$cm^3mol^{-1}bar^{-1}$

25°C 1-atm

I $molkg^{-1}$	$-10^4\phi_k$(exp -0.6(+-)) eq 6.22.17	$\Delta\phi_v$ eq6.22.13	$-10^4\Delta\phi_v\frac{\partial\alpha}{\partial p}$ eq6.22.6a	$-10^4\phi_k$(IP) eq6.22.3	$-10^4\phi_k$(IP) eq6.22.18
0.36	45.70	8.49	1.34	110.9	110.8
0.50	44.99	8.74	2.47	106.3	105.5
0.70	44.11	9.04	4.04	100.2	99.1
1.00	43.00	9.42	6.57	91.1	91.1
1.20	42.35	9.65	7.67	86.7	86.4
1.50	41.47	9.95	9.54	79.8	80.0
2.00	40.18	10.41	11.72	71.2	70.7

The value of Δ_k^o is easily calculated, since, from eqs 6.22.3a, 6.22.6a and with $\Delta\phi_v^o$ from eq 6.22.10,

$$\Delta_k^o = -\frac{\Delta\phi_v^o}{RT}\left((\bar{v}^o - \bar{v}^o)_{MgSO_4} - \Delta\phi_v^o \right)$$

$$= \frac{\Delta\phi_v^o}{RT} = (7.2)^2/RT$$

$$= 21.2 \times 10^{-4} cm^3 mol^{-1} bar^{-1} \qquad (6.22.20)$$

So that, if Δ_k is neglected $10^4\Delta\phi_k^o$ would be about 24 $cm^3mol^{-1}bar^{-1}$.

Millero, Ward et al (1974) showed that the linear correlation of $\Delta\bar{v}^o$ versus $\Delta\bar{K}^o$, shown for weak acids and bases (Lown, Thirsk et al 1968) also applied for ion pairs. A value of about 12 $cm^3mol^{-1}bar^{-1}$ for $10^4\Delta\phi_k^o$ (or $10^4\Delta\bar{K}^o$) is predicted from this correlation, at 25°C.

The limiting value for α can be obtained from eq 4.21.4 and 6.22.2, viz.

$$\alpha_L = 1 + \frac{\phi_v^o(+-) - \phi_v^o(exp)}{\Delta\phi_v^o} \qquad (6.22.21)$$

Normally it is assumed that $\phi_v^o(+-) = \phi_v^o(exp)$. However, the indications are, from accurate ϕ_v^o data, that some ion-pairing persists even in very dilute solutions. For example, the value of $\phi_v^o(+-)$ for $MgSO_4$ is estimated as follows, using the data tabulated in Table 6.2e, and eq 6.22.10

$$\phi_v^o(+-) = \phi_v^o(MgCl_2) + \phi_v^o(Na_2SO_4) - 2\phi_v^o(NaCl) = -7.80 \ cm^3mol^{-1}$$

or

$$\phi_v^o(+-) = \phi_v^o(MgCl_2) + \phi_v^o(K_2SO_4) - 2\phi_v^o(KCl)$$

$$= -7.70$$

so that, from eq 6.22.21, α_L is 0.9.

An estimate of α_L can also be made using ϕ_k^o data, since, from eq 6.22.3a,

$$\alpha_L = 1 + \frac{\phi_k^o(+-) - \phi_k^o(exp)}{(\Delta_k^o + \Delta\phi_k^o)} \qquad (6.22.22)$$

TABLE 6.2e Partial molal volumes of electrolytes at infinite dilution. 25°C 1-atm $cm^3 mol^{-1}$

Electrolyte	ϕ_v^o	source	Table listed in
$MgCl_2$	14.02	Perron et al (1974)	4.1a
Na_2SO_4	11.42	Lee (1966)	4A.3
K_2SO_4	32.00	Millero and Masterton (1974)	4A.4
KCl	26.85	" "	4A.3
$MgSO_4$	-7.18	" "	4A.4
NaCl	16.62	Lee (1966)	4A.3

Values of $10^4\phi_k^o(+-)$, in $cm^3mol^{-1}bar^{-1}$, for LC and HC extrapolations, respectively, are - 133.8 and -143.2 (Table 6.2c and eq 6.22.14). The corresponding values for $10^4\phi_k^o(exp)$ are -134.6 and -135.7. Since Δ_k^o and $\Delta\phi_k^o$ are both positive $(\phi_k^o(+-) - \phi_k^o(exp))$ must be negative, otherwise α_L will be greater than 1. That is, $\phi_k^o(exp)$ must be numerically smaller than $\phi_k^o(+-)$, since they are negative quantities. Within the experimental errors, the LC values predict α_L as 1. The HC results, however, give α_L as

$$\alpha_L = 1 - \frac{7.5}{21.2+10^4\Delta\phi_k^o}$$

Even if $10^4\Delta\phi_k^o$ is taken as 16.3 $cm^3mol^{-1}bar^{-1}$, this gives α_L as 0.8, which is too low.

With α_L taken as 0.9 and $10^4\Delta\phi_k^o$ as 12.8$cm^3mol^{-1}bar^{-1}$, then $(\phi_k^o(+-) - \phi_k^o(exp))10^4$ has a value of -3.4, and using the LC value for $10^4\phi_k^o(exp)$, -134.6, this gives $10^4\phi_k^o(+-)$ as -138.0 in the usual units. This value is almost equivalent to the mean of the LC and HC values. On the other hand, from the work of Fisher (1962) $\Delta\bar{K}^o$ (or $\Delta\phi_k^o$) might be expected to be close to zero.

If $\Delta\phi_k^o$ is taken as zero, then on the same basis as above,

$$10^4\phi_k^o(+-) = -2.1 + \phi_k^o(\exp)$$
$$= -136.7$$

This is near the value for $10^4\phi_k^o(\exp)$ estimated from the HC data (i.e. all the experimental data, $0.2\leq I\leq 2$).

Combining this value with $\Delta\phi_k^o$ taken as zero, from eq 6.22.4, $10^4\phi_k^o(IP)$ also equals $-136.7 cm^3 mol^{-1} bar^{-1}$. If $10^4\Delta\phi_k^o$ is taken as 12.8, $\phi_k^o(IP)$ becomes $-125.2 cm^3 mol^{-1} bar^{-1}$.

This latter value is rather low, compared to the $-140.3 cm^3 mol^{-1} bar^{-1}$ predicted from eq 6.22.18.

6.3 THE TAIT AND TAIT-GIBSON EQUATIONS.

The partial molal compressibilities of the electrolytes in seawater may be computed from the general equation for $\bar{v}_i^{sw}$, after differentiating with respect to pressure. Comparison with the values derived via Young's Rule (Millero 1973) can then be made. However, before this is done, the equations for seasalt (the bulk solute, a mixture of salts) will be developed. The effect of pressure on ϕ_{k_i} will also be considered and the direct calculation of the partial molal volume of seasalt up to 1000 bars will be presented.

A number of equations derived from the Tait-Gibson equation of state have already been presented (eqs 4.61.6, 4.61.10, 4.62.8, 4.71.1 and 5.72.11). Excellent agreement has been shown between quantities calculated from these TG equations (with Ekman's data) and the direct measurements, e.g. $\bar{V}_s$ (the mean partial molal volume of seasalt) and α_{sw} (the coefficient of thermal expansion of seawater). Such results indicate the reliability of this relatively simple equation of state. In addition, the TG equation features parameters that may be related to solution structure, and that are capable of being resolved into the component (salt) contributions. This is in contrast to other equations of state (e.g. the Tumlirz) which are basically mathematical-statistical devices for summarising data, so that they can be accurately interpolated, as required.

Since the TG equation features in most of the following sections, this seems an appropriate point to consider it in more detail, particularly the B and B* parameters which characterise it.

The Tait and Tait-Gibson equations have already been used for systems at 1-atm pressure. In this section the general equations for any pressure will be given together with more details of the parameters.

The Tait-Gibson equation is an extension of the Tait equation, which is an empirical P-V relationship originally used to summarise the compressibility behavior of water (Tait 1888). The Tait equation, in a slightly modified form, was later applied to many other liquids and was found to fit closely the experimental data for non-volatile liquids up to 10,000-12,000 bars (Harned and Owen 1959). Hirschfelder et al (1964) have described the Tait equation as the best empirical representation of the equation of state of liquids. However, there has been criticism of its suitability from time to time (e.g. Hayward 1967 McDonald 1969).

For water, the commonly used form of the equation is

$$\beta_w^{(p)} v_w^{(p)} = -(\partial v_w/\partial p)_T^{(p)} = 0.4343C/(B+p) \qquad (6.3.1)$$

or, in integrated form, from an initial pressure p_o (usually taken as 1 bar) to a final pressure p bars.

$$k_w v_w^{(p_o)} = -(v_w^{(p)} - v_w^{(p_o)}) = C\log((B+p)/(B+p_o)) \qquad (6.3.2)$$

where C = $0.315v_w^{(1)}$, v_w is the specific volume, β_w the isothermal coefficient of compression, k_w is the bulk compression of the water (the change in volume relative to the original volume). For a given temperature B is a positive constant, as noted in Chapter 4, and represents the difference between the expansive pressure (due to thermal energy) and the cohesive pressure (due to molecular interactions that are independent of temperature). By means of eq 6.3.2 the properties of liquids at 1 bar can be estimated from compression measurements at high pressures. The application of the Tait equation to compressibility data will be deferred until Section 6.7.

At present, the main concern is to introduce it as the basis of the Tait-Gibson (TG) equation, and to outline one of its important features, that is, its applicability to solution theory. Ginell (1961) derived the T equation from association theory and used it to interpret the general structure of liquids. He has shown how insights into the liquid structure can be derived from the functional relationships of the two coefficients B and C.

In brief, a liquid is pictured as similar to a polycrystalline or mosaic solid, but differing in the type of structural units, which are associated with holes, and surrounded by certain defects (areas bounded by weak bond-links to the main units). A pressure increase decreases the average size of the main units, creating new holes and extruding some of the defect holes. Ginell found that the coefficient B (or L in his terminology) was given by

$$B = RT/A\Sigma_i b_i N_i \tag{6.3.3}$$

where A is the weight fraction of the hole-defect species (or atom groups), RT has its usual meaning (gas constant x temperature) and the remaining term is one of the structural parameters of liquids. It is the sum of the excluded volumes of each unit (b) multiplied by the number of particles (N) in each unit. These terms are not expected to vary at constant temperature. The other Tait parameter, C', or 0.4343C ($0.1368v_w^{(1)}$ for water), is given by

$$0.4343C = (V-b)/\Sigma_i b_i N_i \Sigma s_x (1/V)^x \qquad (6.3.4)$$

where V-b is the difference between the total volume and the excluded volume, and equals the volume of holes in the liquid. The number of holes increase with temperature. Over the temperature range 20-100°C and pressure range 1-1000 bars, (V-b) is estimated as approximately 0.135 to 0.183 of the volume. The quantity s_x is a function related to the concentration of the various types of unit groupings present, and to A. This quantity probably decreases very slightly with temperature.

The theoretical study of solutions is, and no doubt will continue to be, of extreme difficulty. The model of liquid water by Weres and Rice (1972), for example, illustrates the problems involved. The tractability of the T equation in terms of theoretical interpretation is therefore of some consequence, even though some of the theoretical details and concepts may have to be restructured.

A number of other models of liquid water are available and several have been reviewed recently by Hagler, Scheraga and Némethy (1973). The interested reader is also referred to the texts on water of R.A. Horne (1972) and F. Franks (1972). As will be shown in Section 6.74 the improved version of the 1972 "cluster" model of liquid water (hereafter referred to as LHSN model) by Lentz, Hagler and Scheraga (1974), provides numerical values for structural parameters of liquid water that should be useful for checking suspected "structural" perturbation errors in P-V-T data for water and seawater. In the LHSN model, liquid water is assumed to consist of clusters of sizes from 1-9 molecules, as compact as possible to give a maximum degree of hydrogen bonding. Specific interactions between clusters are not considered. In order to treat their translational degrees of freedom, the clusters are considered as hard spheres in a constant background potential. In the revised treatment, the relative stabilities of hydrogen bonded rings and noncyclic structures are considered, as well as the effects of nonadditive (cooperative) hydrogen bonding in the internal partition function of a cluster. In addition, the dependence of the intermolecular normal mode vibrational frequencies on cluster size and geometry are taken into account. An important feature of the recent treatment is the degree of structural definition that it allows. The physical picture given by Erlander's (1968) theory, that was outlined in Chapter 5, is compatible (ignoring size concepts) with this much more refined and rigorous treatment. Erlander's cluster "aggregates" that break down into clusters at higher temperatures can be compared with the fairly rigid hydrogen-bonded rings of the LHSN model. As the temperature rises, these rings break down into smaller, less completely hydrogen-bonded structures. The two cluster concepts differ in that the LHSN stresses the conversion from rigid to less rigid structures as the temperature rises, whilst the other stresses changes in size of hydrogen bonded regions; these are large at low temperatures compared to the LHSN clusters (of less than 7 water molecules). The LHSN theory will be referred to again in Sections 6.74, 6.75.

The interest in the structure of water can be gauged by noting that for the period 1969-1971 nearly 300 papers were published on the structure of water (see the compilation of Hawkins, 1975).

Two other papers on the Tait equation that may be of interest are those of Winnick and Cho (1971) and Dayantis (1972).

Dayantis showed that the Tait equation (in the form originally given by Tait) may be derived from the Van der Waals equation, assuming that the internal (cohesive) pressure is a constant.

Winnick and Cho (1971) showed that the Tait equation could be used to predict the volume change of water at negative pressures (to -100 atm), within their experimental errors (7.5×10^{-5} cm^3/cm^3). The Tait-Gibson equation, which is an extension of the Tait equation to aqueous salt solutions (see next Section) may be useful for predicting the volume changes of seawater under negative pressures (tensile stresses within the liquid column).

For example, using the equation (see Section 6.81),

$$v_{sw}^{(p)} - v_{sw}^{(1)} = -(1-c)0.315v_w^{(1)}\log\frac{(B^*+p)}{(B^*+1)} \qquad (6.81.1)^*$$

with

$$B^* = 2670.8 + 6.8966S + (19.39 - 0.0703S)t - 0.223t^2 \qquad \text{(bars)}$$

the volume change of seawater of salinity, S, equal to 35 ‰ and temperature, t, equal to $0^{\circ}C$, and under a pressure, p, of -100 bars is 0.004661 cm^3.

6.31 The Tait-Gibson (TG) equation

Gibson (1934, 1935) extended the Tait equation to aqueous solutions by applying Tammann's principle, namely, that a given amount of water in a solution at atmospheric pressure has the properties of the same amount of pure water under a constant pressure greater than atmospheric (i.e. p_e bars greater). This principle was used to compute the compression of aqueous solutions up to 10,000 bars, from the compressibility data on water, the specific volumes of the solutions at atmospheric pressure, and the "effective" volume (ψ_2) of the solute in the solutions.

At any pressure, p, the volume of one gram of solution (i.e. the specific volume) containing x_1 g of solvent and x_2 g of solute may be represented by (see eq 4.62.3):

$$v^{(p)} = x_1\psi_1^{(p)} + x_2\psi_2^{(p)} \tag{6.31.1}$$

where ψ_1 is the apparent specific volume of the solvent in solution (i.e. the volume of 1 g of water under an additional pressure P_e bars), ψ_2 is the volume that 1 g of solute (as a liquid) contributes to the solution.

The addition of the solute (as a liquid) causes the water-contraction $(v_w^{(p)} - \psi_1^{(p)})$, given by the Tait-Gibson equation (see eq 4.62.5):

$$v_w^{(p)} - \psi_1^{(p)} = C\log((B+p_e+p)/(B+p)) \tag{6.31.2}$$

p_e is the change in the internal pressure of the water produced by the solute, just as B is the internal pressure of the water, so that $(B+p_e)$ or B* is the resultant internal pressure. p_e is, from Tammann's principle, the hydrostatic pressure producing the same changes to the properties of water as a given concentration of solute does. Combining these two equations

$$-v^{(p)} = x_1C\log(B^*+p) - x_2\psi_2^{(p)} - x_1(v_w^{(p)} + C\log(B+p)) \tag{6.31.3}$$

Thus, for the pressure range p_o to p, from eqs 6.31.3 and 6.3.2:

$$kv^{(p_o)} = -(v^{(p)} - v^{(p_o)})$$

$$= x_1C\log\left[(B^*+p)/(B^*+p_o)\right] - x_2\left[\psi_2^{(p)} - \psi_2^{(p_o)}\right] \tag{6.31.4}$$

and differentiating at constant temperature and composition, gives

$$\beta^{(p)}v^{(p)} = 0.4343x_1C/(B^*+p) - x_2\partial\psi_2^{(p)}/\partial p \tag{6.31.5}$$

for the compressibility of the solution at p bars. The terms containing ψ_2 are negligible at moderate pressures and concentrations. The values of B and C for water are known so that p_e can be evaluated from the measurement of a single compression over a suitable pressure range.

Gibson (1934, 1935) measured the bulk compression, k, of some thirty electrolytes (including sulfates, halides, nitrates and carbonates), over the pressure range 1-1000 bars. He found that k was largely dependent on the anion, the only cations having significant effect, at moderate concentrations, being H(+), NH_4(+), Be(2+) and Ce(3+) and possibly Cs(+) and Li(+). These cations offset the effect of the anions. Gibson analysed his results via a function A, defined as

$$\underline{A} = -(\Delta pv_w - \Delta pv/x_1) \tag{6.31.6}$$

where $-\Delta pv_w$ is the decrease in volume of 1 g of water over the pressure range 1-1000 bars and $-\Delta pv$ the decrease in volume per g of solution (containing x_1 g of water). A is, essentially, the difference between the specific compression of pure water and the specific compression of water in the solution. A therefore summarises the effect of the solute on the water. The contribution made by the solute to the compression is assumed to be negligible (as long as x_1 is greater than about 0.7g). For a given pressure range the specific compression of pure water is constant at constant temperature so that the relative effect of different salts can be shown via $-\Delta pv/x_1$. From eq 6.31.4 this quantity is given by

$$kv^{(1)}/x_1 \quad \text{or} \quad C\log(B^*+p)/(B^*+1), \text{ since}$$

$(x_2/x_1)(\psi_2^{(p)} - \psi_2^{(1)})$ is negligible.

Values of $-\Delta pv/x_1$, given by Gibson, are plotted in Figure 6.9 as a function of the molality of the anion (upper set of curves). At these moderate concentrations the dominant effect of the anion is clearly shown. The sulfates H_2SO_4, $(NH_4)_2SO_4$, and $BeSO_4$ do not conform to the average trends, as noted by Gibson. The $-\Delta pv/x_1$ values for these salts are much closer to the value of $-\Delta pv_w$ (39.45×10^{-3} ml/g was used by Gibson) which indicates that the cation in these cases is counteracting the specific effect of the anion. The line for H_2SO_4 hardly falls below 39×10^{-3}ml/g while values for $(NH_4)_2SO_4$ and $BeSO_4$ fall near the $Fe(CN)_6^{3-}$ values.

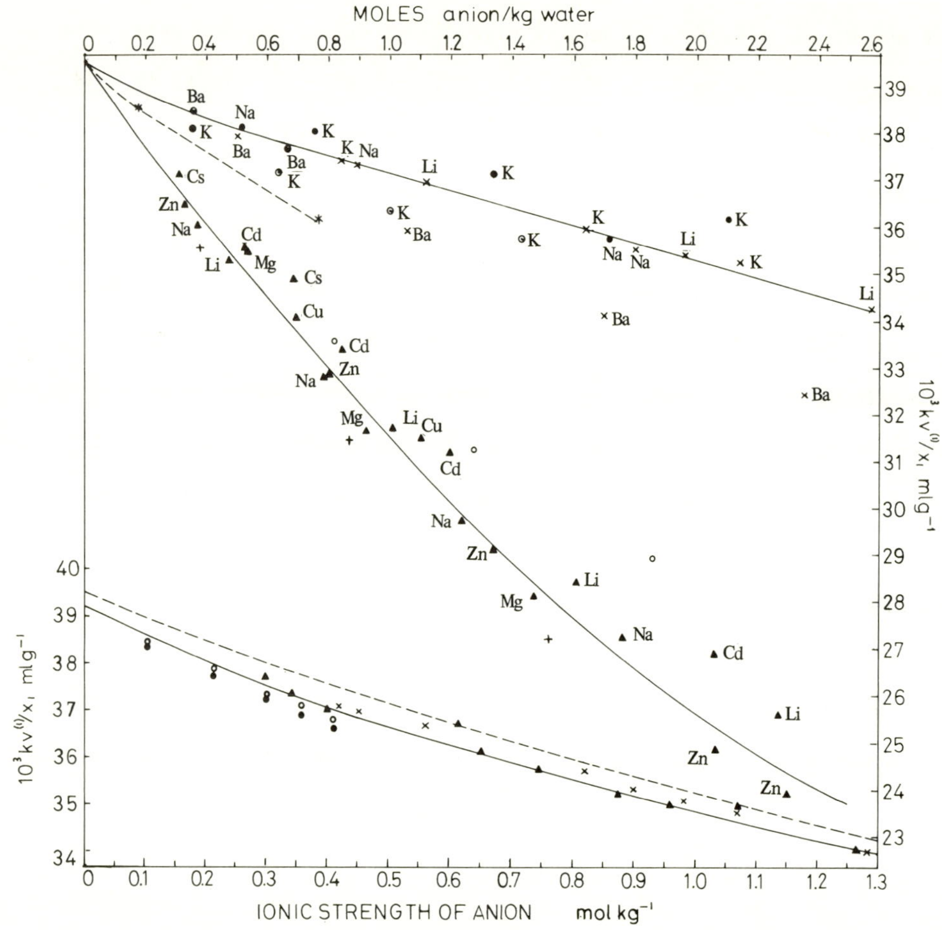

FIGURE 6.9

The change in volume of electrolyte solutions per gram of solution (containing x_1 g water) when pressure changes from 1 to 1000 bars.

k represents the bulk compression for the 1000 bar pressure range.

$$kv^{(1)}/x_1 = -(v^{(p)}-v^{(1)})/x_1 \text{ (or} -\Delta pv/x_1)$$

$$= C\log((B^*+1000)/(B^*+1)) \text{ Data of Gibson (1934,1935).}$$

Solution code for Figure 6.9

Upper curves

symbol	●	⊙	×
anion group	Br^- (Na,K)	NO_3^- (Ba,K)	Cl^- (Ba,Li,K,Na)

symbol	○	▲	+	✳
anion group	CrO_4^{2-} (K)	SO_4^{2-} (Na,Mg,Zn,Cu, Li,Cd,Cs)	CO_3^{2-} (Na)	$Fe(CN)_6^{3-}$ (K)

Lower curves × chlorides, ▲ sulfates

○ Ekman, ● Wilson (B* as given by Li 1967).

Barium chloride shows deviations in the opposite direction at molalities greater than about 0.6. For clarity, these points are not shown but are listed in Table 6.3.

The effect of the anion increases with the charge density and if the ionic strength of the anion is used as the concentration variable the functions for the sulfates and chlorides appear identical (lower curve, Figure 6.9).

This curve has been shifted down to adjust to the more recent k_o value (see section 6.7). The dashed line indicates the position with k_o = 393.5×10^{-4} as used by Gibson. Values of $-\Delta pv/x_1$ for seawater, calculated from $C\log(B^*+10^3)/(B^*+1)$ are shown as a function of $(I_{SO_4^{2-}} + I_{Cl^-})$.

The B* parameters as a function of salinity and temperature are given by:

Ekman's data $B^* = 2670.8 + 6.8966S + (19.39-0.0703S)t - 0.223t^2$ (4.61.7)*

Wilson's 1-atm sound velocity data as analysed by Li(1967) gives

TABLE 6.3 The specific compression of water* in aqueous solutions at 25°C over the pressure range 1-1000 bars.

Salt	$\underline{m}$ molkg^{-1}	$-\Delta pv/x_1(10^4)$ mlg^{-1}	Salt	$\underline{m}$ mol kg^{-1}	$-\Delta pv/x_1(10^4)$ mlg^{-1}
H_2SO_4	0.578	396.9	$(NH_4)_2SO_4$	0.500	368.2
	1.478	391.6		0.841	354.0
	3.708	380.6		1.232	340.3
				1.335	336.7
$BeSO_4$	0.501	366.4		1.892	325.4
	1.057	340.2		2.522	312.5
	1.679	312.7			
	2.379	288.5	$Ce_2(SO_4)_3$	0.153	369.1

* Gibson (1934), values for other salts are shown in Figure 6.9
$-\Delta pv$ is the decrease in volume per gram of solution (containing x_1 grams of water)

$$\underline{B^*_W} = B + (S/37)(252.59 - 2.883t + 4.365\text{x}10^{-2}t^2 - 0.4017\text{x}10^{-3}t^3)$$

with $B = 2689.81 + 20.233t - 0.3081t^2 + 1.38\text{x}10^{-3}t^3$ (6.31.7)

The B* values of Newton and Kennedy, as given by Li (1967) are listed in Table 6.4.

The results are consistent with the results for aqueous solutions of the single salts. The values calculated from the B* of Wilson are too low; the values from the B* of Newton and Kennedy give the closest agreement. This is because Wilson's B* values were estimated from 1-atm data whereas the other data were from averages over pressure ranges 200-600 bars (Ekman) and 100-1300 bars (Newton and Kennedy). As pointed out by Li (1967), the same error in the specific volume at 100 and 1000 bars gives a much larger

TABLE 6.4 Average B* of seawater from the data of Newton and Kennedy (1965)[a]

	400-1300 bars					
$t^{o}C$ / S(‰)	0	5	10	15	20	25
30.52	2858.9	2937.5	3002.0	3061.6	3092.6	3138.7
$\pm\sigma$	10.1	6.8	5.9	6.6	6.2	6.5
34.99	2918.7	2994.1	3051.4	3093.2	3136.4	3163.8
$\pm\sigma$	5.1	10.0	2.3	4.4	4.6	3.8
41.03	2958.1	3026.6	3084.3	3124.1	3176.6	3201.2
$\pm\sigma$	3.4	4.1	8.1	5.2	9.1	6.8

[a] Li (1967)

error in B or B* at the lower pressure. As will be shown in Section 6.8, B* calculated from sound velocity data at higher pressures (200-1000 bars) shows excellent agreement with Ekman's values, so that the high values at 1 bar are anomalous. This is discussed in Section 6.84.

The values of B* for some chlorides (K, Li and Na) and sulfates (Na and Mg) were calculated from Gibson's data and compared with the seawater values (Figure 6.10)

In the concentration range of interest here (up to an ionic strength of about 1.3) the B* values for chlorides and sulfates are linear functions of the molalities. As in Figure 6.9 the ionic strength functions (i.e. I_{anion}) for the sulfates and chlorides were very similar and serve as a base-line to compare the B* values derived from different data sets. Again, the results show that Wilson's B* values are too high.

The relationships of B* to the amv, ϕ_v, and the pmv, $\bar{v}$, have already been given. For a single electrolyte solution at any pressure p, and concentration m (mol kg^{-1}),

$$\phi_v^{(p)} = M_2\psi_2^{(p)} - 315v_w^{(1)}\log[(B^*+p)/(B+p)]/m \qquad (4.62.8)^*$$

and

$$\bar{v}_2^{(p)} = M_2\psi_2^{(p)} - (434.3C\ (\partial B^*/\partial m))/(B^*+p) \qquad (4.61.6)^*$$

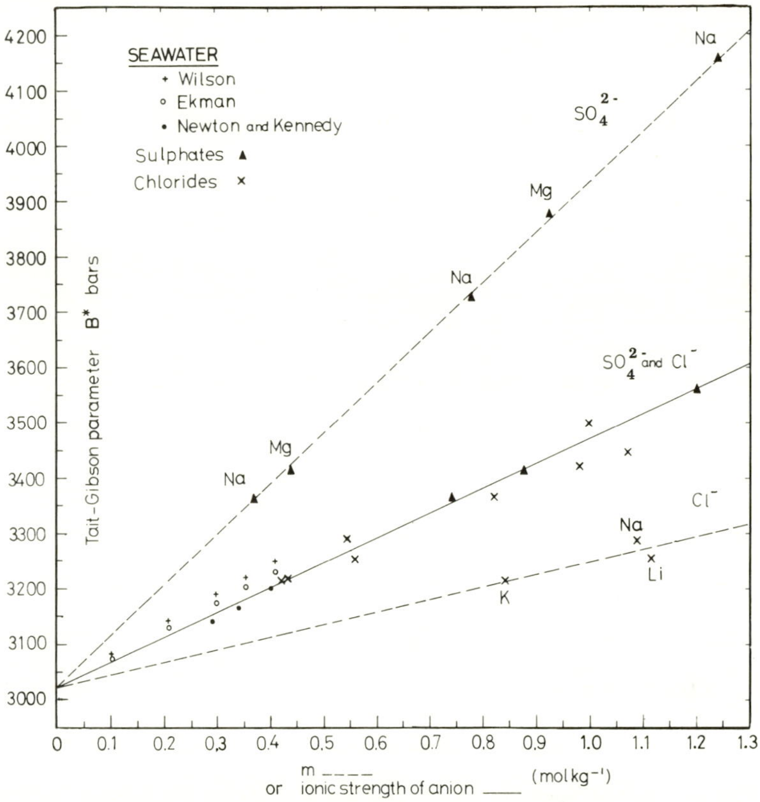

FIGURE 6.10

The Tait-Gibson parameter for sulfates, chlorides and seawater as a function of the ionic strength of the anion. Wilson and Ekman data as given by Li (1967). Newton and Kennedy (1965).

TABLE 6.5 Specific volume of pure melted solute in solution
1 bar $25^{o}C$

Salt	Ionic strength ($molkg^{-1}$)	$\psi_2 (mlg^{-1})$
Li_2SO_4	1.437	0.486
	3.033	0.482
	9.096	0.488
Na_2SO_4	1.113	0.392
	2.346	0.405
	5.28	0.413
Cs_2SO_4	0.921	0.266
	2.073	0.263
	8.295	0.258
$MgSO_4$	1.748	0.326
	3.692	0.344
	11.076	0.372
$(NH_4)_2SO_4$	2.523	0.587
	5.676	0.593
	15.135	0.601
KCl	0.838	0.510
	1.643	0.502
	2.146	0.503
NaCl	0.5-3	0.509
Seasalt	0.64-0.80	0.48625*

* Ekman's data as calculated by Li (1967)

The quantity ψ_2 (the specific volume of the pure melted solute) is the volume that 1 g of solute contributes to the solution on the assumption of Tammann's principle. Gibson estimated ψ_2 from the specific compressions to 1000 bar and the specific volumes of the solutions at 1 bar, eqs 6.31.1 and 6.31.4, neglecting $(\psi_2^{(p)}-\psi_2^{(1)})(x_2/x_1)$.

These values are shown in Table 6.5, together with the value of seawater estimated by Li (1967) from the data of Ekman. Values of ψ_2 (Note that ψ_t is used equivalently, usually to denote the total solute) were also calculated from Dorsey's Table 108 (1940) which lists the data of Knudsen *et al.* for 1-atm. The results support the conclusion of Li, that over the range 30-40‰ salinity, the value of ψ_2 (or ψ_t) is effectively constant (Figure 6.11). The values change very little down to about 15‰ salinity. ($I = 0.3037 molkg^{-1}$).

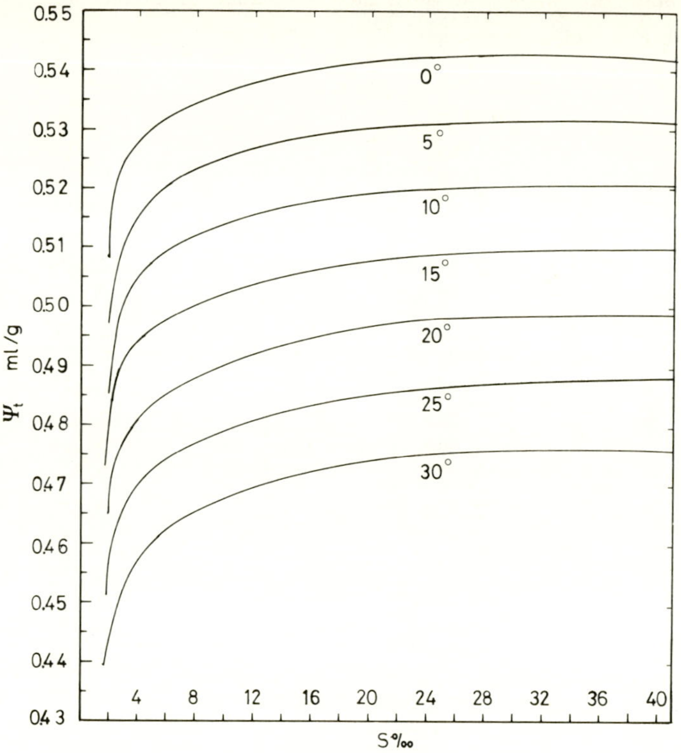

FIGURE 6.11

The volume that one gram of "seasalt" (as a liquid) contributes to the volume of a seawater solution. Calculated from eqs 6.31.1, 6.31.2, (4.61.7)* data from Dorsey (1940, Table 108)

Gibson also made estimates of ψ_2 by assuming that the solute just melts on going into solution and expands about 10%. Values of ψ_1 and p_e and hence $\Delta pv/x_1$ calculated from these values of ψ_2 (and the specific volumes) agreed with the experimental values of Δpv to within 4%. These calculations were, in effect, estimates of the compressibility from specific volumes at atmospheric pressure, compressibility data on pure water, and assumed values of ψ_2 (from the expected volumes of the pure liquified salts), i.e. without any data on the compressibility of the solution. Thus values of $v^{(1)}$, $\psi_2^{(1)}$ in eq 6.31.1 give $\psi_1^{(1)}$; B* is then calculated from eq 6.31.2 and hence k from eq 6.31.4 and β from eq 6.31.5.

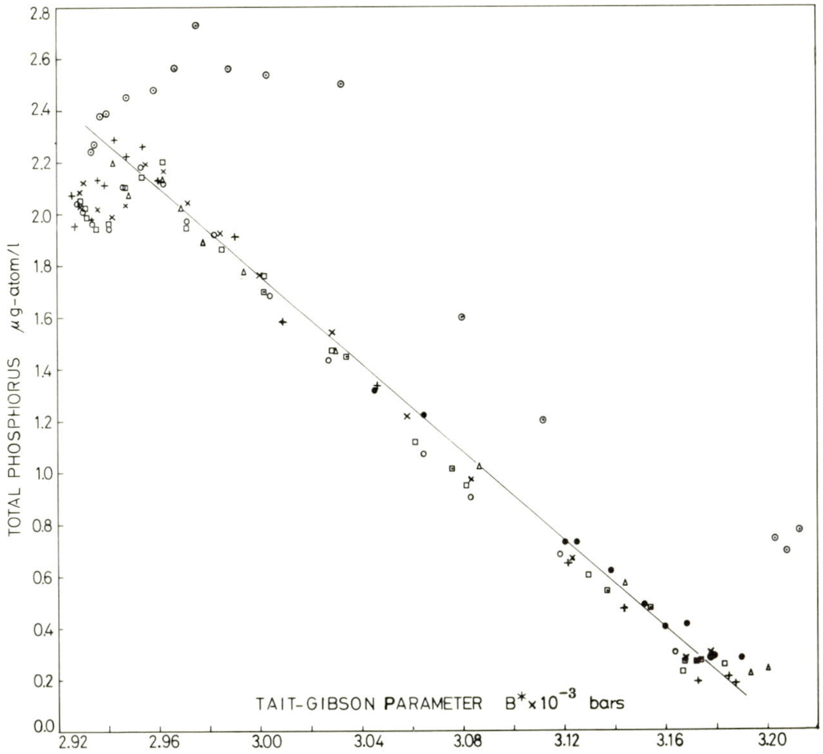

FIGURE 6.12

Total phosphorus as a function of the "internal" pressure of seawater, as represented by the Tait-Gibson parameter B* (eq (4.61.7)*)

Positions of the stations are listed in Table 6.6.

All are in the south-west Pacific region.
N.B. ⊙ is most northerly station

●	G3/49/65	+	G3/95/63
⊡	G3/57/65	⊙	G3/109/63
o	G3/62/65	Δ	G3/121/63
X	G3/76/65	□	G3/126/63

If $\bar{v}_2$ and B* are known, ψ_2 can also be calculated from eq 4.61.6. In this case, it is found that ψ_2 diminishes slightly with concentration.

TABLE 6.6 Positions in time and space of Oceanographic Stations featuring in Fig. 6.12

Station	Date	Maximum depth m	Latitude	Longitude	Source
G3/49/65	22/2/65	300	34 05S	151 35E	CSIRO (1965)
G3/57/65	23/2/65	4489	33 50S	153 25E	
G3/62/65	24/2/65	4583	33 52S	153 27E	
G3/76/65	25/2/65	4479	33 53S	153 29E	
G3/95/63	24/7/63	5005	23 52S	164 55W	CSIRO (1963)
G3/109/63	4/8/63	4000	0 00S	165 02W	
G3/121/63	15/8/63	2320	20 23S	174 51E	
G3/126/63	17/8/63	3125	26 12S	165 49E	

6.32 B* as an oceanographic parameter.

Since B* (or B + p_e) represents the internal "chemical" pressure of seawater it might be expected to be useful as a correlating parameter in biological or chemical studies, especially since it is a function of salinity and temperature and independent of pressure. As an example, the "total phosphorus" of south-west Pacific waters has been plotted as a function of B*. The relationship is practically linear, unlike that for total P versus temperature only, Figure 6.12.

The most northerly station obviously exhibits different phosphorus versus B* characteristics.

6.4 Estimates of the partial molal volume changes with pressure from bulk compression data.

Gibson (1934) found that his function A was simply related to the molality of the solution, m, by

$$A = am + bm^{3/2}$$

where a and b are constants (Table 6.7)

By differentiating A with respect to m, the change in the partial molal volume of the salt for a pressure rise 1-1000 bars can be found. This can

TABLE 6.7 Parameters of the equation

$A = a\underline{m} + b\underline{m}^{3/2}$ of Gibson* (mlg^{-1})

Solute	ax10^4	-bx10^4	av.dev. x 10^4
Na_2SO_4	115.0	33.9	0.5
$MgSO_4$	114.5	30.4	0.4
Li_2SO_4	109.5	32.1	0.4
K_2SO_4	104	29	-
$ZnSO_4$	104.5	24.0	0.6
$CuSO_4$	99	26	-
$CdSO_4$	94.9	24.3	1.0
Cs_2SO_4	89.1	28.0	0.4
$BeSO_4$	69.6	16.3	0.8
$(NH_4)_2SO_4$	68.1	22	0.6
NaCl	35	6.6	
NaBr	30	6.1	

* Gibson (1934)

be shown as follows:

Since $$A = \Delta pv/x_w - \Delta pv_w \qquad (6.31.6)^*$$

$$\partial A/\partial m_i = \partial(\Delta pv/x_w)/\partial m_i \qquad (6.4.1)$$

The volume containing 1000 g of water and m_i moles of salt is given by

$$V = (10^3 + m_i M_i)v$$
$$= 10^3 v/x_w \qquad (6.4.2)$$

where v is the specific volume, x_w the mass of water per g of solution, and

$$\partial V/\partial m_i = \bar{v}_i = 10^3\ \partial(v/x_w)/\partial m_i$$

hence $$\bar{v}_i^{(p)} - \bar{v}_i^{(1)} = 10^3\ \partial A/\partial m_i$$
$$= 10^3(a + 3/2bm^{\frac{1}{2}}) \qquad (6.4.3)$$

and $$\bar{v}_i^{o(p)} - \bar{v}_i^{o(1)} = 10^3 a \qquad (6.4.4)$$

Values of $\bar{v}_i^{(p)} - \bar{v}_i^{(1)}$ have been estimated for a number of electrolytes, with the ionic strength at 0.723, equivalent to a salinity of 35‰. The concentration range was too high to give accurate values for $\bar{v}^o$ so that eq 6.4.4 was not used. The molality of the salt, corresponding to I, is $m_i = 2I/z_+z_-\nu_i$, as usual.

The values are compared with those calculated previously by Leyendekkers (1974) for the electrolytes in a seawater medium (Table 6.8). The seawater values, as anticipated by the author, are somewhat higher than might be expected relative to the binary solutions. The calculation for the seawater medium is given in Section 6.11.

An estimate of changes in $\bar{K}_i$ with pressure, for the binary solutions can be made from the results for $\bar{v}_i^{(1000)} - \bar{v}_i^{(1)}$ as follows, assuming the function

$$\bar{K}_i^{(p)} = \bar{K}_i^{(1)} + B_k(p-1) \qquad (6.4.5)$$

TABLE 6.8 The effect of 1000 bars pressure on the partial molal volumes of electrolytes in the binary solutions at an ionic strength 0.723 $molkg^{-1}$ (equivalent to S=35‰) at 25°C

Salt	$(\bar{v}_i^{1000} - \bar{v}_i^{1})$	$(\bar{v}_i^{sw\,1000} - \bar{v}_i^{sw\,1})$	$B_k x10^6$	$-\bar{K}^1 x10^4$	$-\bar{K}^{1000}$ $x10^4$	$-\phi_k^{1000}$ x 10^4	$-\phi_k^{1000}$ x 10^4
	$mlmol^{-1}$			$mlmol^{-1}bar^{-1}$			
			eq 6.4.5			eq 6.4.7	a
$CaCl_2$		5.5					
KBr	(1.85) b						
KCl	(2.29)	3.0					
NaCl	2.66	3.80	2.07	36.9	16.2	20.6	21.1
NaBr	2.22	3.60	2.87	36.6	7.9		
Na_2SO_4	9.00	11.07	6.51	124.0	58.9	76.8	68.9
$MgSO_4$	9.51	10.54	7.07	130.6	59.9	73.3	67.3
$MgCl_2$	(5.83)	6.98	(4.70)				
Li_2SO_4	8.59		4.36	107.7	64.1		
Cs_2SO_4	6.85		1.47	75.9	61.2		
$(NH_4)_2SO_4$	5.19		1.69	60.4	43.5		
K_2SO_4	8.26	9.47					
$ZnSO_4$	8.92						
$CuSO_4$	8.24						
$CdSO_4$	7.94						
$BeSO_4$	5.92						

a Neglecting the change in s_k' with pressure results indicate that s_k' at 1000 bars versus $s_k'^{(1)}$ are approximately: NaCl (10.3 vs 11.4), Na_2SO_4 (42 vs 23.6) and $MgSO_4$ (31.5 vs 17.6). b from additivity principle.

and since $(\partial\bar{v}_i/\partial p)_T = \bar{K}_i$

$$\bar{v}_i^{(1000)} - \bar{v}_i^{(1)} = \bar{K}_i^{(1)}999 - \tfrac{1}{2}B_k(999)^2 \qquad (6.4.6)$$

Values of B_k for a few electrolytes are shown in Table 6.8.

TABLE 6.9 The effect of concentration* on the coefficient B_k of the equation $\bar{K}_i^{(p)} = \bar{K}_i^{(1)} + B_k(p-1)$ at 25°C

		$B_k x10^6$ mlmol^{-1}bar^{-2}				
Salt	S‰	20	25	30	35	40
NaCl		2.38	2.27	2.16	2.07	1.98
NaBr		2.99	2.95	2.91	2.87	2.84
Na_2SO_4		6.77	6.67	6.59	6.51	6.43
$MgSO_4$		7.23	7.17	7.12	7.07	7.02
Li_2SO_4		5.09	4.83	4.59	4.36	4.15
Cs_2SO_4		2.59	2.19	1.81	1.47	1.15
$(NH_4)_2SO_4$		2.41	2.15	1.91	1.69	1.48

* Ionic strength has been given in terms of equivalent salinity, i.e. $I = 0.01994S/(1-10^{-3}S)$

If changes in the slope, s_k, with pressure are ignored, values of ϕ_k^{1000} can be obtained from

$$\phi_{k_i}^{1000} = \bar{K}_i^{1000} - (3/2)s_k'^{(1)} I^{\frac{1}{2}} \tag{6.4.7}$$

Values of ϕ_k^{1000} can be estimated also from

$$\phi_{k_i}^{(p)} = \frac{(434.3x0.315)v_w^{(1)}(B-B_i^*)}{m_i(B_i^*+p)(B+p)} \tag{6.4.8}$$

which is derived from eq 4.62.8 by differentiating w.r.t. p (changes in ψ_2 with pressure are negligible).

Estimates for a few electrolytes are shown in Table 6.8. The results indicate that s_k' decreases slightly for NaCl, but almost doubles for the sulfates.

Values of $\bar{K}_i^{(p)}$ can also be estimated from this equation by differentiating with respect to I, provided B_i^* is known as a function of solution composition and temperature. B_k is a function of salinity and the small changes for the range 20-40‰ are shown in Table 6.9.

The values of $\bar{v}_i^{sw}$ for the major components of seawater, at 1 bar, were not very different from the values in the binary solutions at the same ionic strength as the seawater solution. The values of $\bar{V}_s$ estimated from the binary solutions were thus not much different than the actual values (derived from the direct measurements of the components), see Figure 4.8. It is of interest, therefore, to calculate the value of $\bar{V}_s^{(1000)}$ using the experimental $\bar{v}_i^{sw}$ values for 1 bar and the volume change for 1000 bars pressure estimated for the binary solutions (Table 6.8). The values of the individual components and the weights and final result are summarised in Table 6.10.

TABLE 6.10 Estimates of the partial molal volumes of seawater components at 1000 bars
$I = 0.723$ molkg^{-1}(water) (S=35‰)
$t = 25^{o}C$

Salt	$\bar{v}_i^{sw(1)}$ (exp)	$(\bar{v}_i^{(1000)} - \bar{v}_i^{(1)})$	$\bar{v}_i^{sw(1000)}$	f_i^*
NaCl	18.90	2.66	21.56	0.796440
Na_2SO_4	21.03	9.00	30.03	0.054895
$NaHCO_3$	27.07	-	29.87	0.004539
KCl	29.20	2.29	31.49	0.017666
$MgCl_2$	19.62	5.82	25.44	0.103630
$CaCl_2$	22.00	5.00	27.00	0.020098
KBr	(37.24)	1.85	39.09	0.001602
NaF	(0.59)	-	6.27	0.000139

$$\bar{V}_s^{A(1000)} = \Sigma_i f_i \bar{v}_i^{sw(1000)} = 22.75 \text{ mlmol}^{-1} \text{(salts)}$$

$$\bar{V}_s^{I(1000)} = \frac{\Sigma_i m_i \bar{v}_i^{sw(1000)}}{\Sigma_i \nu_i m_i} = 10.44 \text{ mlmol}^{-1} \text{ (ions)}$$

$$\bar{V}_s^{sw(1000)} = \frac{\Sigma_i m_i \bar{v}_i^{sw(1000)}}{(10^3 + m_s M_t)} = 11.69\text{x}10^{-3} \text{ mlg}^{-1} \text{(seawater)}$$

* solution A (Table 4.16)
Bracketed $\bar{v}_i^{sw}$ values are estimates (Section 4.4)

It will be of interest to compare this value with those derived from the seawater data via the TG and WM equations. The calculation of the partial molal volume of seasalt over the range of pressures 1-1000 bars will now be considered.

6.5 The partial molal volume of seasalt at pressures up to 1000 bars

Values of $\bar{V}_s^{(p)}$ for any pressure p, can be estimated directly from:

General.

$$\bar{V}_s^{(p)} = D\partial v_{sw}^{(p)}/\partial S + M_t v_{sw}^{(p)} \quad \text{ml mol}^{-1}\text{(salts)} \qquad (6.5.1)$$

where $D = S(1-c)(M_t+(1-c)M_t'/c)$, $c = 10^{-3}S$

from eq 4.61.14

$$\bar{V}_s^{I(p)} = \bar{V}_s^{(p)}/\Sigma\nu_i f_i \quad \text{ml mol}^{-1}\text{(ions)} \qquad (6.5.2)$$

(M_t', M_t and $\Sigma\nu_i f_i$ are given in Table 4.16)

$$\bar{V}_s^{sw(p)} = S(1-c)\partial v_{sw}^{(p)}/\partial S + x_2 v_{sw}^{(p)} \quad \text{ml g}^{-1}\text{seawater} \qquad (6.5.3)$$

where $x_2 = w_s/(w_s+10^3)$, with w_s representing the mass of salt per kg water, $w_s = 1.0049S/(1-c)$. These equations are now applied to various equations of state.

6.51 TG equation of state.

The simple relationship for the partial molal volume of seasalt has been given previously (Section 4.6)

$$\bar{V}_s^{(p)} = M_t\psi_t - (A(1-c)^2 v_w^{(1)} \partial B^*/\partial S)/(B^*+p) \qquad (6.51.1)$$

where $A = 434.3 \times 0.315 M_t'$

$$B^* = 2670.8+6.8966S+19.39t-0.0703St-0.223t^2 \qquad (4.61.7)^*$$

$$\partial B^*/\partial S = 6.8966 - 0.0703t$$

For solution (A) $M_t' = 68.0811$

$$M_t \psi_t = 37.09 - 0.1498t \qquad (4.61.9)*$$

alternatively, $$\bar{V}_s^{(p)} = \bar{V}_s^{(1)} + (Av_w^{(1)}(1-c)^2 \partial B^*/\partial S)(p-1)/(B^*+1)(B^*+p) \qquad (6.51.2)$$

which is useful when $\bar{V}_s^{(1)}$ is accurately known.

6.52 WM equation of state.

This equation is much more complicated and will only be given in summary, the coefficients and various functions are listed in Table 4.18.

Using eq 6.5.1,

$$\bar{V}_s^{(p)} = D\,(\partial v_{sw}^{(1)}/\partial S + ((p-1)v_w^{(1)} \; \partial R_{sw}/\partial S)/R_{sw}^2 + M_t(v_{sw}^{(1)} - (p-1)v_w^{(1)}/R_{sw}) \qquad (6.52.1)$$

with $v_{sw}^{(1)} = 1.000027/(1+\sigma_t 10^{-3})$

$\sigma_t = f(S,t)$ 16 coefficients

$\partial v_{sw}^{(1)}/\partial S = -(1.000027\text{x}10^{-3}\partial\sigma_t/\partial S)/(1+\sigma_t 10^{-3})^2$

$R_{sw} = B_{sw}(S,t) + A_{1_{sw}}(S,t)(p-1) + A_{2_{sw}}(S,t)(p-1)^2$

22 coefficients

$v_w^{(1)} = f(t)$ 8 coefficients

Values of $\bar{V}_s^{(p)}$ were calculated from eqs 6.51.1 and 6.52.1. The TG values are given in Figure 6.13 as a function of pressure and temperature for a salinity of 35‰. A typical P-T line for south west Pacific water is also shown. As for $\bar{E}_s$ the value remains fairly constant from about 500-2000 metres depth, the "thermodynamic stability region".

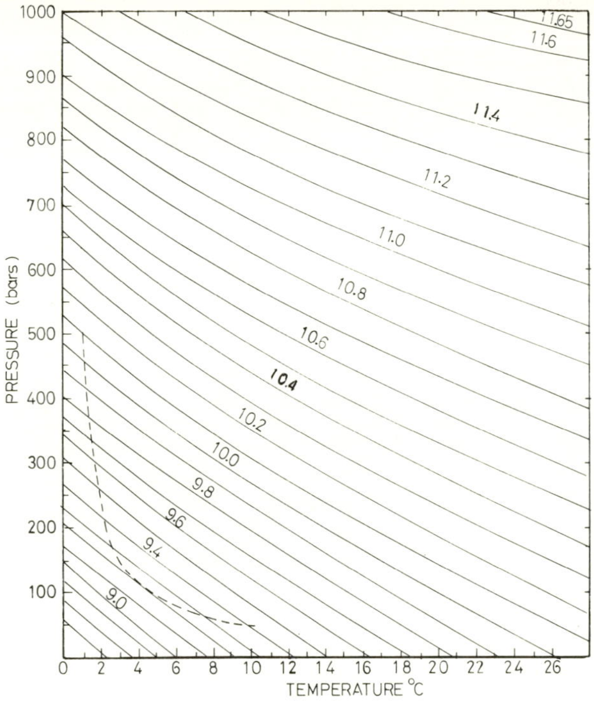

FIGURE 6.13

The mean partial molal volume of seasalt

$\Sigma m_i \bar{v}_i^{sw} / (10^3 + \Sigma(m_i M_i))$ ml per g seawater, $(x10^3)$

as a function of temperature and pressure.

From the data of Ekman (as given by Li 1967) eq 6.51.1

S = 35‰

The P-T characteristics of a south-west

Pacific station are superimposed,

------ Stn G3/93/63, lat. 27 45S,

long. 164 57W (23rd July 1963), salinity

range 34.4-34.7‰. CSIRO (1963).

The values from the WM equation are higher, the deviations increasing with pressure. The general trends are shown in Figure 6.14, together

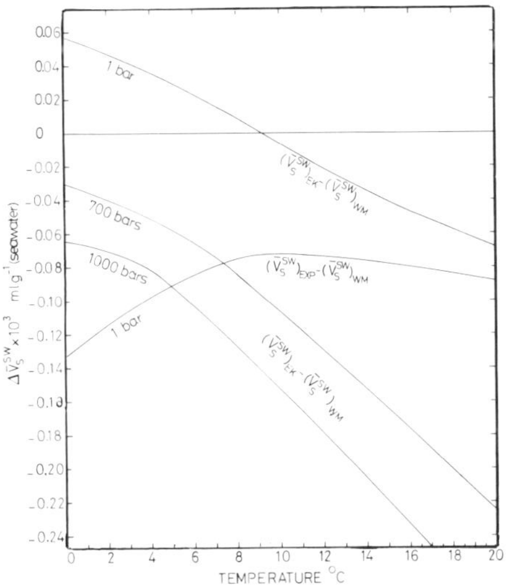

Differences between the mean partial molal volume of seasalt, calculated from the Tait-Gibson equation of state (Ekman's data, coded EK) and the Wang-Millero equation of state (coded WM)

Exp
experimental value
(Duedall 1968)

FIGURE 6.14

with the deviations of the WM results from the direct experimental results at 1 bar. At the higher temperatures, where the direct experimental results are more reliable, the deviations of the WM values from those of Ekman are in the same direction, but larger (increasing with pressure) than those of the corresponding experimental ones at 1 bar. In addition, the value of $\bar{V}_s$ at 25°, estimated in section 6.4, is in excellent agreement with the value from the TG equation (22.75mlmol^{-1}(salts) versus 22.69 mlmol^{-1}(salts) for the TG result). The WM value is 0.64 mlmol^{-1} higher.

On present evidence, the TG equation is expected to give better results than the WM, probably as a result of the better functional form for the concentration variable, salinity. In general, it might be expected, on the basis of the binary solution results, that a polynomial in $(S/(1-10^{-3}S))^{\frac{1}{2}}$ rather than S, would give better results for a "least-squares fitted" equation of state, i.e. one not based on any theoretical model.

As well, the small perturbations of the water structure, that appear to be associated with the acoustic data (Section 6.74) probably contribute additional errors to derivatives of the WM equation.

6.53 The apparent specific volume, asv, and the partial molal volume, pmv, of water in seawater.

The equations for calculating the asv ($\psi_w^{(p)}$), and pmv ($\bar{V}_w^{(p)}$), of water in seawater have been derived in Section 4.7. In summary,

$$\psi_w^{(p)} = v_w^{(1)} \left[1 - 0.315\log\frac{(B+p_e+p)}{(B+1)}\right] \quad (4.62.5)^*$$

(see, also eq 6.3.2), $B+p_e = B^*$

$$\bar{V}_w^{(p)} = (v_w^{(1)}/0.05551) \left[1-0.315\log\frac{(B^*+p)}{(B+1)} + ((0.4343\times0.315)S(1-c)\partial B^*/\partial S)/(B^*+p)\right] \quad (4.72.4)^*$$

Values of B* functions are given in section 6.51, B =B*(S=0), $v_w^{(1)}$ is given by eq 5.71.1, Table 5.18,

$c=10^{-3}S(‰)$, as usual.

More generally,

$$55.51\bar{V}_w^{(p)} = 10^3 v_w^{(p)} - (I^2/W)(\partial\Phi_v/\partial I) \quad (4.72.5)^*$$

or

$$55.51\bar{V}_w^{(p)} = 10^3 v_w^{(p)} - (S^2/M_t')\partial\Phi_v/\partial S$$

with $M_t' = 68.0811$ for Φ_v^A, solution A (Table 4.16).

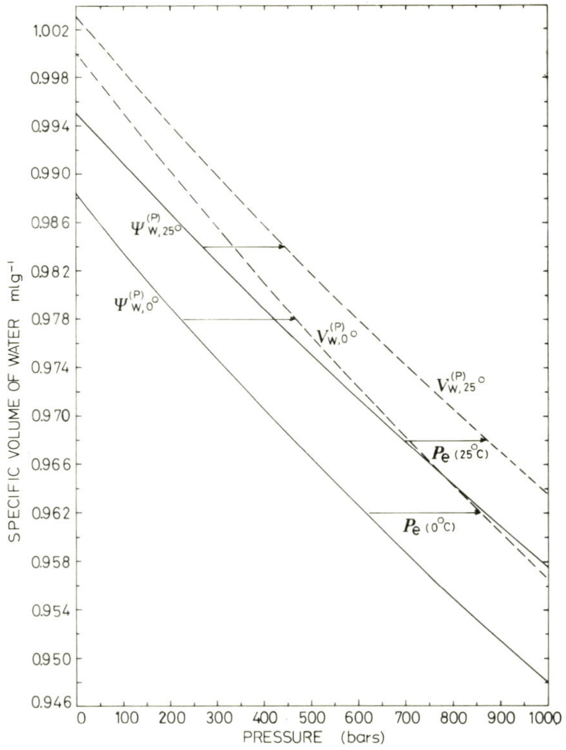

FIGURE 6.14a

The specific volume of water ($v_w^{(p)}$) and the apparent specific volume of water in seawater ($\psi_w^{(p)}$), at S = 35‰ as a function of pressure. P_e represents the internal pressure due to the salt.

The value of $\psi_w^{(p)}$ is equivalent to the specific volume of pure water under an additional applied pressure of p_e bars, (as indicated by eq 4.62.5). This is illustrated in Figure 6.14a where values of $v_w^{(p)}$ (from the tabulation of Fine and Millero (1973), based on acoustic data) are shown as a function of pressure, at 0° and 25°C. These values contrast with those of $\psi_w^{(p)}$ (calculated from 4.62.5 with S = 35‰ and using Ekman's B*, Li 1967).

The latter are shifted back by p_e bars along the pressure co-ordinate; because of the salt pressure, a lower external pressure is needed to achieve the same decrease in volume. An interesting feature of Figure 6.14a is that the values of ψ_w at 25°C, 700-900 bars, are very close to the values of v_w at 0°C at the same pressures. Apparently, at these pressures, a reduction of 25°C in temperature has the same effect on 1 g of water as the addition of the salt.

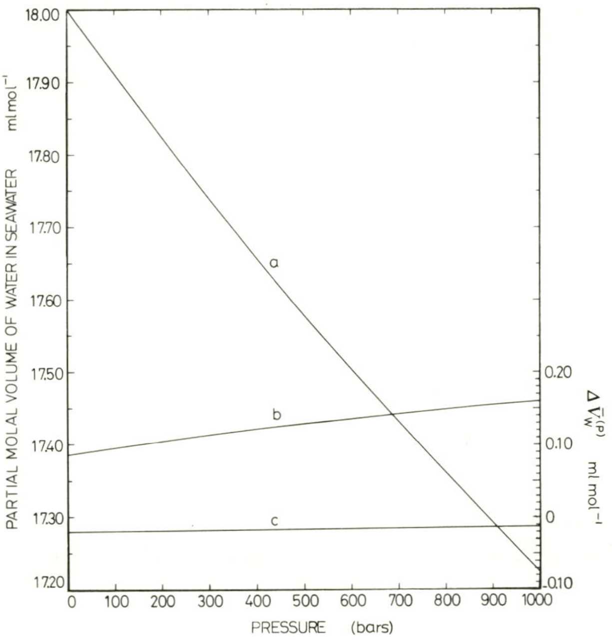

FIGURE 6.14b

Effect of temperature, salinity and pressure on the partial molal volume of water in seawater.

a $\bar{V}_w^{(p)}$ at 0°C and S = 35‰

b $\bar{V}_w^{(p)}$ (30°C) - $\bar{V}_w^{(p)}$(0°C) at S=35‰

c $\bar{V}_w^{(p)}$ (40‰) - $\bar{V}_w^{(p)}$ (5‰) at 0°C

The pmv of water in seawater is practically insensitive to change in salinity, from 5-40‰ the change is only about $0.02 ml mol^{-1}$. The temperature change (per ^{o}C versus per (‰)) is about 4 times larger and doubles with an increase of 1000 bars pressure. The salinity effect decreases with pressure. These trends are illustrated in Fig. 6.14b, R.H.S. ordinate ($\Delta\bar{V}_w^{(p)}$). The value of $\bar{V}_w^{(p)}$ decreases with pressure. At 1000 bars it has decreased by about 5%, line a in Fig. 6.14b. This is largely due to the decrease in v_w, the specific volume of pure water, as the gradient $\partial\Phi_v/\partial S$ does not change much with pressure (see eq 4.72.5).

6.6 Compressibilities from sound velocity measurements.

An important application of the limiting value of the partial molal compressibility, $\bar{K}_i^o$, is in determining the effect of pressure on equilibrium constants, via the relationships

$$RT(\partial \ln \underline{K}/\partial p)_{T,m} = -\left[(\Sigma\bar{v}_i^o)_{\tilde{p}} - (\Sigma\bar{v}_i^o)_r\right] \tag{6.6.1}$$

and
$$-((\Sigma\bar{K}_i^o)_{\tilde{p}} - (\Sigma\bar{K}_i^o)_r) = (\partial\Delta\bar{v}^o/\partial p)_{T,m} \tag{6.6.2}$$

where i, represents any component of the system, and $\tilde{p}$ and r denote the products and reactants, respectively. $\underline{K}$ represents the thermodynamic equilibrium constant. These equations and the application of them to solution equilibria are illustrated by the recent work of Millero and Masterton (1974), Ward and Millero (1974), Millero, Ward et al (1974), Read (1975), Olofsson and Hepler (1975), Fisher and Fox (1975) and Fisher (1975). These authors cite much of the relevant literature. Leyendekkers (1975) has used these equations to calculate the effect of pressure on the chemical potential of seawater components.

In the present section, the method of calculating $\bar{K}_i^o$ from sound velocity will be briefly outlined. The calculation is based on Laplace's equation, as is the WM equation of state, and those compressibility equations developed from sound velocity data (e.g. Kell's equation for water at 1-atm 1970). It therefore seems appropriate to note here the restraints of this equation, and to outline the extension, via thermodynamics, to the isothermal compressibility.

6.61 Isentropic and Isothermal Compressibilities.

The Laplace equation has been known for over 160 years, and can be deduced from the principles of hydrodynamics. The periodic compressions of the solution, due to the propagation of the sound waves, take place adiabatically (without heat exchange) and (presumably) reversibly, up to very high frequencies. The equation relates the adiabatic coefficient of compression $\beta_{\$}$ $(-(1/V)(\partial V/\partial p))$ in bar^{-1}, to the sound velocity, u, $(cmsec^{-1})$,

$$u^2 = 10^6/\beta_{\$}d \qquad (6.61.1)$$

where d is the density (gcm^{-3}), $\$$ represents the entropy.

The adiabatic (or isentropic) coefficient is given the subscript $\$$ to denote constant entropy. For, in the case of reversible adiabatic changes, which are assumed for the sound propagation, the entropy change is zero (i.e. $\Delta\$=0$). If, however, $\Delta\$\neq 0$, then the relationship is not strictly valid. Because the speed of sound can now be measured to very high accuracy (for water it is measured routinely to 5ppm with a sensitivity better than 1 ppm (0.0015m/sec)) it is generally assumed that compressibility values derived from sound velocity data are the most accurate. However, because of the constraint that $\Delta\$=0$, the adiabitic values may exhibit larger errors than anticipated, under certain conditions. It may be significant, for example, that anomalies have been reported in the temperature dependence of the speed of sound in water at 1 bar and low temperatures (around 0-10°C). An important structural change may occur in water around 4°C (maximum density), and it is of note that the speed of sound is very nearly linear (with respect to temperature) in that region (Horne, 1969). While Del Grosso and Mader (1973) concluded (from over 1000 relative measurements of the speed of sound in water between 0° and 10°C) that no anomaly larger than 2 ppm (0.003m/sec) exists, there may be entropy changes which introduce errors in the derived quantity $\beta_{\mathcal{S}}$. Other errors will be introduced via the auxiliary quantities which are needed to convert to the isothermal coefficient.

The relationship between the isentropic $(-(1/V)(\partial V/\partial p)_{\mathcal{S}})$ and isothermal $(-(1/V)(\partial V/\partial p)_T)$ compressibilities can be found using the principles of eq 2.2.7 i.e.

$$(\partial p/\partial V)_{\mathcal{S}} = (\partial p/\partial V)_T + (\partial p/\partial T)_V(\partial T/\partial V)_{\mathcal{S}} \quad (6.61.2)$$

The last term can be converted to quantities for which data are generally available, as follows. From the first and second laws of thermodynamics

$$C_p - C_V = T(\partial p/\partial T)_V(\partial V/\partial T)_p \quad (6.61.3)$$

where C_p and C_V are the heat capacities at constant pressure and volume, respectively.

Also

$$(\partial T/\partial V)_{\mathcal{S}} = -(\partial \mathcal{S}/\partial V)_T/(\partial \mathcal{S}/\partial T)_V$$

$$= -(T/C_V)(\partial p/\partial T)_V \quad (6.61.4)$$

so that, from 6.61.3

$$(\partial T/\partial V)_{\mathcal{S}} = -(C_p/C_V - 1)/\ (\partial V/\partial T)_p \quad (6.61.5)$$

Also,

$$(\partial p/\partial T)_V = -(\partial V/\partial T)_p/\ (\partial V/\partial p)_T \quad (6.61.6)$$

Substituting 6.61.5 and 6.61.6 into 6.61.4 gives

$$\beta_{\mathcal{S}} = \beta C_V/C_p \quad (6.61.7)$$

where β is the isothermal compressibility coefficient (the subscript T is omitted for convenience, as this quantity is to be frequently referred to in the text). Since data for C_V are not usually available, a more useful relationship can be found by eliminating this quantity.

From 6.61.3, 6.61.6, $C_V = C_p - T\alpha^2 V/\beta$

so that 6.61.7 becomes

$$\beta_S = \beta - T\alpha^2 V/C_p \qquad (6.61.8)$$

with α in deg^{-1}, C_p in $cm^3 bardeg^{-1}$, β in bar^{-1}, T in absolute temperature units (273.15+t°C), and the volume V in cm^3.

or

$$\beta_S = \beta - 0.0239\alpha^2 TV/C_p$$

with C_p in $caldeg^{-1}$ (if joule deg^{-1} units used then 0.0239 is replaced by 0.1). More commonly, the specific heat (c_p) is used (i.e. C_p per gram)

$$\beta_S = \beta - 0.0239\alpha^2 T/(c_p)d$$

where d is the density. The isothermal compressibility is thus obtained from the sound velocity via

$$\beta = 10^6/u^2 d + 0.0239\alpha^2 T/(c_p)d \qquad (6.61.9)$$

The last term is zero at the temperature of maximum density, since $\alpha = (-1/d)(\partial d/\partial T)_p$ is then zero. At 25°C the difference between β and β_S is about $10^{-2}\beta$.

Table 6.11 lists values of C_p/C_V (equivalent to β/β_S - eq 6.61.7) given by Vedam and Holton (1968) for distilled water. These ratios have been derived from the sound velocity data of Wilson (1959).

Vedam and Holton constructed the PVT surface of water from the ultrasonic velocity data, using eq 6.61.9 as a base, and derived the specific volume and various other thermodynamic quantities. Crease (1962) had earlier calculated the specific volume of seawater similarly. More details of this are given below where the WM equation of state is outlined.

TABLE 6.11 Comparison of the isothermal and adiabatic compressibilities of water , β/β_{ϕ}*

t^oC Pressure $kgcm^{-2}$	0	10	20	30	40
1	1.001	1.001	1.006	1.015	1.026
100	1.000	1.002	1.008	1.017	1.027
200	1.000	1.002	1.009	1.018	1.029
300	1.000	1.003	1.010	1.019	1.030
400	1.001	1.005	1.011	1.021	1.031
500	1.001	1.006	1.013	1.022	1.032
600	1.002	1.007	1.014	1.023	1.034
700	1.003	1.008	1.016	1.025	1.035
800	1.005	1.010	1.017	1.026	1.036
900	1.006	1.012	1.017	1.028	1.038
1000	1.007	1.013	1.021	1.029	1.039

* taken as equivalent to the ratio C_p/C_v, given by Vedam and Holton (1968, Table VII)

6.62 Limiting compressibility ϕ_k^o

The application of Laplace's eq 6.61.1 to the determination of ϕ_k^o ($=\bar{K}^o$) has only been fairly recent, Barnartt(1952) used the equation to derive the concentration-dependence of u, viz.

$$u = u_w + A_u c + B_u c^{3/2} \qquad (6.62.1)$$

where u_w is the velocity of sound in water. The relationship for $\beta-\beta_{\phi}$, usually designated δ is of the same form

$$\delta = \delta_w + A_{\delta} c + B_{\delta} c^{3/2} \qquad (6.62.2)$$

The partial molal compressibility of an electrolyte at infinite dilution, $\bar{K}_i^o$ is obtained from the coefficients A_u and A_{σ}, data for pure water and the limiting apparent molal volume ϕ_v^o, via

$$-\bar{K}_i^o = (2000A_u/u_w + M_i d_w - 2\phi_{v_i}^o) - (10^3 A_{\delta} + \delta_w \phi_{v_i}^o) \qquad (6.62.3)$$

Owen and Simons (1957) give the derivation of this relationship in detail. See also, Harned and Owen (1958, p387) and the thesis of Kearns (1966) for further details.

Values of $\bar{K}^o$ derived in this way, even allowing for unreliabilities in the auxiliary data, are generally considered more accurate than the direct method which involves two long and somewhat uncertain extrapolations.

6.63 Compressibility equations from Acoustic data

Equation 6.61.9 in the form

$$(\partial d/\partial p)_{T,S} = 1/u^2 + (T/c_p)(\ (1/d)(\partial d/\partial t))^2_{P,S} \qquad (6.63.1)$$

(with $\underline{S}$ representing salinity), when integrated at constant temperature and salinity over the pressure range 1 to p bars, becomes

$$d^{(p)} - d^{(1)} = \int_1^p dp/u^2$$

$$+ \quad T\int_1^p \frac{((1/d)(\partial d/\partial T))^2}{c_p} dp \qquad (6.63.2)$$

with $$c_p^{(p)} = c_p^{(1)} - \int_1^p (T\partial^2(1/d)/\partial T^2) \quad dp \qquad (6.63.3)$$

These three equations form the basis of Crease's calculations of the specific volume of seawater at pressures above atmospheric. Crease pointed out the constraint of equation 6.63.1 but accepted Del Grosso's (1952) conclusion that the effect of non-equilibrium conditions are negligible. He used an iterative technique for calculating the second term of equation 6.63.2, covering temperature blocks of up to 16°C in steps of 2°C. The reliability of the temperature derivatives so obtained can be seen from Table 5.22, where they are compared with the direct experimental values.

Crease's results showed that Ekman's specific volume values, and those derived from the sound velocity agreed to better than 30ppm (cm^3/g) over conditions prevailing for most ocean basins.

Vedam and Holton (1968) adopted a similar iterative technique in their calculation of thermodynamic properties of distilled water using sound velocity data (Wilson's (1959) at pressures 1-1000 $kgcm^{-2}$ and their own measurements to 10,000 $kgcm^{-2}$). The adiabatic compressibility was found at each point of a P-T grid (10°C strip in 1°C steps) from velocity data. Integration of the compressibilities over pressure then gave estimates of the volume. The pressure and temperature coefficients, C_p, C_V, and the ratio of C_p/C_V (Table 6.11) were also calculated from the constructed P-V-T surface.

Wang and Millero (1973) have questioned the reliability of Crease's method of computation, preferring to work directly from eq 6.61.9 with $\Delta\Theta$ representing the adiabatic correction

$$(- \partial V/\partial p) = V^2/u^2 + \Delta\Theta \qquad (6.63.4)$$

to generate iteratively an equation of state from the sound velocity data. However, it was necessary to predict in advance the most suitable relationship between specific volume and pressure. McDonald's (1969) detailed statistical analysis of the P-V water data of Kell and Whalley, using many forms of equations, indicates some of the difficulties involved in choosing the best equation. Unless the model is applicable (and there does not appear to be any unequivocal way of deciding this) the parameters and standard deviations obtained must be discounted as biased. McDonald found that a general polynomial equation (3rd degree Slater equation, designated 3SE) gave the best fit to Kell and Whalley's data (0°C and 50°C sets only were considered, the latter set appearing less accurate). Whereas, another polynomial (the 3rd degree David-Gordon, 3DGE) was more appropriate for the sound velocity data (Wilson's plus Vedam and Holton's at high pressures,

50°C set only). The 3SE gave a very poor fit to this data. The Murnaghan equation was judged the best of the non-linear equations in this case, with the quadratic secant-modulus running a poor fourth.

This latter equation was the one chosen by Wang and Millero since it appeared to give the best fit for the compressibilities derived from Wilson's sound velocity data over the 1-1000 bar range. The linear form of this equation is equivalent to the Tumlirz equation used by Eckart (1958) and by Wilson and Bradley (1968). Both Bradshaw and Schleicher (1970) and Caldwell and Tucker (1970) have noted the apparent unsuitability of this form of equation for seawater. See also Section 4.61. However, the extended form (including the p^2 term for higher pressures) appears to be satisfactory for the sound velocity data.

The P-V relationship chosen was thus

$$v^{(1)} - v^{(p)} = p'v_w^{(1)}/(B + A_1p' + A_2p'^2) \qquad (6.63.5)$$

A summary of this equation and the coefficients have been given previously (Section 4.6 Table 4.18) $v^{(1)}$ and $v^{(p)}$ are the specific volumes of the solution (subscripted sw for seawater, w for water) at applied pressures of zero (atmospheric pressure) or p' (bars), p' = p-1. A_1 and A_2 are empirical constants, being functions of temperature and salinity. They were determined by an iterative method after substituting equation 6.63.5 and its derivative with respect to pressure (eq 6.72.5) into eq 6.63.4.

The values of the coefficient B (for water and seawater) were actually $B_w = 1/\beta_w^{(1)}$ and $B_{sw} = v_w^{(1)}/v_{sw}^{(1)}\beta_{sw}^{(1)}$. These coefficients are thus not fitted, except in terms of the temperature variations, expressed in the form of polynomials. The compressibility values were derived from the sound velocity data of Wilson (1959, 1960), together with the accessory data needed to convert to the isothermal values. The latter were derived from Cox and Smith (1959, specific heats seawater), Knudsen's (1901) σ_t equation as given by Fofonoff (1962) and Sweers (1971), to derive specific volumes and expansibilities. The water compressibility values $\beta^{(1)}$ were those derived by Kell (1970) from recent sound velocity data.

The coefficient A_1 was determined in the low-pressure range (not specified) where the nonlinear effects are assumed to be negligible. The value of A_2 was then determined iteratively using Bradshaw and Schleicher's (1970) expansibilities as a first estimate. The values of A_1 and A_2 were then fitted to polynomials in salinity and temperature (Table 4.18). Fine and Millero (1973) used the same form of equation as that outlined by Wang and Millero (1973) for calculating isothermal compressibilities, specific volumes and expansibilities of pure water. The results of such calculations will be discussed in the following sections.

6.7 The isothermal compressibility of water.

The results to date are considered ambiguous in certain respects, and will therefore be analysed in some detail. Apart from defining more clearly the areas of uncertainty in the compressibility values, it is hoped that such an analysis will provide some insights into the peculiar structural properties of water.

Until recently (Kell 1970, Fine and Millero 1973) the P-V-T data of Kell and Whalley (1965) for distilled water was accepted as the most reliable available. As shown by McDonald (1969), different equations, chosen to represent these data, can produce variations of $0.1\mathrm{x}10^{-6}\mathrm{bar}^{-1}$ in $\beta^{(1)}$. However, the polynomials chosen by Kell and Whalley should result in a maximum error of less than half this value (on the basis of McDonald's analysis). Nevertheless, the KW values at 1-atm show deviations as large as $0.11 \times 10^{-6} \text{bar}^{-1}$ from the corresponding values estimated from acoustic data (Kell 1970). These deviations occur at low (0-10°C) and high (90-120°C) temperatures. Other reliable direct data, Millero et al. (1969, 2-55°C, to 18 bars) and Diaz Pena and McGlashan (1959, 5-60°C to 30 bars) show larger deviations, but, significantly, with the same temperature trends. The statistical analysis by McDonald (1969) also showed the incompatibility of these two groups of data (derived from the acoustic versus direct measurements). Comparison of the KW and acoustic data was made for the 50°C sets (at this temperature the acoustic data (Vedam and Holton 1968) covered a pressure range about ten times wider than that of the KW data). As will be seen later, this temperature would be expected to show the smallest differences between the two data types. However, the acoustic data did not fit the same equation as the KW data. This was interpreted as systematic errors in the sound velocity data, possibly deriving from the considerable smoothing of the data. It would be interesting to see a similar analysis for a temperature of 90-100°C, say, or 0-10°C.

The recent comparison of the KW results and sound derived values by Fine and Millero (1973) show that the 1-atm discrepancies extend up to 1000 bars.

It should be noted, that an inherent assumption in these results (both of McDonald and of Fine and Millero) is that the models used are valid for the data employed.

6.71 One atmosphere results.

Kell (1970) has developed an equation for the isothermal compressibility of water at 1-atm, from acoustic and direct measurements (weighted by 1/100), covering the temperature range 0-150°C, viz.

$$10^6\beta_w^{(1)}(\text{bar}^{-1}) = (50.88630 + 0.717582t + 0.7819867\text{x}10^{-3}t^2 + 31.62214\text{x}10^{-6}t^3 - 0.1323594\text{x}10^{-6}t^4 + 0.6345750\text{x}10^{-9}t^5)/(1 + 21.65928\text{x}10^{-3}t)$$

(standard error estimated as $0.0021\text{x}10^{-6}\text{bar}^{-1}$) (6.71.1)

Kell has also given equations for the isothermal compressibility from sound velocity data alone, the isentropic compressibility coefficient $(\beta_w^{(1)})_s$, the sound velocity u, and $1/u^2T$ as a function of t.

All take the same form as 6.71.1, viz.

$$(a_o+a_1t + a_2t^2 + \ldots + a_5t^5)/(1+b_1t) \qquad (6.71.1\underline{a})$$

The respective coefficients are shown in Table 6.12 together with those of the polynomial of Kell and Whalley for 1-atm ($b_1 = 0$).

TABLE 6.12 Coefficients of equations for compressibility and sound velocity at 1 atm

Coeff.	$10^6(\beta_w^{(1)})_s$	$10^6(\beta_w^{(1)})_T$		$10^{12}/u^2T$	u
	bar^{-1}	bar^{-1}		$\text{cm}^{-2}\text{sec}^2$	metres sec^{-1}
		Sound velocity	Direct		
a_o	50.85541	50.88539	50.9804	50.84744	1402.385
a_1	0.2962265	0.5200294	-0.374957	0.2816838	5.03522
10^3a_2	1.864987	2.140826	7.21324	1.566732	-58.3087
10^6a_3	-1.71515	7.37010	-64.1785	-2.04020	345.300
10^9a_4	37.2105	44.0991	343.024	28.1355	-1645.13
10^9a_5	-	-	-0.684212	-	3.9625
10^3b_1	13.07477	17.77306	-	12.72426	-
std error of function	0.0016	0.0014	0.04	0.0015	0.038

Kell (1970)

6.72 Pressures to 1000 bars.

Acoustic data.

As noted in the previous section 6.6 Fine and Millero (1973) have developed an equation of state for water using the acoustic data of Wilson (0-90°C and 1-1000 bars). Wilson's data were refitted (with an average deviation $\pm$0.22m/sec from experimental values) by normalising his 1-atm data to that given by Kell (1970) using

$$u^{(p)} = u^{(1)} + a_1 p + a_2 p^2 \qquad (6.72.1)$$

where $u^{(1)}$ is given in Table 6.12 (last column), with a_1, a_2 given by

$$a_1 = 1.0987\text{x}10^{-2} + 3.4233\text{x}10^{-5}t - 1.6352\text{x}10^{-7}t^2 + 2.1158\text{x}10^{-9}t^3$$

$$a_2 = 7.900\text{x}10^{-8} - 5.2177\text{x}10^{-9}t + 6.6051\text{x}10^{-11}t^2 - 3.4284\text{x}10^{-13}t^3$$

The 2SM (2nd degree secant-modulus) equation for water,

$$v_w^{(p)} = v_w^{(1)} - v_w^{(1)}p'/(B_w + A_{w_1}p' + A_{w_2}p'^2) \qquad (6.63.5)^*$$

was used as the equation of state, with p' representing the applied pressure (i.e. p-1). The B and A coefficients being derived as explained above.

B_w was obtained from the reciprocal of $\beta_w^{(1)}$ as given by equation 6.71.1 above (Table 6.12 column 3) and refitted as a function of temperature to give

$$B_w = 19654.320 + 147.037t - 2.21554t^2 + 1.0478\text{x}10^{-2}t^3 - 2.2789\text{x}10^{-5}t^4 \qquad (6.72.2)$$

The other derived coefficients are given by

$$A_{w_1} = 3.2891 - 2.3910\text{x}10^{-3}t + 2.8446\text{x}10^{-4}t^2 - 2.8200\text{x}10^{-6}t^3 + 8.477\text{x}10^{-9}t^4 \qquad (6.72.3)$$

$$A_{w_2} = 6.245\text{x}10^{-5} - 3.913\text{x}10^{-6}t - 3.499\text{x}10^{-8}t^2 + 7.942\text{x}10^{-10}t^3 - 3.299\text{x}10^{-12}t^4 \qquad (6.72.4)$$

Differentiation of eq 6.63.5 with respect to pressure at constant temperature gives the compressibility

$$\beta = -(1/v^{(p)})\,(\partial v^{(p)}/\partial p)_T$$

$$= v_w^{(1)}(B_w - A_{w_2} p'^2)/v_w^{(p)}\,(B_w + A_{w_1} p' + A_{w_2} p'^2)^2 \qquad (6.72.5)$$

Fine and Millero (1973) have tabulated values of v, β and expansibilities (derived from the temperature derivation) for the ranges: applied pressure 1-1000 bars (100 bar intervals), and temperature 0-100° (at 5° intervals). Some representative values are given in Tables 6.13 and 6.14 for pressures in $kgcm^{-2}$.

TABLE 6.13 Comparison of specific volumes derived from sound velocity data via different analytical methods

$v\ cm^3g^{-1}$

t^oC / p $kgcm^{-2}$		0	10	20	30	40	50
1	FM	1.000160	1.000300	1.001797	1.004369	1.007842	1.012107
	VH	1.0002	1.0003	1.0016	1.0043	1.0078	1.0121
100		0.995299	0.995726	0.997399	1.000066	1.003577	1.007831
		0.9954	0.9957	0.9974	1.0000	1.0036	1.0078
200		0.990545	0.991244	0.993085	0.995846	0.999393	1.003640
		0.9906	0.9913	0.9931	0.9958	0.9994	1.0036
300		0.985941	0.986895	0.988894	0.991744	0.995329	0.999570
		0.9860	0.9869	0.9889	0.9917	0.9953	0.9996
400		0.981481	0.982673	0.984821	0.987757	0.991378	0.995615
		0.9815	0.9827	0.9848	0.9877	0.9914	0.9956
500		0.977160	0.978572	0.980860	0.983877	0.987534	0.991769
		0.9772	0.9786	0.9808	0.9838	0.9875	0.9918
600		0.972971	0.974589	0.977008	0.980102	0.983791	0.988026
		(0.9739)	0.9746	(0.9779)	0.9801	0.9838	(0.9889)
700		0.968909	0.970718	0.973259	0.976424	0.980146	0.984381
		0.9690	0.9707	0.9732	0.9764	0.9801	0.9844
800		0.964970	0.966954	0.969609	0.972841	0.976593	0.980827
		0.9650	0.9670	0.9696	0.9728	0.9767	0.9808
900		0.961149	0.963294	0.966054	0.969348	0.973128	0.977362
		0.9612	0.9633	0.9660	0.9693	0.9731	0.9773
1000		0.957441	0.959733	0.962590	0.965941	0.969747	0.973980
		0.9575	0.9598	0.9626	0.9659	0.9697	0.9740

TABLE 6.13 (continued)

p kgcm^{-2} \ t°C		60	70	80	90	100
1	FM	1.017089	1.022740	1.029027	1.035934	1.043453
	VH	1.017	1.0226	1.0289	1.0358	1.0436
100		1.012763	1.018327	1.024494	1.031247	1.038578
		1.0127	1.0182	1.0243	1.0311	1.0387
200		1.008526	1.014010	1.020065	1.026675	1.033829
		1.0085	1.0139	1.0199	1.0265	1.0339
300		1.004415	1.009826	1.015778	1.022254	1.029244
		1.0044	1.0097	1.0156	1.0221	1.0293
400		1.000423	1.005766	1.011622	1.017975	1.024813
		1.0004	1.0057	1.0115	1.0178	1.0248
500		0.996543	1.001824	1.007591	1.013828	1.020524
		0.9965	1.0017	1.0075	1.0137	1.0205
600		0.992769	0.997992	1.003676	1.009806	1.016369
		0.9927	0.9979	1.0035	1.0097	1.0163
700		0.989095	0.994264	0.999871	1.005901	1.012340
		0.9891	0.9942	0.9997	1.0057	1.0123
800		0.985515	0.990635	0.996170	1.002105	1.008428
		(0.9859)	0.9906	0.9960	1.0020	1.0084
900		0.982025	0.987098	0.992565	0.998412	1.004626
		0.9820	0.9870	0.9924	0.9983	1.0045
1000		0.978620	0.983648	0.989052	0.994816	1.000927
		0.9786	0.9836	0.9889	0.9947	1.0008

FM Fine and Millero (1973)

VH Vedam and Holton (1968)

Bracketed values - excessive deviations

Underlined values - deviations $\geq$ 100ppm

The specific volume results showed deviations from the KW values up to 169 ppm (average about 115ppm) at the higher pressures. The compressibility results showed a parabolic shift in the KW values. The maximum deviation was about 0.205×10^{-6} bar^{-1} at 400 bar, 5°C. Around this region, 0-10°C, 140-300 bars, the sound velocities calculated from the

KW data are in better agreement with measured velocities than elsewhere (except 10°C, 800-960 bars - see section 6.75). The maximum deviations in the sound velocities (calculated versus experimental) occurs at 400 bars, but at 90° rather than at 5°C. These results seem somewhat incompatible and suggest overlapping errors.

TABLE 6.14 Comparison of pressure differential of specific volume, from sound velocity data using different analytical methods $-(\partial v/\partial p)_T$ $cm^3g^{-1}/kgcm^{-2}$

P $kgcm^{-2}$ \ t^oC		0	10	20	30	40	50	60	70	80	90	100
1	FM	50.89	47.83	45.98	44.97	44.59	44.72	45.27	46.20	47.49	49.13	51.15
	VH	50.9	47.8	45.9	44.9	44.6	44.7	45.2	46.1	47.5	49.3	51.9
100		49.27	46.40	44.64	43.67	43.29	43.38	43.87	44.73	45.92	47.44	49.30
		49.3	46.4	44.6	43.7	43.3	43.4	43.8	44.7	45.9	47.6	49.9
200		47.70	45.02	43.36	42.42	42.04	42.11	42.55	43.33	44.43	45.84	47.57
		47.7	45.0	43.3	42.4	42.0	42.1	42.5	43.3	44.4	45.9	48.1
300		46.20	43.69	42.13	41.23	40.86	40.90	41.30	42.02	43.03	44.34	45.95
		46.2	43.7	42.1	41.2	40.9	40.9	41.3	42.0	43.0	44.4	46.4
400		44.76	42.43	40.95	40.10	39.74	39.76	40.13	40.79	41.73	42.94	44.44
		44.8	42.4	40.9	40.1	39.8	39.8	40.1	40.7	41.7	43.0	44.8
500		43.38	41.21	39.83	39.02	38.67	38.69	39.01	39.63	40.50	41.63	43.03
		43.4	41.2	39.8	39.0	38.7	38.7	39.0	39.6	40.5	41.7	43.3
600		42.06	40.04	38.75	37.99	37.66	37.66	37.97	38.53	39.35	40.41	41.71
		42.1	40.0	38.7	38.0	37.7	37.7	38.0	38.5	39.3	40.5	41.9
700		40.78	38.92	37.72	37.01	36.69	36.70	36.97	37.50	38.26	39.25	40.48
		40.8	38.9	37.7	37.0	36.7	36.7	37.0	37.5	38.2	39.3	40.6
800		39.56	37.84	36.73	36.07	35.77	35.78	36.04	36.53	37.24	38.17	39.32
		39.6	37.8	36.7	36.1	35.8	35.8	36.0	36.5	37.2	38.2	39.4
900		38.38	36.81	35.78	35.17	34.90	34.90	35.15	35.61	36.28	37.15	38.23
		38.4	36.8	35.8	35.2	34.9	34.9	35.1	35.6	36.2	37.2	38.3
1000		37.25	35.81	34.87	34.31	34.07	34.08	34.31	34.75	35.38	36.20	37.21
		37.3	35.8	34.9	34.3	34.0	34.1	34.3	34.7	35.3	36.2	37.2

FM Fine and Millero (1973)

VH Vedam and Holton (1968)

Fine and Millero concluded that the values derived by them, for the specific volume and its derivatives, are the most accurate available (as at November 1973). Vedam and Holton (1968) had earlier used the same sound velocity data as Fine and Millero (i.e. Wilson's 1959 data) to calculate various thermodynamic properties over 1-1000 bar. The method of analysing the data was different, being essentially equivalent to that used by Crease (1962) as outlined in Section 6.63.

Although the precision was less (specific volume rounded to 100ppm and compressibility to $0.1 \times 10^{-6} bar^{-1}$) it is of interest to compare these results with those of Fine and Millero (Table 6.13). McDonald (1969) found, for the 50°C data, a standard deviation of 3×10^{-5} for the specific volume results, an order of magnitude smaller than the maximum errors expected by Vedam and Holton. Evidence of very considerable smoothing was found. The VH values are in close agreement with the FM values, generally within the rounding errors, for the temperatures below about 70°C. Some notable discrepancies (bracketed in Table 6.13) occur at $600 kgcm^{-2}$ (0°, 20° and 50°) and at $800 kgcm^{-2}$ (60°C). For the 80° and 90° sets at all pressures the VH values are $1 \times 10^{-4} cm^3 g^{-1}$, or more, lower than the corresponding FM values. The agreement improves appreciably at 100°C. At these higher temperatures the auxiliary data makes a larger contribution (Table 6.11) and this may be the cause of the discrepancies. Values of $\beta v, (-\partial v/\partial p)_T$, are compared in Table 6.14. The close agreement, mostly to $0.03 \times 10^{-6} bar^{-1}$ shows that the different types of analysis give comparable results. An exception is 100°C, for which the VH values are much larger up to $600 kgcm^{-2}$. This is the grid-edge area of VH's analysis so that the FM values are probably more accurate.

Crease has also calculated the specific volume of distilled water from Wilson's sound velocity data, and using the 1-atm data given by Tilton and Taylor (1937); the revised Chappuis Table. Comparison with Amagat's values (from Dorsey 1940) showed agreement to 50ppm or better up to about 900 bars. At this pressure errors in sound velocity of +1 m/sec give a corresponding error of about 40ppm. Fine et al (1974) found discrepancies higher than this for Wilson's (1960) seawater sound velocity measurements, relative to the measurements of Del Grosso and Mader (1972); see also Millero and Kubinski (1975). A maximum deviation of 1.77m/sec was found at 37‰, 5°C and 1000 bars. Similar errors may apply to the pure water measurements of Wilson (1959). However, as will be shown in Section 6.75, Amagat's data are not as reliable at these higher pressures.

6.73 Tait Equation

McDonald (1969) concluded from his analysis of the KW data (0° and 50°C sets only) that certain equations of state, including the Tait equation are unsuitable for water. Actually, what he showed was that they are not suitable for the KW data sets at these two temperatures. Since the KW data may have errors as large as 169 ppm at higher pressures, the conclusion regarding the best equation for water must be treated with reserve. Especially since the best equation for the KW data differed from that of the Wilson-Vedam-Holton acoustic data, and Fine and Millero found yet another equation to be more suitable for the Wilson data. Obviously, there are serious limitations in analyses of this type, however statistically elegant and sound they may be. The choice of mathematical model (generally with little regard for any physical reality), weighting effects, and reliability of the data all introduce their own distortions.

Li (1967) chose the Tait equation (6.3.1) as a suitable theoretical model for water. He fitted a whole range of distilled water data to this equation in order to derive the two parameters C and B, which are both functions of temperature, but independent of pressure. $C = 0.315v_w^{(1)}$ where $v_w^{(1)}$ is given by a rational polynomial in t (Table 5.18) and B is given by a 2nd or 3rd degree polynomial in t. McDonald (1969) has criticised Li's approach and the techniques. However, a lot of this criticism is invalid and somewhat biased (to the statistical approach). Ekman's data, for example, is referred to as "ancient" and the implication is that it is not very accurate because of Eckart's (1958) criticism. Apparently, the observation by Li regarding Eckart's mistake in pressure conversion was overlooked, together with other work indicating the reliability of Ekman's results (Crease 1962, Caldwell and Tucker 1970, Bradshaw and Schleicher 1970).

Li found that the Tait equation gave excellent representation of the data sets of Amagat (1893) and Ekman (1908), the two most internally consistent sets analysed. An interesting feature of his results was that the B parameter derived from 1-atm sound velocities was about 20 bars higher than that derived from the direct measurements. An error of 10 ppm in the specific volume would give an error in B of about 20 bars at 1-atm. However, this positive deviation remained nearly constant from 0-25°C and was a characteristic of the sound velocity data, not being shown by the direct measurements.

Since the B parameter, Section 6.3, is related to the structural properties of water, and in view of other evidence (given in this and following sections) regarding the possibility of some peculiarity in the acoustic derived values, this anomaly will now be examined in more detail.

The analyses of Vedam and Holton and of Fine and Millero enable the B parameter to be calculated from sound velocity data at higher pressures. Errors in the specific volume of 10ppm will only introduce errors of a few bars in B over the range 400-1000 bars. Table 6.15 gives a summary of the results. The values of some direct measurements are shown together with 1-atm sound velocity values from various sources.

The B parameter derived from the "acoustic" specific volumes are generally in close agreement with the values from direct measurement. The values of Wilson and Bradley exhibit a skew effect (too high at 0° and too low at 30°). It is significant that Wilson and Bradley found difficulty in fitting the low temperature, low pressure data for distilled water to their equation of state. They point out that the thermal expansion for distilled water is predicted as zero at 3°C rather than 4°C and the specific volume at 0°C is predicted as $1.0000 cm^3 g^{-1}$. These peculiarities are probably due to the constraint imposed that $\beta_w^{(1)}$ at 4°C has the value $49.46 \times 10^{-6} bar^{-1}$ (sound velocity data). This gives the B values a positive bias around 0°. In terms of temperature, the effect corresponds to a shift of -1°C, i.e. the B versus t curve is shifted back by about 1°C, as for the sound velocity results. This shift is illustrated by the values of E3 in Table 6.15 which were derived from the equation for B based on Ekman's data (Li 1967). An explanation for this temperature shift will be given in Section 6.75.

For sodium chloride solutions a drop in the observed temperature of maximum density of 1°C (from 4 to 3°C) occurs around a solute concentration of 0.06 moles/liter (Foster and Finch 1974). In terms of ionic strength this is equivalent to a salinity of 3‰ Via the Tait Gibson equation, for seawater, this solute concentration corresponds to an increase in the effective pressure of $B^* - B = 6.8966S - 0.0703St$ or 20 bars, with $t = 3°C$. This is consistent with the depression of the TMD noted for water data forced to fit the acoustic values, and the error in B (20 bars too high for the acoustic data).

TABLE 6.15 The B parameter of the Tait equation*
as a function of temperature

B (bars) ± std.dev.

t°C	FM	A	EK1	EK2	WB	DMc	(LM')
0	2671.8	2668.1	2671.4	2670.8	2683.5		2675.6
	1.2	1.0	0.3		6.5		
5	2762.5	2762.8	2763.0	2762.2	2765.5	2746.7	2766.6
	1.8	6.1	0.6		6.0		
10	2840.3	2836.4	2843.6	2842.4	2835.5	(2836)	2846.4
	2.6	1.4	0.5		5.1		
15	2906.1	2902.4	2909.7	2911.5	2895.9	2900.3	2910.7
	2.9	1.5	0.9		4.7		
20	2960.6	2955.2	2964.8	2969.4	2946.7	(2949.5)	2965.0
	3.3	2.1	0.4		3.8		
25	3004.5		(3005.8)		2987.2	2998.2	3007.2
	3.4				3.2		
30	3038.5	3030.8			3016.4	(3040.0)	3041.6
	3.2	2.3			2.6		
35	3063.7						
	2.7						
40	3080.3	3081.4					
	2.4	2.6					
45	3089.1						
	1.7						
50	3090.8	3094.8					
	1.1	1.8					
55	3085.8						
	0.4						
60	3074.7	3081.3					
	0.5	2.5					
65	3058.0						
	1.3						
70	3035.9	3038.1					
	2.1	7.5					
75	3009.1						
	2.9						
80	2977.8	2967.2					
	3.7	13.8					
85	2942.3						
	4.5						
90	2903.0	2901.3					
	5.3	14.2					

Sound velocity results at 1 bar

t°C	W	K	E3
0	2689.8	2687.4	2690.4
5	2783.5	2781.1	2780.8
10	2862.7	2860.3	2858.3
15	2928.6	2926.1	2922.6
20	2982.3	2979.8	2974.0

TABLE 6.15 (continued)

t^oC	FM
95	2860.1
	5.9
100	2813.9
	6.5

FM calculated from "acoustic" specific volumes, Fine and Millero (1973)

A Amagat as listed by Li (1967, Table 1)

EK1 Ekman as listed by Li (1967, Table 5)

EK2 From $B^*_{S=0}$ (Li 1967)

WB Wilson and Bradley (1968)

These 4 sources give averaged values for the range 400-1000 bars.

DMc Diaz Pena and McGlashan (1-30 bars) 1959 .

LM Lepple and Millero (1971) with compressibility of glass assumed to be 0.07×10^{-6} bar^{-1} higher.

W Wilson as calculated by Li (1967)

K Kell (1970) E3 Ekman's equation $B = 2670.8 + 19.9t - 0.26t^2$ with t = (t+1).

Amagat's B values tend to be slightly lower than the FM values, equivalent to somewhat larger compressibility values (Figure 6.15).

These results illustrate that the B parameter is essentially independent of pressure up to 1000 bars. The corresponding Tait-Gibson coefficient will also be shown to be independent of pressure in Section 6.8.

The 1-bar sound velocity results are all about 20 bars higher (or 1°C lower). These results could be interpreted to mean that the direct measurements and the sound velocity measurements are associated with different structural effects. These differences, in practice, give rise to different solutions, so that different compressibilities apply. However, the effective constancy of B suggests that the deviations of the sound velocity values are due to small entropy changes associated with the measurements. This means that the measurements are not strictly isentropic and eq 6.61.9 does not apply without the necessary correction factor.

The specific volumes are only affected to a small extent because of the functional relationship with sound velocity (eq 6.63.2). However, even an error of 10ppm in v will give an error of about 0.3×10^{-6} bar^{-1} in the compressibility. While the deviations are most obvious at the low pressures, they also occur at higher pressures (as indicated by the small deviations of B). These errors can be smoothed out by using the average B value

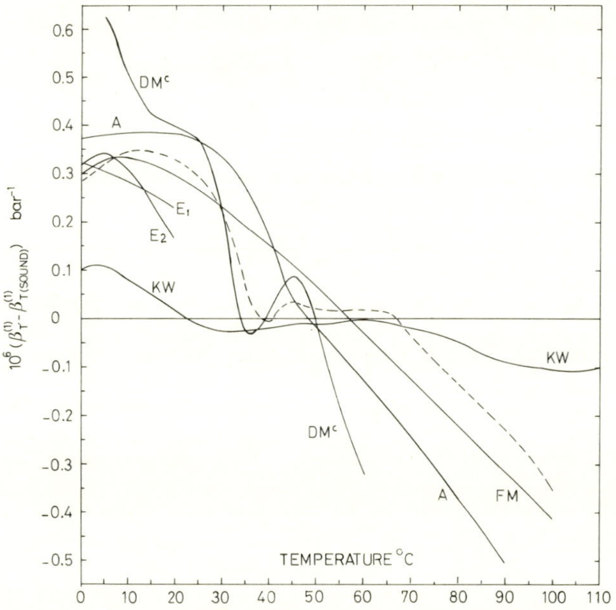

FIGURE 6.15

Deviations of isothermal compressibilities of water (direct - acoustic) at 1 bar, as a function of temperature.
DMc Diaz Pena and McGlashan (1959), A Amagat, E1 Ekman (water data), E2 Ekman (seawater data extrapolated to zero salinity) - from the equations given by Li (1967). KW Kell and Whalley (1965)
FM calculated (via Tait eq) from "acoustic" specific volumes of Fine and Millero (1973 Table IV)
----- predicted from $\Delta\mu$ values, see Figure 6.21.

at the higher pressure to recalculate B over the whole pressure range. For this reason, the $\beta^{(1)}$ values for the sound velocity results were re-calculated from the average B value at the higher pressures via

$$\beta^{(p)} = Av_w^{(1)}/(B+p)v_w^{(p)} = A/(B+1) \quad \text{with } p=1 \qquad (6.3.1)*$$

and A = 0.315x0.4343

At temperatures around 55-60°C the value of B had a standard deviation of less than 1 bar over the whole pressure range (1-1000 bar) so that high and low pressure values of B gave much the same compressibility.

The deviations from the values predicted by eq 6.71.1 (sound velocity equation of Kell, 1970), that is, assuming no entropy changes, are shown in Figure 6.15 as a function of temperature. The corresponding deviations for direct compressibilities are shown for comparison. The similarity in the temperature trends of all the data is quite striking, i.e. positive deviations at low temperatures (0-30°C), zero deviations at mid temperatures and negative deviations around 100°C. Such trends are particularly interesting in view of the anomalies in sound velocity that appear to have shown up in highly accurate measurements (e.g. the intimations at 4°, 40°, and 90°C - Del Grosso and Mader 1972a). These anomalies are, however, very small (probably less than 0.003m/sec - Del Grosso and Mader 1972b). They would not show up in the compressibilities unless they were associated with some non-equilibrium effects such as entropy changes during the adiabatic compression. The high B values at low pressure are believed to be linked to these sound velocity anomalies. The associated deviations in β might be expected to be associated with structural properties of water. This will now be considered.

6.74 <u>Deviations of β as a function of structural parameters of water</u>.

Structural factors are often important determinants in aqueous solutions.

The parameters describing the limiting apparent molal volumes (ϕ_v^o) of electrolytes, viz. A and B' in

$$\phi_v^o = Ar^3 - B'z^2/r \quad (4.5.2)*$$

have increments attributable to the water and its structural properties. A structural factor is absorbed into A (up to temperatures around 100°C) so that its temperature dependence should correspond to that of the structural factor since A itself should be independent of temperature (Curthoys and Mathieson 1970). A as a function of temperature is shown in Figure 6.16.

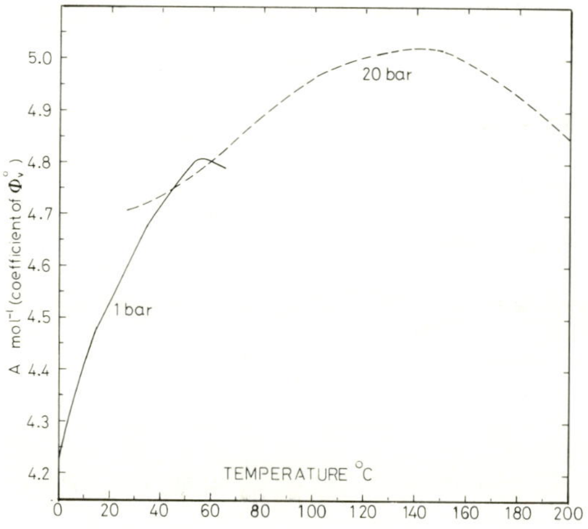

FIGURE 6.16

The coefficient A of the equation

$\phi_v^o = \underline{A}r^3 + B'z^2/r$ (Chapter 4 eq 4.5.2)

as a function of temperature

(Curthoys and Mathieson 1970).

The compressibility deviations ($\beta^{(1)}$ predicted from the B values (via the Tait equation) minus the corresponding value of $\beta^{(1)}$ calculated from sound velocity (Kell 1970), via eq 6.71.1a - Table 6.12, column 3) were plotted against values of A at corresponding temperatures (Figure 6.17).

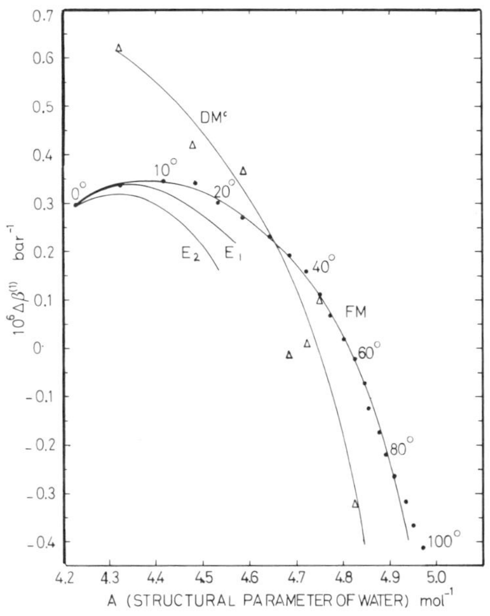

FIGURE 6.17

Isothermal compressibility deviations for water (direct minus acoustic) as a function of the A parameter in the equation

$$\phi_v^o = \underline{A}r^3 + B'z^2/r \quad \text{(Section 4.5)}$$

A values taken from the compilation given by Curthoys and Mathieson (1970) and from Figure 6.16.

It can be seen that a distinct correlation exists. The deviations are also correlated with the entropy of solvation parameter, $166x10^3(1/D^2)(\partial D/\partial T)$ $cal deg^{-1}$, where D is the dielectric constant of water (see eq 4.5.4 and Figure 6.18).

It seems, indeed, that the deviations are related to perturbations in the water structure. The deviations of the direct measurements at low pressures (1-30 bar) are similarly correlated. Thus, the results are not

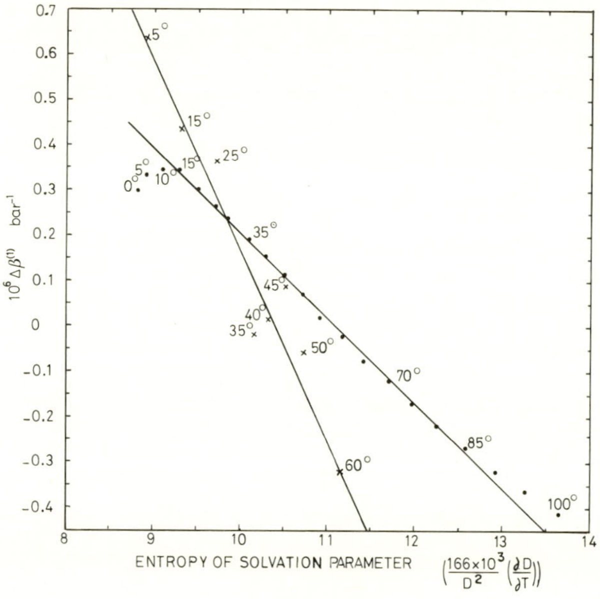

FIGURE 6.18

Isothermal compressibility deviations (direct-acoustic) as a function of the dielectric constant of water (D). n.b. Entropy of solvation of an ion (of charge z and radius r) equals $(166x10^3/D^2)(\partial D/\partial T))z^2/r$. cal $deg^{-1}mol^{-1}$

Direct values from:

x Diaz Pena and McGlashan (1959)

● (via the Tait equation) from "acoustic" specific volumes, as given by Fine and Millero (1973), at 5°C intervals, see text.

peculiar to the B parameter (which, however, is a more precise representative of it, when calculated from very reliable data) but reflect structural anomalies during the velocity measurements. As will be shown later, these perturbations occur at the higher pressures also. The associated errors in the B parameter only show up clearly at the low pressures where errors of 10ppm in the specific volume can cause a 20 bar error in B. As will be shown in Section 6.8 the same results apply to seawater.

In their recent improved version of the HSN "cluster" model of liquid water (see Section 6.3), Lentz, Hagler and Scheraga (1974) give a table of various quantities that describe the temperature-induced structural changes computed for liquid water. These include the quantities $(E_I)_{aq}$ and $(E_U)_{aq}$. The former is "the average energy per mole of bond required to break $(1-X_{HB})$ of the bonds in ice to obtain the equilibrium distribution of water clusters at a given temperature." X_{HB} is the mole fraction of unbroken hydrogen bonds. The quantity $(E_U)_{aq}$ is "the average energy per mole of bond needed to break all the bonds in the **equilibrium** distribution of water clusters at a given temperature and produce all unbonded water molecules in the liquid". Both these quantities include the shifts in the intramolecular zero-point energy on breaking a hydrogen bond. The two energies would be equal if the energies of all the hydrogen bonds in all the clusters present in water were the same as the energy of a hydrogen bond in ice. The difference in the two quantities occur because the model takes into account the effect of cluster size and geometry on the hydrogen bond energies (i.e. the model covers co-operative effects, second-neighbour interactions, and hydrogen randomness in the development of the internal energies of clusters).

The differences between the "acoustic" and "direct" isothermal compressibilities are simply related to the difference in $(E_I)_{aq}$ and

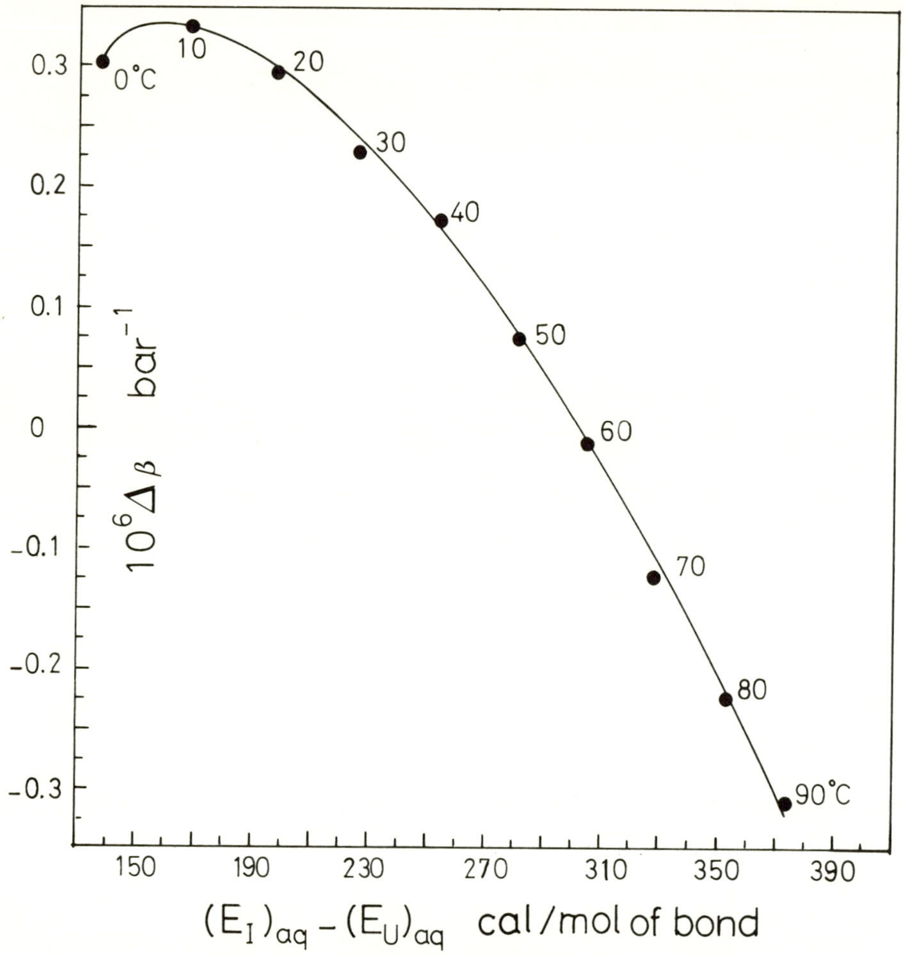

FIGURE 6.18a

Differences in the "direct" and "acoustic" isothermal compressibilities of water (β_{exp} - β(sound), see Figure 6.15) versus a structural parameter of water from the "cluster" model of Lentz, Hagler and Scheraga (1974), see text.

$(E_U)_{aq}$, as illustrated by Figure 6.18<u>a</u>. This correlation supports the idea that the acoustic measurements are associated with small entropy changes. Perhaps, small irreversible changes in cluster size and/or geometry are involved.

The temperature features of Figure 6.15 show up in the pressure coefficients of expansibility, and of the specific heat, as can be seen from Figures 6.19 and 6.20.

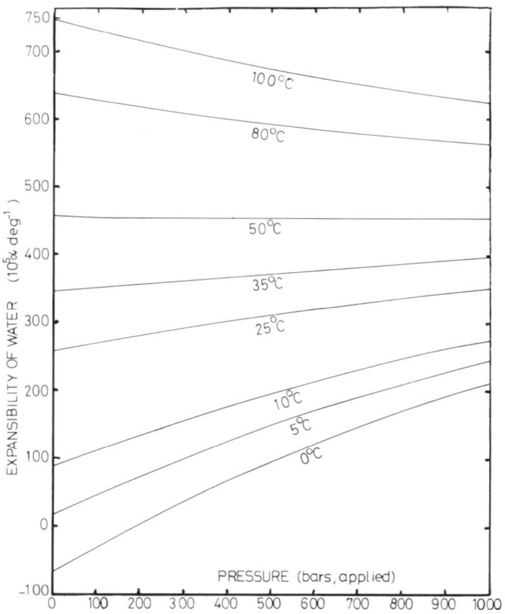

The expansibility of pure water as a function of pressure, calculated from acoustic data (Fine and Millero 1973). The gradient $(\partial\alpha/\partial p)_T$ is equivalent to $-(\partial\beta/\partial T)_p$.

FIGURE 6.19

The mid-temperatures are characterised by much flatter slopes, and the high versus low temperatures by slopes of opposite sign. The thermal properties of water are linked closely to its structural properties. The large heat capacity of water is partly due to the absorption of heat necessary to dissociate the molecular aggregates characteristic of water.

The addition of an electrolyte increases the internal pressure and lowers the heat capacity of water, diminishing the number of aggregates. Increasing the external pressure (to a moderate degree, ca. 1000 bar) has a similar effect. Near the freezing point the number of aggregates is very large and dissociation of the larger aggregates causes a decrease in the volume. The effect of pressure is to enhance the temperature effect on the volume at low temperature, and inhibit it as the boiling point region is approached. In view of the above anomalies between "sound-velocity" compressibility and directly measured compressibility, and the link with water structure, it seems worthwhile to look at the thermal effects of adiabatic compression of water.

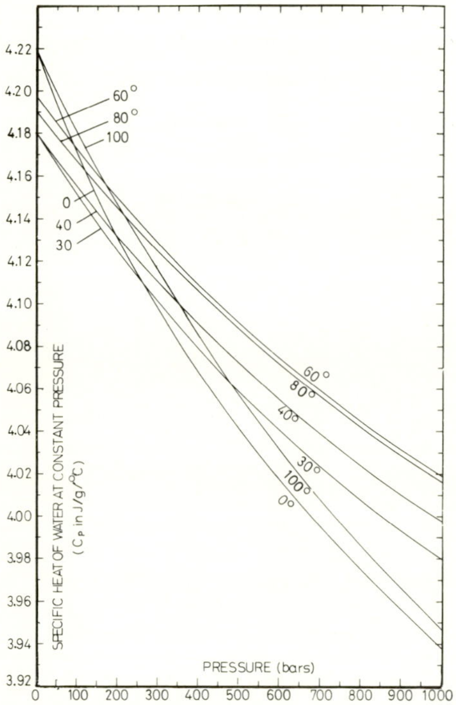

The specific heat of water (at constant pressure) calculated from the sound velocity data of Wilson by Vedam and Holton (1968, Table VII (e)).

FIGURE 6.20

6.75 Effect of adiabatic compression on the thermal energy of water. The specific heat of water at constant volume, C_V, the ratio of specific heats C_p/C_V, or their differences (C_p-C_v) can be computed from direct measurements of compressibility, thermal expansion and specific heat at constant pressure. The computed values are called static values. When computed from sound velocity they may be referred to as dynamic, although in most calculations they are assumed as static (Kell 1970). Similarly, the change in temperature on adiabatic compression is termed either static (from the thermal expansion) or dynamic (from direct measurement of the temperature change associated with a sudden release of pressure). No differences between the static and dynamic values would be expected, provided dynamic changes produce no finite changes (in terms of time equilibrium) in the internal state of the molecules of the liquid. For water there is a large disparity between the periods of intermolecular vibration and the time required for molecular reorientation and translations (about 50:1) and it might be expected that the dynamic and static values of these thermal properties would differ if equilibrium could not be re-established fast enough. There is certainly ample evidence for the absence of rapid molecular reorientations in water (Weres and Rice 1972).

It has long been known that the increase in temperature on adiabatic compression differs according to whether it is observed directly (dynamic) or is computed from the specific heat and the coefficient of thermal expansion (Dorsey 1940). The results are given in terms of the coefficient μ (the decrease in temperature per unit drop in pressure on adiabatic expansion). Dorsey has incorrectly designated this quantity as the Joule-Thomson coefficient (the distinction between the two coefficients and some interesting thermodynamic relationships featuring them, will be discussed in the next section, 6.76, before going on to the compressibility of seawater). For the present purposes we note that the μ coefficient is associated with reversible adiabatic compression or expansion and is defined by

$$\mu = (\partial t/\partial p)_s \qquad (6.75.1)$$

i.e. the temperature change associated with isentropic adiabatic pressure changes (cf. eq 6.61.2). Using the same notation as Dorsey, μ_c is the computed or static value and μ_o the observed or dynamic value. The values listed by Dorsey (Table 126, II and III) and derived from Pushin

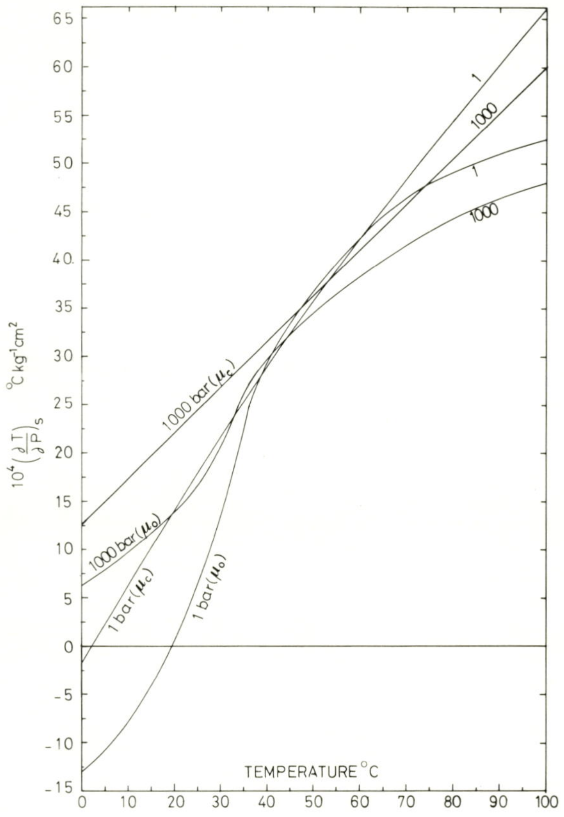

FIGURE 6.21
The dynamic (μ_o) and static (μ_c) values of $(\partial T/\partial p)_s$ as a function of temperature, from the data given by Dorsey (1940 Table 126, II and III).

and Grebenshchikov (1923) are shown in Figures 6.21 and 6.22 as functions of temperature and pressure. The values for $500 kgcm^{-2}$ are listed in Table 6.16.

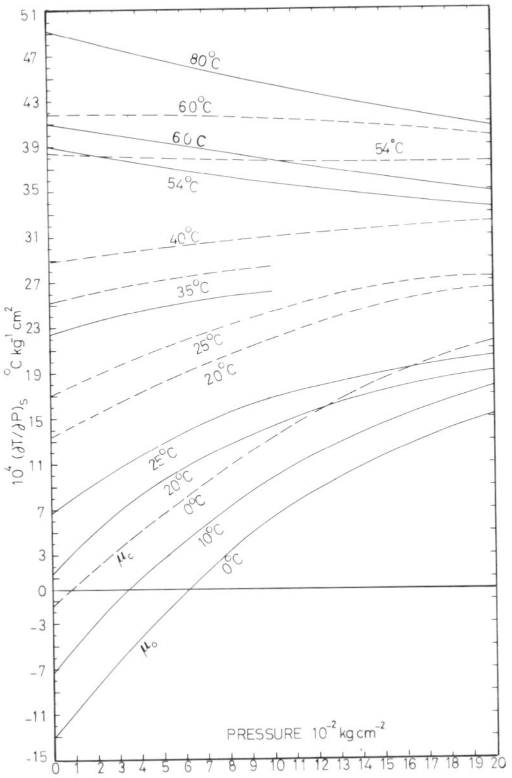

FIGURE 6.22

The change in temperature of water during reversible adiabatic compression.

_______ "dynamic" (μ_o)

------- static (μ_c)

(notation and data from Dorsey 1940, Table 126, II and III).

The temperature trends of $\mu_c - \mu_o$, ($\Delta\mu$), are very close to those of $\Delta\beta$ (Figure 6.15), i.e. large at high and low temperatures, and negligible in the middle temperature range (45-65°C). The values of μ_o at 600-1000 bars are close to the value of μ_c at 1 bar. The movement of μ_o up to these pressures would represent a change in temperature around 1°C. This agrees with the "compensating" temperature change estimated for correcting the B parameter from 1-atm sound velocity data (Table 6.15).

These results reinforce the above conclusions that structural perturbations associated with the acoustic measurements are the main cause of the deviations between the compressibilities derived from the direct measurements and the acoustic (SV hereafter) data.

In Figure 6.23 the change in magnitude of β (direct - SV) with temperature, at 1-bar pressure, is compared with the corresponding change

TABLE 6.16 The Dynamic and Static Coefficients[a] of Reversible[b] Adiabatic Expansion of Water at 500 $kgcm^{-2}$

	$10^4\ (\partial t/\partial p)_S$ $^{\circ}C/kgcm^{-2}$		
$t^{\circ}C$	μ_c	μ_o	$\mu_c - \mu_o$
0	(6.8)[c]	(-2.0)	8.8
5	9.5	0.4	9.1
10	12.2	3.0	9.2
15	15.2	6.2	9.0
20	(17.5)	9.5	8.0
25	21.0	(13.0)	7.0
30	23.9	20.0	3.9
35	26.5	25.2	1.3
37	28.0	(27.3)	0.7
40	(30.0)	29.5	0.5
54	38.0	(37.1)	0.9
60	(41.7)	40.0	1.7
70	47.5	44.2	3.5
80	(50.0)	(46.8)	
	54.8		8.0
90	59.0	48.5	10.5

a From the compilation given by Dorsey (1940, Table 126, II and III) and derived from Pushin and Grebenshchikov (1923)

b Assuming no significant pressure gradients developed during the measurements for the dynamic values. N.B. the Joule-Thomson coefficient is the corresponding quantity for irreversible conditions (Table 6.18)

c Direct values bracketed, others are interpolated.

in $\Delta\mu$. The similarity is very close. An interesting feature occurs at 35°C. In this region, the $\Delta\mu$ versus temperature curve resembles a titration curve, with heat as the titrant, the end point occurring around 35°C. More formally, the behaviour may be compared with a second order phase transition. Is there some clear-cut change in the species (or phase space regions) at this temperature? On the basis of the Weres-Rice (1972) model of liquid water, the configurational component of the heat capacity receives contributions from a number of sources: intra-molecular and intermolecular zero-point energy changes, basic-cell structural-configurational probability changes (BCSCP changes) and long-range electrostatic interaction changes. The relative contributions no doubt vary with temperature. The BCSCP changes are possibly involved predominantly. The cell structural variations with temperature other than simple thermal expansion are important (Weres and Rice 1972). The average energy, $(E_U)_{aq}$, (per mol of bond) to produce all unbonded water molecules in the liquid (on the basis of the equilibrium distribution of water clusters at a given temperature) has been calculated by Lentz, Hagler and Scheraga (1974) - see previous Section. The values are listed in Table 6.16a. As can be seen, the minimum energy occurs just near 35°C.

TABLE 6.16a The average energy needed to break all the hydrogen in the equilibrium distribution of water clusters at a given temperature (includes shifts in intramolecular zero point energies)

$(E_U)_{aq}$ cal/mol of hydrogen bond

t°C	0	10	20	30	35	40	50	60	70	80	90	100
	2250	2233	2224	2220	2219	2221	2227	2237	2250	2266	2286	2309

It is well known that transitional changes occur around 30-40°C in many other properties of water, viz. viscosity, refractive index, surface tension and thermal expansion. There is a minimum in the viscosity of water at 34°C, near which temperature the pressure dependence of the viscosity is eliminated. The isothermal compressibility of water has a minimum value around 46°C. The heat capacity of water reaches a minimum about 35°C, and then continually increases as for other liquids and solids (e.g. ice). This minimum at 35°C has been interpreted by

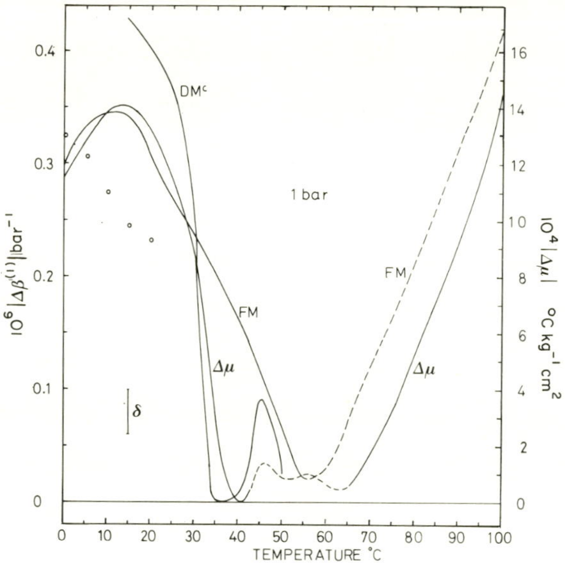

<u>FIGURE 6.23</u>

Comparison of the absolute deviations of isothermal compressibilities (direct-acoustic) and differences in $(\partial T/\partial p)_s$, or μ, (static - dynamic) for water. Dashed sections indicate negative values.

O Ekman, as given by Li (1967).

DMc Diaz Pena and McGlashan (1959), interpolated curve to 50°, see Figure 6.15.

FM Fine and Millero (1973), see text

$\Delta\mu$ predicted from Figure 6.22 and Dorsey (1940 Table 126, II and III)

δ error associated with $0.5 msec^{-1}$ error in sound velocity measurements.

Erlander (1968*a*) as being due to two opposing specific heat factors. The destruction of "cluster aggregates" of water decreases the average distance between water molecules because of the loss of the empty pockets associated with such aggregates. On the other hand, the destruction of water "clusters" increases the average distance between molecules. The "cluster aggregates" predominate in the temperature region 0-18°C, whereas, from 60-100°C noncompressible water "clusters" are produced exclusively. Both processes coexist in the transition region of approximately 18-60°C. See, also, the discussion on structural types of water aggregates outlined by Millero *et al* (1974). These authors also cite the earlier references dealing with these concepts. In principle, the ideas are similar to those of Erlander (1968*a*), in that two main forms of water types exist. One, the "structured" form, dominates at low temperatures (cf Erlander's "cluster aggregates") and the second type, the non (or less-) structured type is characteristic of water at high temperatures (60-100°C), (cf Erlander's "cluster"). The effect of the solute is interpreted in terms of the ions enhancing the "nonstructured aggregate". We have pointed out, Section 5.73, the close parallel between the effect of an applied pressure, and the effect of the solute. That is, the solute exerts an "effective" pressure p_e bars. The effect of pressure is to increase the "clusters". That is, the result is the same as indicated by the interpretation of Millero *et al* (1974). However, our discussion is on a quantitative basis because the value of p_e as a function of temperature and concentration of solute is known.

As pointed out in Section 6.3, the statistical thermodynamic treatment of the LHSN "cluster" model of liquid water (Lentz, Hagler and Scheraga 1974) gives a similar picture of the effects of temperature on water structure. As the temperature increases from 0-100°C the rigid-hydrogen bonded ring structures break down into smaller, less completely hydrogen-bonded structures. Although, the main temperature effects in this model are associated predominantly with structural (rigid to less rigid structures) rather than size changes.

The "direct" measurements in Figure 6.23 are represented by Ekman's results, those of Diaz Pena and McGlashan, Amagat, and by the values calculated from the Tait equation using the acoustic B parameter values. The latter are close to the Ekman data (0-20°C experimental range) and to Amagat's data up to higher temperatures (as illustrated by the B values in Table 6.15). Possibly, at 35°C, the B parameter smooths over the subtle structural change, or the specific volumes have larger errors in this region. The other data versus the sound velocity data indicate its existence (Figure 6.15). The correlation of the B derived deviations with the simple structural parameters (Figures 6.17 and 6.18) suggests that the change at 35°C is of a different type.

The magnitude of the $\Delta\mu$ changes with temperature at higher pressures are compared with the $\Delta\beta$ changes in Figure 6.24. Here the agreement is not as close, no doubt due to compensating errors of different types being present. For example, the $\Delta\mu$ values suggest an error in $\Delta\beta$ of about 0.19×10^{-6} bar^{-1} at 10°C, 1000 bar pressure, whilst the actual deviation is about 0.01×10^{-6}bar^{-1}. This is in agreement with the other data (Ekman and Amagat) so that the discrepancy must come from the SV derived compressibility. In Table 6.17 are listed values of the sound velocity calculated from the KW data for 964.25 bars. The abnormally low value at 10.2°C suggests that at this temperature and pressure the sound velocity results are in error. If this is the case, then it indicates the danger of simply assuming that small $\Delta\beta$ means more accurate data from both sources.

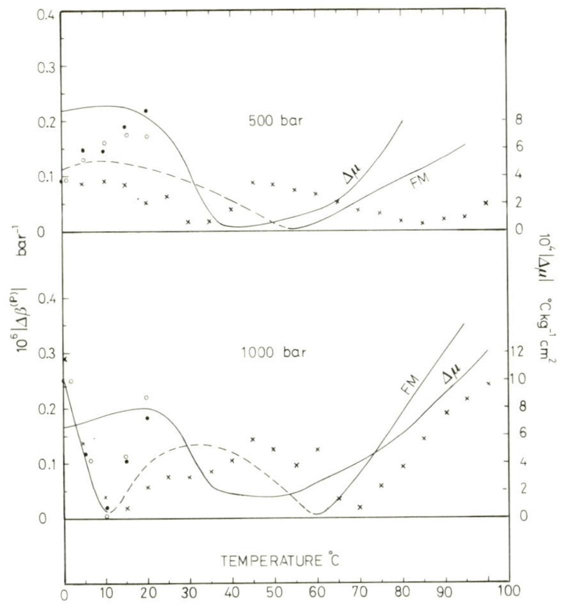

FIGURE 6.24

Comparison of isothermal compressibility deviations (direct - acoustic) and differences in $(\partial T/\partial p)_{s}$ or μ (static - dynamic) - absolute values for water.

● Ekman, B*(S=0) as given by Li(1967)

O Ekman, B (pure water data)

X Amagat, as given by Li (1967).

FM: "direct" value of $\beta^{(1)}$ from acoustic specific volumes via the Tait model, Fine and Millero (1973)

$\Delta\mu$ predicted from Figure 6.22 same scale as in Figure 6.15.

Dashed sections indicate negative regions (except Amagat's deviations at 500 bar which are all negative).

TABLE 6.17 The differences between measured and calculated sound speeds (from the KW data) in pure water, as calculated by Fine and Millero (1973), at a pressure of 964.25 bars

t^oC	Δu(m/sec)	t^oC	Δu(m/sec)
0.91	1.54	49.57	2.44
2.77	1.03	59.59	2.45
10.20	0.19	69.66	2.33
19.66	0.59	78.70	2.20
29.95	1.49	91.27	2.13
39.42	2.12		

Values of the specific volume were calculated from the Tait equation (6.3.1) using the B values derived from the acoustic data. These values are given as a function of temperature by

$$B = 2671.8 + 19.454t - 0.27028t^2 + 9.798\times10^{-4}t^3 \quad \text{(bar)}$$

for temperature range 0 to 45°C

(std.dev. $\pm$0.06) (6.75.2)

and $$B = 2740.9 + 15.568t - 0.19437t^2 + 4.5994\times10^{-4}t^3 \quad \text{(bar)}$$

$45°C < t \leq 100°C$

(std.dev. $\pm$0.09) (6.75.3)

Compare: $B(\text{Ekman}) = 2670.8 + 19.90t - 0.260t^2$ (bar)
for $0 \leq t \leq 20°C$ (Li 1967) (6.75.2a)

The values are compared with the values given by Fine and Millero and with the corresponding values of Amagat and Ekman in Figure 6.25 (0° and 25°C) and Figure 6.26 (60° and 90°C). The deviations of the Kell and Whalley values are also shown. For the lower temperatures the deviations for the "acoustic-B" - derived values, (AcB) values, and for Ekman's data are generally less than 30ppm. The Amagat deviations are all positive and somewhat higher. As already noted by Fine and Millero, the deviations of Kell and Whalley increase with pressure to relatively large values, up to nearly 160ppm at 0°C, 1000 bars. At 60°C the AcB deviations are negligible, since, at this temperature, the scatter of B is less than 1 bar for the whole pressure range. Even at 90°C the maximum deviations are only around 50ppm up to 900 bars. The deviations of Amagat's data are excessive at 90°C. The standard deviation of B (Table 6.15, column 3) indicates that Amagat's data are unreliable at temperatures near and above 80°C.

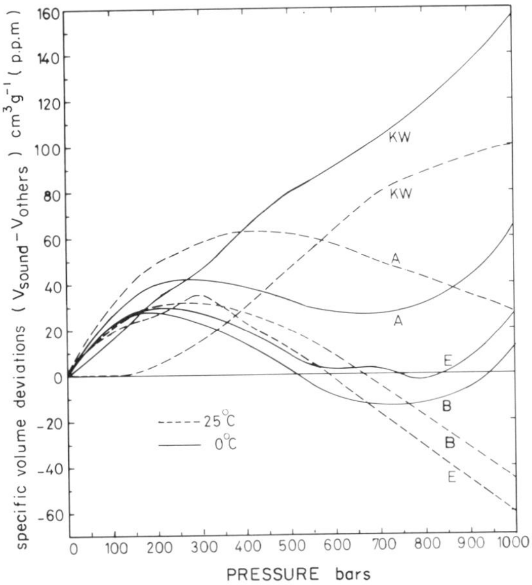

FIGURE 6.25

Differences (ppm) between the specific volumes of water, calculated from acoustic data (Fine and Millero 1973) and those from:

KW Kell and Whalley(1965)
A Amagat (1893)
E Ekman (1908)
B Tait equation with the "acoustic" B parameter (Table 6.15)

for temperatures 0 and 25°C.

The final conclusions are that the Tait equation (6.3.1) with B given by 6.75.2 and 6.75.3 should give values of the specific volumes very close to the true values. The compressibility values (from eqs 6.3.1, 6.75.2 and 6.75.3) should be unequivocally reliable to better than 0.1×10^{-6} bar^{-1} for pressures 200-1000 bars, and probably of the same reliability at low pressures. The FM equation (or any "acoustic" derived equation), apart from being necessarily more complex, will carry errors due to the structural perturbations associated with the sound velocity measurements. These errors are most serious for the compressibilities at low pressures where they may be as large as 0.3×10^{-6} bar^{-1}.

Recently, Kell and Whalley (1975) have revised their 1965 data on the compression of water in the range 0-150°C and 0-1kbar. They used

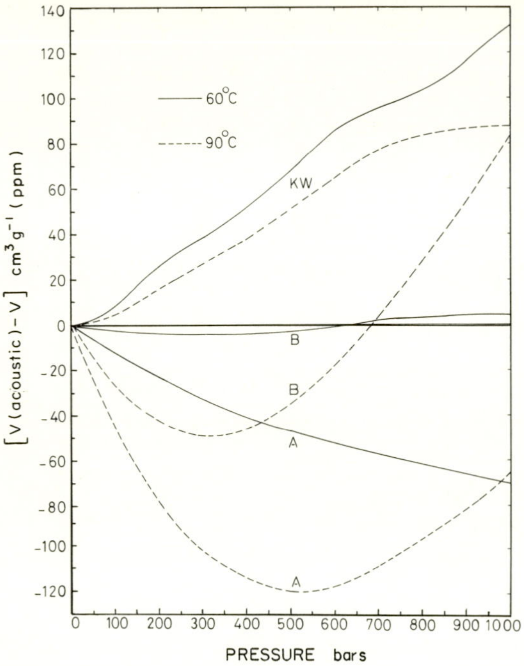

FIGURE 6.26
Differences (ppm) between the specific volumes of water, calculated from acoustic data (Fine and Millero 1973) and from direct measurements, code as in Figure 6.25 for temperatures 60° and 90°C.

sound velocity data and eq (6.63.2) to calculate the density along the lines used by Crease (1962) and by Vedam and Holton (1968) - see Section 6.63.

Kell and Whalley found variations of 20-40ppm at 1000 bars between the results from different sound velocity data (Wilson (1959) and Barlow and Yazgan (1967)), and from different methods of analysis of the same data (e.g. their results and those of Fine and Millero (1973)).

They confirmed that their directly measured values, relative to a stainless steel vessel, were in error. They attributed this to an error in the value used for the compressibility of the stainless steel. They adjusted the original densities relative to the sound velocity results, reassessing the compressibility of their measuring vessel. The new values agreed with the acoustic-derived values within ±20 ppm. The differences were roughly sinusoidal in the pressure with one cycle over 1-1000 bar, in agreement with B in our Fig. 6.25, i.e., the acoustic values are higher than the direct values at low pressures and vice versa for the high pressures. The reason for such differences may be explained as above; although, the results at the temperatures above $60^{o}C$ should be reversed (see Fig. 6.26).

6.76 Adiabatic Compression of water - The Joule Thomson (JT) and "reversible JT" coefficients.

The JT coefficient (μ_{JT}) is the change in temperature per unit drop in pressure, under adiabatic conditions, and measures the internal latent heat of expansion. At ordinary temperatures and pressures gases cool during such an expansion (except hydrogen and helium). The most common apparatus for the production of liquid air depends on this process, which applies to any fluid substance. The JT coefficient is formally defined by

$$\mu_{JT} \quad = \quad (\partial T/\partial p)_H \qquad (6.76.1)$$

where H represents the enthalpy. If the expansion occurred, not in a thermally insulated system, but in a system at constant temperature, then the heat which would be absorbed per unit difference in pressure is given by

$$(\partial H/\partial p)_T \quad = \quad - \quad \mu_{JT} C_p \qquad (6.76.2)$$

which follows from

$$(\partial T/\partial p)_H \quad = \quad -(\partial H/\partial p)_T/(\partial H/\partial T)_p \qquad (6.76.3)$$

(see Chapter 2)

A thermodynamic requirement to which any empirical equation of state must conform is given by

$$V \quad = \quad T(\partial V/\partial T)_p \quad + \quad (\partial H/\partial p)_T \qquad (6.76.4)$$

(derived from the equations

$$dH \quad = \quad TdS \quad + \quad VdP \quad \text{and}$$

$$(\partial S/\partial p)_T \quad = \quad - \quad (\partial V/\partial T)_p \quad)$$

TABLE 6.18 The Joule-Thomson Coefficient[a] for water,

$\mu_{JT} = (\partial T/\partial p)_H$

10^4 $^oC/kgcm^{-2}$

t^oC / p $kgcm^{-2}$	0	10	20	30	40	50
100	-236.3 (-235)	-227.8	-220.6	-213.9	-207.6	-202.3
200	-234.6 (-236)	-227.6	-220.4	-213.7	-208.2	-202.7
300	-233.8 (-237)	-226.8	-219.7	-213.9	-208.2	-203.0
400	-232.5	-226.3	-219.6	-213.6	-208.3	-203.3
500	-231.4	-225.3	-219.1	-213.5	-207.8	-203.1
600	-230.0	-224.2	-218.5	-213.3	-208.0	-203.4
700	-228.5	-224.1	-218.2	-212.7	-208.2	-203.3
800	-227.3	-223.4	-217.9	-212.7	-207.8	-203.7
900	-226.3	-222.2	-217.6	-212.1	-208.0	-204.1
1000	-225.4	-221.7	-216.9	-212.2	-207.7	-203.4
t^oC	60	70	80	90	100	
p $kgcm^{-2}$						
100	-196.9	-191.7	-186.1	-179.0	-170.8 (-178)	
200	-197.5	-192.6	-187.3	-181.1	-173.6 (-179)	
300	-198.0	-193.5	-188.5	-183.1	-176.2 (-181)	
400	-198.6	-194.3	-189.6	-184.7	-178.7	
500	-199.2	-194.6	-190.7	-186.0	-181.2	
600	-199.3	-195.4	-191.8	-187.4	-183.4	
700	-199.8	-195.7	-192.3	-188.6	-185.1	
800	-200.0	-195.9	-193.3	-189.8	-186.7	
900	-199.9	-195.7	-193.7	-190.5	-188.1	
1000	-200.5	-195.9	-194.2	-191.1	-189.6	

a calculated from $(V(T\alpha-1)/C_p)$ using data of Vedam and Holton (1968).
Bracketed values from W. Koch (as given by Dorsey 1940, Table 126,IV)

Thus,

$$\mu_{JT}C_p = V(T\alpha - 1) \qquad (6.76.5)$$

(since $\alpha V = (\partial V/\partial T)_p$)

The values for water listed by Dorsey (1940, Table 126, IV) from W. Koch (1934) are the only correct or nearly correct values of μ_{JT}. Values calculated from eq 6.76.5, using the data given by Vedam and Holton (1968), are shown in Table 6.18.

Some interesting equations involving μ_{JT} are

$$C_V = C_p(1-\mu_{JT}(\partial p/\partial T)_v) - V(\partial p/\partial T)_v \qquad (6.76.6)$$

where C_V is the specific heat at constant volume, with the alternatives, obtained by substituting for $(\partial p/\partial T)_V$ any of the following:

$$(\partial p/\partial T)_v = (\partial \underline{S}/\partial V)_T \qquad (6.76.6\underline{a})$$

$$(\partial p/\partial T)_v = (C_p - C_V)/\alpha VT \qquad (6.76.6\underline{b})$$

$$(\partial p/\partial T)_v = \alpha/\beta \qquad (6.76.6\underline{c})$$

The latter, combined with eq 6.76.6, gives

$$\beta = \alpha(\mu_{JT}C_p + V)/(C_p - C_V) \qquad (6.76.7)$$

The differentiation of $(\partial H/\partial p)_T = -\mu_{JT}C_p$ (6.76.8)

with temperature gives

$$(\partial C_p/\partial p)_T = -\mu_{JT}(\partial C_p/\partial T)_p - C_p(\partial \mu_{JT}/\partial T)_p \qquad (6.76.9)$$

(since $C_p = (\partial H/\partial T)_p$)

When μ_{JT} is zero,

from eq 6.76.7 $(C_p - C_V = \alpha V/\beta)_{\mu_{JT}=0}$ (6.76.7a)

from eqs 6.76.6 (b and c) $(\alpha = 1/T)_{\mu_{JT}=0}$ (6.76.6d)

and from eq 6.76.9, $((\partial C_p/\partial p)_T = -C_p(\partial \mu_{JT}/\partial T)_p)_{\mu_{JT}=0}$ (6.76.9a)

The temperature at which u_{JT} passes through zero is called the inversion temperature. For water, this occurs around 240-250°C at low pressures (up to 1-300 bars).

The Joule-Thomson effect represents an irreversible adiabatic process. The corresponding reversible process is one in which the adiabatic compression or expansion causes no sensible pressure gradients to develop within the system. For such a process the entropy change is zero, i.e. $\Delta\underline{S} = 0$, and the process is isentropic. An example is presumed to be that of sound propogation as discussed in Section 6.6.

For an isentropic compression there is normally a change in temperature. The measured ratio between this temperature change and the pressure change gives

$$(\partial T/\partial p)_{\underline{S}} = \mu_c' \qquad (6.76.10)$$

and since

$$(\partial T/\partial p)_{\underline{S}} = -(\partial \underline{S}/\partial p)_T/(\partial \underline{S}/\partial T)_p \qquad (6.76.11)$$

with

$$(\partial \underline{S}/\partial p)_T = -(\partial V/\partial T)_p \qquad (6.76.12)$$

and

$$(\partial \underline{S}/\partial T)_p = C_p/T \qquad (6.76.13)$$

$$(\partial T/\partial P)_{\underline{S}} = T(\partial V/\partial T)_p/C_p \qquad (6.76.14)$$

or

$$\mu_c = T\alpha V/C_p \qquad (6.76.15)$$

Thus, it can be seen that the difference between the reversible and irreversible coefficients is given by

$$\mu_c - \mu_{JT} = T\alpha V/C_p - V(T\alpha - 1)/C_p = V/C_p \qquad (6.76.16)$$

Values of μ_c calculated from the data of Vedam and Holton (1968) are shown in Table 6.19.

Similar relationships may be derived for μ_c as for μ_{JT}. Differentiation of eq 6.76.13 with pressure, and using eq 6.76.12 gives

$$(\partial C_p/\partial p)_T = -T(\partial \alpha V/\partial T)_p \qquad (6.76.17)$$

Eq 6.76.15 differentiated with temperature gives

$$(\partial \mu_c/\partial T)_p = (\alpha V/C_p) + T(\partial \alpha V/\partial T)_p/C_p - T\alpha V(\partial C_p/\partial T)_p/C_p^2 \qquad (6.76.18)$$

or

$$-T(\partial \alpha V/\partial T)_p = \alpha V - C_p(\partial \mu_c/\partial T)_p - \mu_c(\partial C_p/\partial T)_p \qquad (6.76.19)$$

and combined with eq 6.76.17 gives

$$(\partial C_p/\partial p)_T = -\mu_c(\partial C_p/\partial T)_p - C_p(\partial \mu_c/\partial T)_p + \alpha V \qquad (6.76.20)$$

TABLE 6.19 The temperature change [a] on reversible adiabatic compression of water

$\mu_c = (\partial T/\partial p)_S$

$10^4 \times$ °C/kgcm^{-2}

p kgcm^{-2} \ t°C	0	10	20	30	40	50
100	-2.2	6.9	15.1	22.4	29.0	35.2
200	0.0	8.4	15.9	22.8	29.1	35.0
300	2.0	9.8	16.8	23.3	29.3	34.9
400	4.0	11.1	17.7	23.8	29.4	34.8
500	5.8	12.2	18.4	24.1	29.5	34.7
	(6.8)		(17.5)		(30.0)	
600	7.5	13.4	19.1	24.6	29.6	34.5
700	9.1	14.5	19.8	24.9	29.7	34.4
800	10.5	15.5	20.4	25.2	29.8	34.4
900	11.7	16.4	21.1	25.5	30.0	34.3
1000	12.9	17.2	21.5	25.8	30.0	34.2
	(13.2)		(22.0)		(30.9)	

p kgcm^{-2} \ t°C	60	70	80	90	100
100	41.3	47.7	54.8	62.9	72.9
200	40.8	46.9	53.7	61.5	70.7
300	40.4	46.2	52.6	60.2	68.8
400	40.1	45.6	51.7	58.8	67.0
500	39.7	45.0	50.8	57.6	65.3
	(41.7)		(50.0)		
600	39.3	44.4	50.0	56.5	63.9
700	39.1	43.9	49.2	55.5	62.5
800	38.7	43.4	48.5	54.6	61.1
900	38.4	42.7	47.8	53.6	60.0
1000	38.2	42.3	47.0	52.8	58.9
	(41.3)		(46.2)		

a calculated from $\alpha VT/C_p$ using data of Vedam and Holton (1968)

bracketed values from Dorsey (1940 Table 126, III Bridgman's data)

The same equation may be obtained from eq 6.76.16 and eq 6.76.9 directly, since

$$C_p(\partial\mu_{JT}/\partial T)_p = C_p(\partial\mu_c/\partial T)_p - \alpha V + V(\partial C_p/\partial T)_p/C_p \qquad (6.76.21)$$

Also, since $C_p(\partial\mu_c/\partial T)_p = \alpha V + T(\partial(\alpha V)/\partial T)_p$

$$- \mu_c(\partial C_p/\partial T)_p \qquad (6.76.19)^*$$

from eq 6.76.15 differentiated with temperature, so that $(\partial C_p/\partial p)_T = -T\partial(\alpha V)/\partial T$, from eq 6.76.20

Similarly, from eqs 6.76.6, 6.76.7 and 6.76.16

$$C_V = C_p(1-(\mu_c - V/C_p)(\partial p/\partial T)_V) - V(\partial p/\partial T)_V \qquad (6.76.22)$$

$$\beta = \alpha C_p \mu_c/(C_p - C_V) \qquad (6.76.23)$$

when $\mu_c = 0$, $\alpha = 0$ and

$$\left[(\partial C_p/\partial p)_T = -C_p(\partial\mu_c/\partial T)_p\right]_{\mu_c=0}$$

or $$\left[T(\partial(\alpha V)/\partial T)_p = C_p(\partial\mu_c/\partial T)_p\right]_{\mu_c=0} \qquad (6.76.20\underline{a})$$

from eq 6.76.20

and $\left[C_V = C_p\right]_{\mu_c=0}$ from eq 6.76.22

6.8 The Isothermal Compressibility of Seawater

Because of the high accuracy of the sound velocity measurements it is generally assumed, as for pure water, that these data can give more accurate values for the compressibility than direct measurements. However, as indicated by the considerations brought forward in the previous section, such assumptions may be incorrect. In this section, the direct measurements of Ekman (1908), and Wilson and Bradley (1968), will be compared with the values calculated from the sound velocity measurements of Wilson. Ekman's measurements only begin at 200 bars. However, these have been combined with the 1-atm Knudsen et al. data (Dorsey 1940, Table 108) via the relationship

$$v_{sw}^{(1)} = (1-c)v_w^{(1)}(1-0.315\log\frac{(B^*+1)}{(B+1)}) + c(0.54214 - 2.1896\times10^{-3}t) \qquad (6.8.1)$$

with $c = 10^{-3}S‰$,(units ml/g)

as given by Li (1967), and as discussed in Section 4.6.

Although most of the equations have been given previously, each set will be summarised again here for convenience.

6.81 Tait-Gibson equation - Ekman's data.

The specific volume of seawater is given by (Li 1967):

$$v_{sw}^{(p)} = v_{sw}^{(1)} - (1-c)0.315v_w^{(1)}\log\frac{(B^*+p)}{(B^*+1)} \qquad (6.81.1)$$

$v_w^{(1)}$ is given by eq 5.71.1 and Table 5.18 in cm^3/g

(normalisation of units is made via 1 ml = 1.000028cm^3)

$$B^* = 2670.8 + 6.8966S + (19.39-0.0703S)t - 0.223t^2 \qquad (4.61.7)^*$$

$$B = B^*(S=0)$$

Differentiating with pressure gives the compressibility

$$\beta^{(p)}, \quad (\text{or } -(1/v_{sw}^{(p)})(\partial v_{sw}^{(p)}/\partial p)$$

$$\beta^{(p)} = 0.4343\times0.315/((B^*+p)(B_1+B_2)) \quad \text{bar}^{-1} \qquad (6.81.2)$$

with $B_1 = (1-0.315\log\frac{(B^*+p)}{(B+1)})$

and $B_2 = (0.54214 - 2.1896\times10^{-3}t)c/(1-c)v_w^{(1)}$

The pressure derivative is simply given by

$$-(\partial v_{sw}^{(p)}/\partial p)_{T.Sal.} = A(1-c)v_w^{(1)}/(B^*+p) \qquad (6.81.3)$$

with A = 0.315x0.4343, Sal. representing salinity

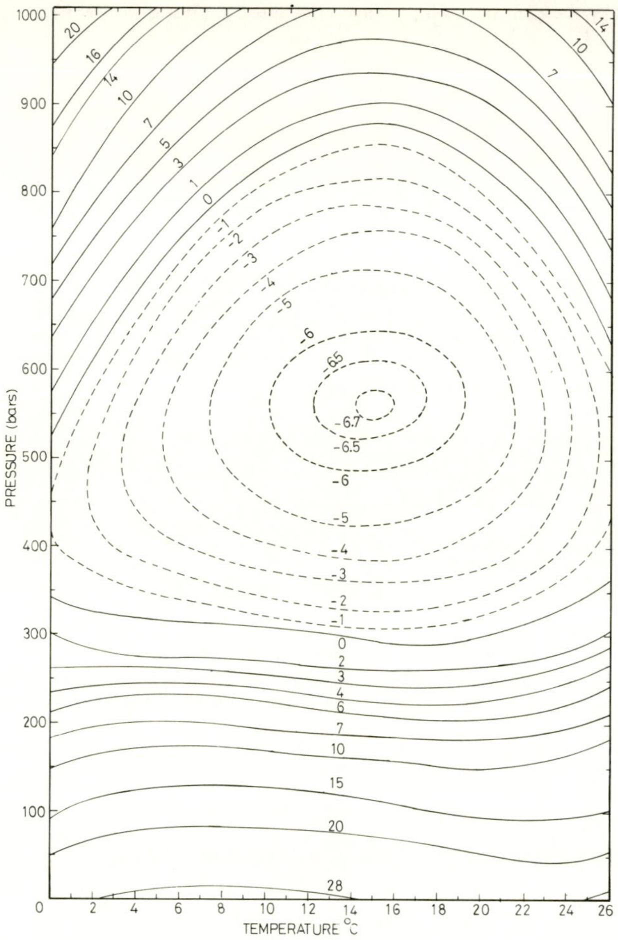

FIGURE 6.27

The deviations of $\beta_{sw}^{(p)}$(Ekman)$-\beta_{sw}^{(p)}$(Wang-Millero) $bar^{-1}x10^{8}$, as a function of temperature and pressure. Ekman values calculated from eq 6.81.2 Wang-Millero values calculated from eq 6.82.3.

6.82 "Acoustic" equation of Wang and Millero.

The development of this equation has been outlined in Sections 6.6, 6.72.

$$v_{sw}^{(p)} = v_{sw}^{(1)} - (p-1)v_w^{(1)}/R_{sw} \quad (cm^3/g) \tag{6.82.1}$$

with $R_{sw} = B_{sw} + A_{1_{sw}} p' + A_{2_{sw}} p'^2 \quad (p' = p-1)$

the coefficients (B and A's) being functions of salinity and temperature (22 coefficients - Table 4.18)

and $v_{sw}^{(1)} = 1.000027/(1+\sigma_t 10^{-3})$

with $\sigma_t = f(S,T)$ (16 coefficients - Table 4.18)

Differentiating with pressure gives

$$-(\partial v_{sw}^{(p)}/\partial p)_{T,Sal.} = v_w^{(1)}(B_{sw}-A_{2_{sw}}(p-1)^2)/R_{sw}^2 \tag{6.82.2}$$

and $$\beta^{(p)} = v_w^{(1)}(B_{sw}-A_{2_{sw}}(p-1)^2)/v_{sw}^{(p)} R_{sw}^2 \tag{6.82.3}$$

with p the absolute pressure

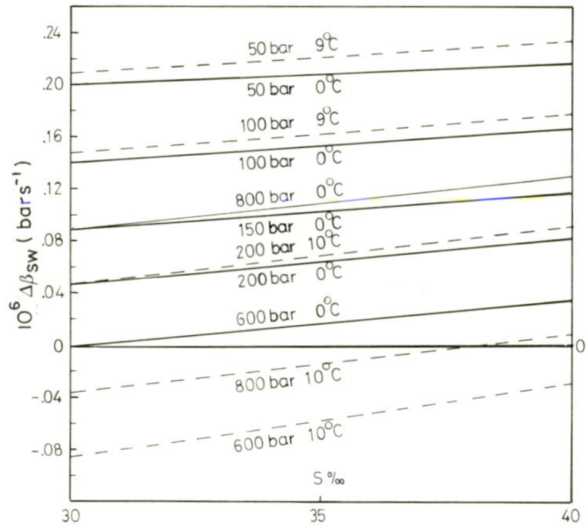

FIGURE 6.28

Differences in the isothermal compressibility of seawater $\beta_{sw}^{(p)}$ (Ekman) - $\beta_{sw}^{(p)}$ (Wang-Millero) as a function of salinity.

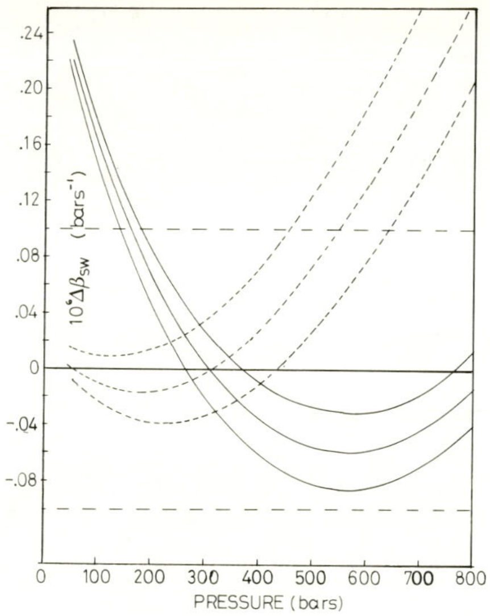

FIGURE 6.29

Differences in the isothermal compressibility of seawater at 9°C, $bar^{-1}x10^6$

———————— $\beta_{sw}^{(p)}$(Ekman)-$\beta_{sw}^{(p)}$(Wang-Millero)

- - - - - - - - $\beta_{sw}^{(p)}$(Ekman)-$\beta_{sw}^{(p)}$(Wilson-Bradley)

6.83 Tumlirz Equation (Wilson and Bradley)

$$v_{sw}^{(p)} = 0.70200 + A/E \qquad (cm^3/g)$$

with $A = 1752.73 + 11.01t - 0.0639t^2 - 3.9986S - 0.0107tS$

and $E = p + 5880.9 + 37.592t - 0.34395t^2 + 2.2524S$

from which

$$\beta_{sw}^{(p)} = A/(E(0.702E + A)) \qquad (6.83.1)$$

$$= A/E^2 v_{sw}^{(p)}$$

or

$$-(\partial v_{sw}^{(p)}/\partial p)_{T,Sal.} = A/E^2 \qquad (6.83.2)$$

6.84 Results. These three equations have been used to calculate the compressibility of seawater for the ranges 1-1000 bars pressure and 0-25°C in temperature.

From 200-1000 bars the values from Ekman's data generally agree with the "acoustic" results to better than $0.1\times10^{-6}\text{bar}^{-1}$. The deviations are higher in two small segments at high pressures, at either end of the temperature range (Figure 6.27). As might be expected, from the pure water results (Section 6.7) the deviations at low pressure are much higher, around $0.3\times10^{-6}\text{bar}^{-1}$ at 1 bar. The overall results are summarised by Figures 6.27, 6.28, 6.29.

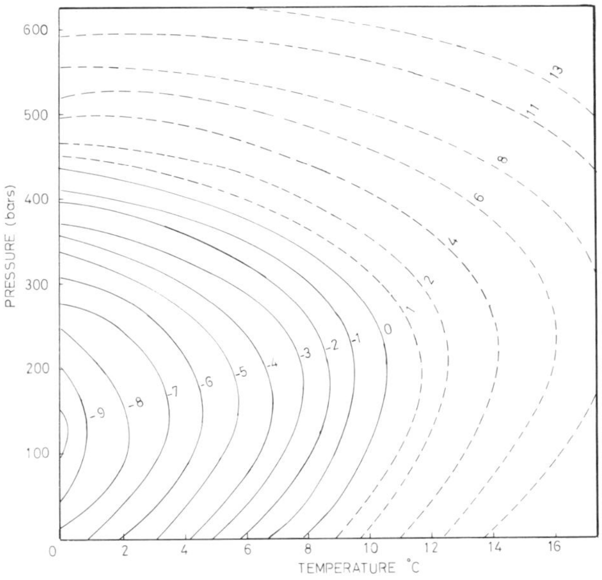

FIGURE 6.30

The deviations of $\beta_{sw}^{(p)}(EK) - \beta_{sw}^{(p)}(WB)$ $\text{bar}^{-1}\times10^{8}$, as a function of temperature and pressure.

Ekman (EK) values calculated from eq 6.81.2.

Wilson-Bradley (WB) values calculated from eq 6.83.1

In contrast to the "acoustic" compressibilities, the direct values of Wilson and Bradley are in close agreement (less than 0.1×10^{-6} bar^{-1} deviations) with the values of Ekman at low pressures and temperatures (up to about 15°C and 1-600 bars), Figure 6.30.

The values of Wilson and Bradley are believed to be less reliable at higher temperatures because of an error in the temperature coefficient of $(\partial v_{sw}/\partial p)_{T,Sal.}$ as illustrated in Figure 6.31. The temperature gradient of the Wilson-Bradley values is too steep. This deficiency in the temperature coefficient shows up in the expansibility results (Bradley and Schleicher 1970, Caldwell and Tucker 1970) and is possibly amplified by the form of the equation of state used (i.e. the Tumlirz equation). At 1 bar, the steepness of the gradient brings the values of $(\partial v_{sw}/\partial p)$ closer to the acoustic values which show an abnormal displacement downwards, relative to the trends shown at the other pressures.

Values of the B* parameter, calculated from the WB data set at 0°C, show very little change with pressure. They are in excellent agreement with the values of Ekman up to 15°C, as expected from the $\beta^{(p)}$ results, Table 6.20.

TABLE 6.20 The B* Parameter of the Tait-Gibson Equation of State for Seawater (in bars) S = 35‰

$t^{o}C$	Ekman 200-600 bars	Wilson and Bradley at 1 bar	Pressure bars	Wilson and Bradley at 0°C
0	2912.2	2908.5	1	2908.5
5	2991.3	2990.7	100	2910.7
10	3059.2	3063.6	200	2909.4
15	3116.0	3120.0	400	2907.2
20	3161.6	3179.4	600	2911.0
25	(3196.0)^{x}	3217.7	800	2915.2

x extrapolated value, Ekman's values from eq (4.61.7)*, Section 6.81.

Wilson and Bradley (1968)

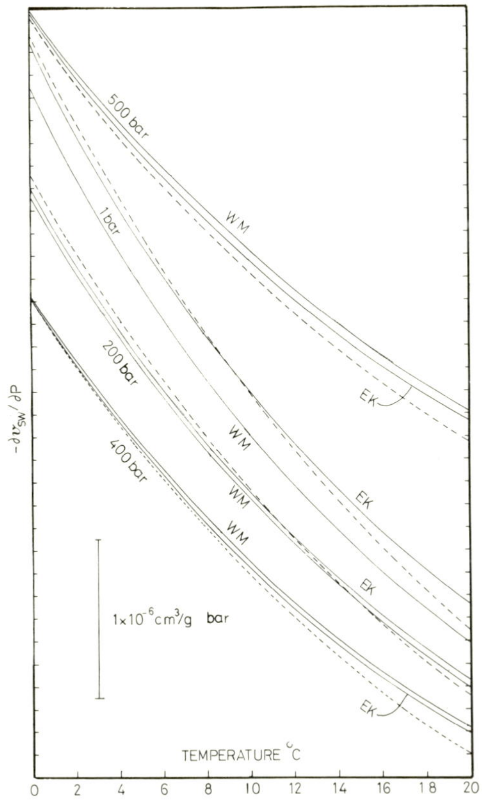

FIGURE 6.31

Comparison of the Pressure gradients
of the specific volume of seawater,
calculated from various data sources,

EK Ekman as given by Li (1967), eq 6.81.3

WM Wang and Millero (1973), eq 6.82.2

--- Wilson and Bradley (1968) eq 6.83.2.

The B* values calculated by Li (1967) from Wilson's sound velocity data at 1-atm, show the same shift as the B values, relative to the direct measurements. That is, the B*-B values (or P_e, the "salt" pressure) from both the direct and sound velocity results agree to within a few bars.

Therefore, on the evidence of the pure water analysis, it might be expected that the compressibilities derived from the sound velocity data, for seawater, are also in error, because of non-equilibrium conditions during the measurements.

Since the errors show up much more at low pressures, it is of interest to compare the "acoustic" values with other direct measurements at 1 bar.

TABLE 6.21 The Isothermal Compressibility of Seawater at S = 35‰, 1-atm

$\beta \times 10^6$ bar^{-1}

$t^{o}C$	0	5	10	15	20	25	30
Data							
LM + E1	46.64	45.39	44.34	43.44	42.79	(42.19)	
LM + E2	46.63	45.39	44.32	43.47	42.86	(42.35)	
E	46.60	45.34	44.32	43.49	42.85	(42.35)	
WB	46.66	45.35	44.25	43.35	42.62	42.10	41.65
LM + A	46.70	45.39	44.46	43.59	42.99	42.47	
NK	46.49	45.30	44.43	43.81	43.20	42.81	42.17
W	46.33	45.06	44.04	43.23	42.60	42.13	41.79
LM + W1	46.33	45.05	44.04	43.21	42.63	42.10	41.82

LM + X — $(\beta_{sw}-\beta_w)$ values of Lepple and Millero (1971) plus β_w values of X as indicated,
E1 — Tait eq with Ekman's B*(S=0) values, eq 6.81.2
A — Amagat (as given by Li 1967)
W1 — Wilson (1959)
E — Ekman (1908)
WB — Wilson and Bradley (1968)
NK — Newton and Kennedy (1965)
W — Wilson (1960)
E2 — Tait eq with Ekman's B values, eqs 6.81.2 and 6.75.2a.

Table 6.21 lists values of $\beta_{sw}^{(1)}$ from various sources. The agreement between the direct measurements at low temperatures is particularly good. The values of Lepple and Millero are relative to pure water so that

$\beta_{sw}^{(1)}$ will naturally depend on $\beta_{w}^{(1)}$. Since the discrepancy in the sound derived B values are associated with water-structure equilibrium effects, it would be expected that $\beta_{sw}-\beta_{w}$ would not show the effect, since it cancels. This is shown in the B*-B results. Thus $\beta_{sw}-\beta_{w}$ results are not so useful in this respect. However, the LM data show that values of B*-B are indeed reliable, that is, the pressure effect of the salt, p_e, on water is accurately known. This is an important result for membrane permeability work, for which the lower external pressures often are of most interest.

As far as the range of oceanographic interest is concerned, the Tait-Gibson equation, which is much simpler than the WM equation, is believed to give reliable results for the compressibility, using the data of Ekman, (1908) and Knudsen *et al*. (Dorsey 1940, Table 108) as summarised in eq 6.81.2.

The cumulative evidence (especially from pure water) suggests that the "acoustic" compressibilities carry errors due to "sustained" perturbations of the water structure during the sound wave propagations. These errors may be somewhat obscured by other compensating errors but are largest at low pressures (ca. 1-150 bars). It is expected that sound velocities cannot be accurately predicted from direct compressibility measurements unless this effect is taken into account. In addition, the errors would be expected to be least at 50-60°C, but no measurements are available in this range, as far as is known. Such errors would, of course, not show up in accurate relative measurements (e.g.seawater data relative to pure water data) provided the two sets were compatible (i.e. both "acoustic" or both direct).

Recently, Emmett and Millero (1974) measured the specific volume of seawater (S = 35‰) relative to the specific volume of pure water ("acoustic" values as given by the equation of Fine and Millero 1973) - see eq 6.63.5, Section 6.72. The temperature and pressure ranges covered were -2° to 40°C and 0-1000 bars. The values of $v_{sw}^{(p)}$(S=35‰) showed an average deviation of ± 7ppm from those calculated from the WM equation of state (Table 4.18); 85% of the measured values agreed with the latter to within ± 15ppm. These results could be expected for accurate measurements, on the basis of the analysis given above. However, if the densimeter used in these experiments had been calibrated on the basis of the E (Ekman) or B (Tait-Acoustic) pure water data sets (eqs 6.75.2a and 6.75.2 respectively, coupled with eq 6.3.1) the agreement would have been biased towards the earlier "direct" data sets as listed in Table 6.21 (e.g. E and WB). In any case, the characteristic compressibility deviations (β, direct - β, acoustic) still show up in a modified form (around 0.1×10^{-6} bar^{-1} at low pressures, p < 100 bars). In short, these more recent results do not change the conclusions reached above. Nor does the paper by Fine, Wang and Millero (1974) on the equation-of-state-of-seawater. Again, the conclusions rest on the premise that the "acoustic" pure water results (Fine and Millero 1973) are the most accurate available. Mention is made in both these 1974 papers of work done by Bradshaw and Schleicher (a personal communication during 1973), and by Duedall and Paulowich (1973). Apparently, Bradshaw and Schleicher determined the compression of seawater at 10°C and combined the results with their expansibility results of 1970 (see Chapter 5). Since this work is not yet published, no details are readily available, so that further comment is not warranted. The measurements of Duedall and Paulowich (1973) were of the quantity $(\beta_{sw}-\beta_w)$ at 15°C. Therefore, their results merely confirm the above conclusion that $(\beta_{sw}-\beta_w)$ is now accurately known (to within ± 0.05×10^{-6} bar^{-1}, or better).

6.9 The Mean apparent molal compressibility of seasalt.

Millero (1973) presents the results of measurements of the ϕ_k of the salts NaCl, $MgCl_2$, Na_2SO_4 and $MgSO_4$ from 0-45°C and the ionic strength range 0.1-1.0 molar. He gives these results in the form of ϕ_k^o and s_k^* for the individual ionic components, together with estimates of the other major ions of seawater (Ca(2+), K(+), Sr(2+), HCO_3(-), Br(-) and F(-)), see Section 6.2. As usual, this author works in terms of equivalents and has estimated the apparent equivalent compressibility of seawater as a function of concentration and temperature (Millero 1974).

The corresponding mean molal quantity can be estimated from

$$\phi_K = \Sigma_i f_i \phi_{k_i} \qquad (6.9.1)$$

in the usual way, with the f_i values listed in Table 4.16. Comparison can then be made with the values of Φ_K calculated from various seawater data sets via

$$\phi_K = (10^3/m_s d_w)(\beta_{sw}-\beta_w) + \beta_{sw}\phi_v \qquad (6.9.2)$$

or the equivalent

$$\phi_K = (10^3/m_s)((v\beta)_{sw} - (v\beta)_w) + (\beta v)_{sw} M_t \qquad (6.9.3)$$

which, combined with the Tait and Tait-Gibson equations, gives for any pressure, p,

$$\phi_K^{(p)} = 434.3 \text{x} 0.315 M_t' v_w^{(1)} (1-c)(B-B^*)/S(B^*+p)(B+p) \quad (6.9.4)$$

so that the value relative to the value at 1 bar is simply given by

$$\phi_K^{(p)} = \phi_K^{(1)} (B^*+1)(B+1)/(B^*+p)(B+p) \qquad (6.9.5)$$

The form of this equation minimises errors in B* and B so that reliable values of $\phi_K^{(p)}$ are obtainable, provided $\phi_K^{(1)}$ is accurate. A check on the accuracy at 1 bar can be made by comparing the values with those from eq 6.9.1.

Values of ϕ_K were calculated from eq 6.9.4, and from eq 6.9.3 using the "acoustic" data, with $m_s = S/(1-c)M_t'$, $M_t = 68.4148$g, $M_t' = 68.0811$g for solution A (Table 4.16).

The relationship for $(v\beta)_{sw}$, (i.e. $(-(\partial v_{sw}/\partial p)$, is given above (Section 6.82).

The values at 1 bar, as a function of temperature, are compared with the values estimated from eq 6.9.1 (using the ϕ_{k_i} values given by Millero (1973)), Figure 6.32.

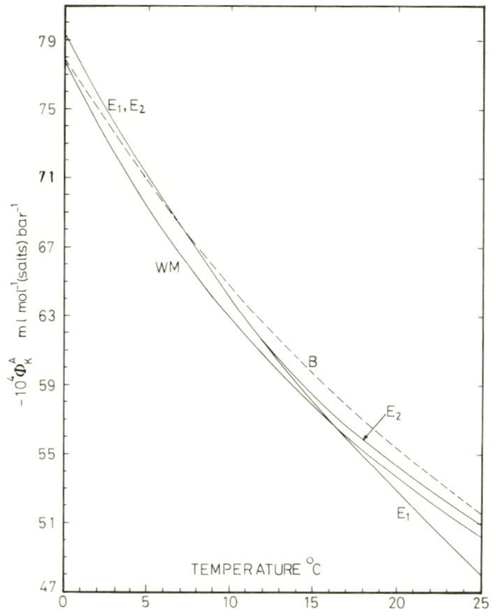

FIGURE 6.32

Mean apparent molal compressibility as a function of temperature.
S = 35‰, 1 bar.
E1 and E2 Ekman's data as given by
Li (1967) with B=B*(S=0) for E1, and B from water data for E2.
B from binary solution data (Millero 1973,
Parsons 1959 - see text)
Solution A Table 4.16 used as a basis. To convert to $\text{mlmol}^{-1}(\text{ions})\text{bar}^{-1}$ divide by 2.178796.

As pointed out in Section 6.2 the sodium sulfate values were weighted by Millero to compensate for "mixing" effects. However, the value of ϕ_K is practically unaffected because the value of $\phi_k^o(Na_2SO_4)$ is the same and the total contribution of the sulfate salt is relatively small, as illustrated in Figure 6.35 for $\bar{K}_s$ (the partial molal quantity).

Up to 15°C the values of Ekman are generally within $1x10^{-4}$ ml$mol^{-1}bar^{-1}$ of the summed electrolyte values (which is about the experimental error for the binary solutions). This agreement is maintained to 25°C (beyond the experimental range of Ekman, viz. 0-20°C) if B from the pure water data (eq 6.75.2a) is used instead of B*(S=0). The latter value has been used for the higher pressures, however, since it is adequate in the temperature range of oceanographic interest, and has been consistently used for other derivatives of the volume, and found to be satisfactory.

The "acoustic" values, except at 0°C are numerically lower. This is due to βv, $((\partial v/\partial p)_{T,Sal.})$ for seawater being too low, as illustrated in Figure 6.31.

The value of $(\beta v)_{sw}-(\beta v)_w$ is somewhat high at 0°C as indicated by the Ekman value. This term is much the same for the Ekman and acoustic values, as discussed in Section 6.84. Thus, there is a compensating effect at 0°C for the WM line. Of course, the "mixing effect" may not be entirely negligible. Measurements of the partial molal compressibilities $(\bar{K}_i^{sw})$ of the components, i, in seawater might solve this problem.

Values of $\phi_K^{(p)}$ in ml per gram of seawater (i.e. $m_s\phi_K/(10^3+ m_sM_t)$, the total solute effect divided by the total mass of the solution in grams) for the P-T ranges 0-1000 bars, 0-26°C are summarised in Figure 6.33.

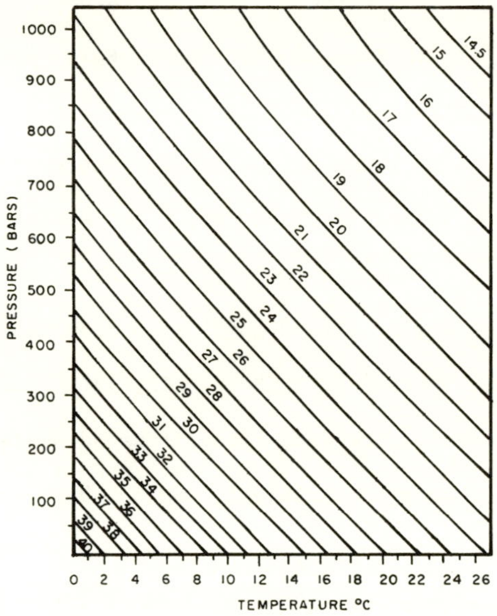

FIGURE 6.33

The mean apparent molal compressibility of seasalt, $-m_s\phi_k/(10^3+m_sM_t)\times10^7$ ml per gram seawater bar^{-1}

Data of Ekman (1908).

eq 6.9.4
S=35‰

(Ekman's data), and Figure 6.34 (acoustic data - in terms of the deviations from Ekman's values). It is concluded that eq 6.9.4 adequately represents the values in the oceanographic P-T range.

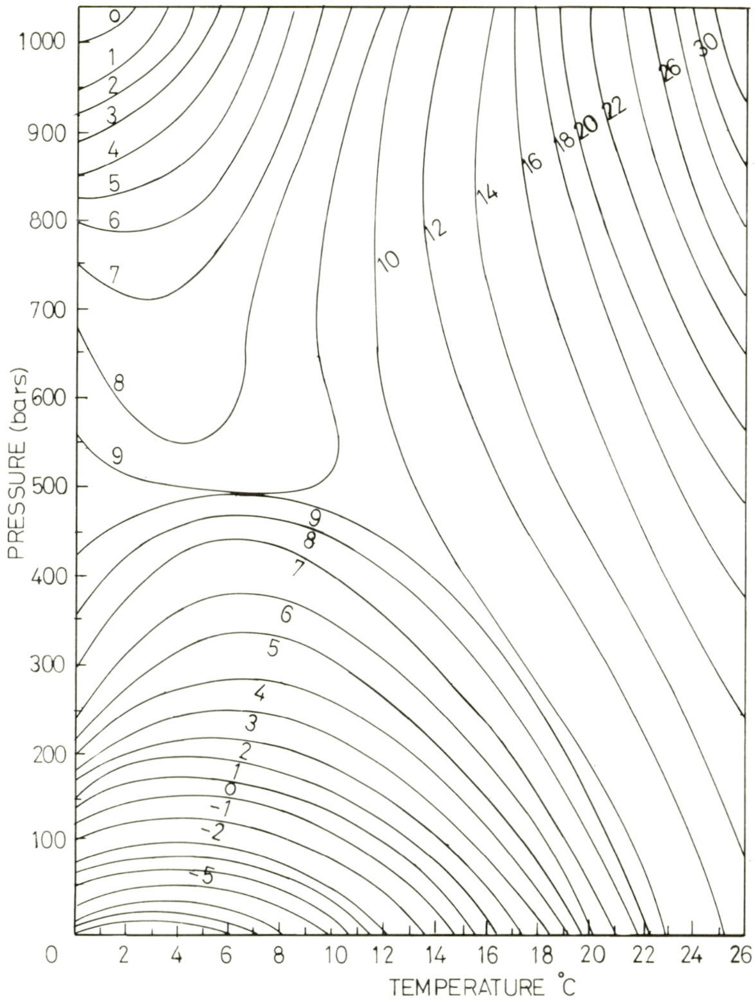

FIGURE 6.34

Deviations between the values of $m_s\phi_K$ from the WM and TG equations of state, $m_s\left[\phi_K(\mathrm{WM}) - \phi_K(\mathrm{EK})\right] / (10^3 + m_s M_t)$, in 10^8(ml per g seawater)bar^{-1}

S=35‰

See equations 6.9.4 and 6.9.3 and Table 6.22

6.10 The Mean Partial Molal Compressibility of Seasalt, $\bar{K}_s$

Values of $\bar{K}_i$ for seawater components in the binary solutions have been given in Section 6.2. An estimate of the total mean effect of the solute (seasalt) may be obtained from

$$\Sigma_i f_i \bar{K}_i \qquad (6.10.1)$$

where the sum covers all the components in the usual way.

Figure 6.35 shows the total effect for a seawater solution containing one kilogram of pure water (i.e. $\Sigma m_i \bar{K}_i$, since there are m_i moles of each component, $\underline{i}$, in such a solution). The contribution of the main components, NaCl, $MgCl_2$ and Na_2SO_4, are shown for comparison.

The value of $\bar{K}_s$ may also be obtained directly from seawater data, in the same way as the corresponding apparent molal quantity, since

$$\bar{K}_s = \phi_K + I(\partial\phi_K/\partial I)_{T,p}$$

$$= \phi_K + S(1-c)(\partial\phi_K/\partial S)_{T,p} \qquad (6.10.2)$$

Thus, from eq 6.9.3,

$$\bar{K}_s = (v\beta)_{sw} M_t + S(1-c)(Q+M_t)(\partial(v\beta)_{sw}/\partial S)_{T,p} \qquad (6.10.3)$$

with $Q = M_t'(1-c)/c = 68.0811(1-c)/c$ g and

$M_t = 68.4148$ g (for solution A, Table 4.16)

For the Tait-Gibson equation this reduces simply to

$$\bar{K}_s = -(A(1-c)^2 \partial B^*/\partial S)/(B^*+p)^2 \qquad (6.10.4)$$

with $A = 434.3 \times 0.315 M_t' v_w^{(1)}$

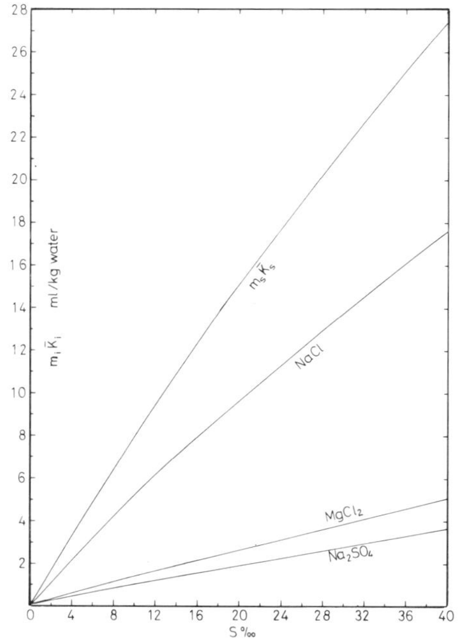

FIGURE 6.35

The contribution of the three major seawater components to the partial molal compressibility of seasalt, $m_s\bar{K}_s$ (or $\Sigma m_i\bar{K}_i$), for a seawater solution containing 1 kg of water and m_i moles of each component, $\underline{i}$, (solution A, Table 4.16, with $M_t' = 68.0811$, $M_t = 68.4148$ g)

$$m_s = \Sigma m_i = S/(1-c)M_t'$$

There is 1 kg water in $(10^3 + m_s M_t)$ g seawater

For Ekman's data

$$B^* = 2670.8 + 6.8966S + (19.39-0.0703S)t -0.223t^2$$

and $\partial B^*/\partial S = 6.8966 -0.0703t$

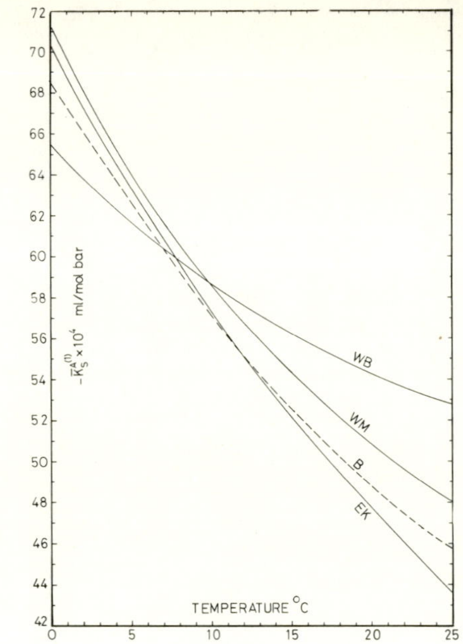

FIGURE 6.36

The mean partial molal compressibility of seasalt $\Sigma m_i \bar{K}_i / \Sigma m_i$ (solution A, Table 4.16) at 1 bar, estimated from various data sets.

-------B from binary solution data, Millero (1973)

EK Ekman (1908), eq 6.10.4

WM Wang and Millero (1973), eqs 6.10.3, 6.10.6.

WB Wilson and Bradley (1968), eqs 6.83.2, 6.10.3,6.10.5

with $c = 10^{-3}S$, $v_w^{(1)}$ from eq 5.71.1, Table 5.18.

For the <u>Tumlirz equation</u> of Wilson and Bradley

$$(\partial(v\beta)_{sw}/\partial S)_{T,p} = (1/E^3)(E(3.9986+0.0107t) + 4.5048A) \quad (6.10.5)$$

with

$$(v\beta)_{sw} = A/E^2 \quad (6.10.5\underline{a})$$

$$A = (1752.73 + 11.01t - 0.0639t^2 - 3.9986S - 0.0107tS)$$

$$E = (p + 5880.9 + 37.592t - 0.34395t^2 + 2.2524S)$$

For the <u>acoustic equation</u> of Wang and Millero

$$(\partial(v\beta)_{sw}/\partial S)_{T,p} = (v_w^{(1)}/R_{sw}^2)\left[(\partial B_{sw}/\partial S) - p'^2(\partial A_{2_{sw}}/\partial S) - (2/R_{sw})(B_{sw} - A_{2_{sw}}p'^2)\partial R_{sw}/\partial S\right] \quad (6.10.6)$$

with $R_{sw} = B_{sw}(S,t) + A_{1_{sw}}(S,t)p' + A_{2_{sw}}(S,t)p'^2$

with p' representing the applied pressure (i.e. p'=p-1). The functions and coefficients of this equation are too many to give in detail (see Table 4.18).

$v_w^{(1)}$ is given by eq 5.71.1, Table 5.18.

The 1-bar values of the mean partial molal compressibility $\bar{K}_s^{(1)}$ (solution A) are shown in Figure 6.36 as a function of temperature.

The EK and WM values are in close agreement at low temperatures, but diverge away from each other above about 10°C. This is due to the apparent error in the WM salinity gradient at the higher temperatures as shown for the mean partial molal volume (Figure 4.8 Section 4.6).

These errors persist to high pressures as shown by Figure 6.37. Even at 0°C, the agreement is less at the higher pressures.

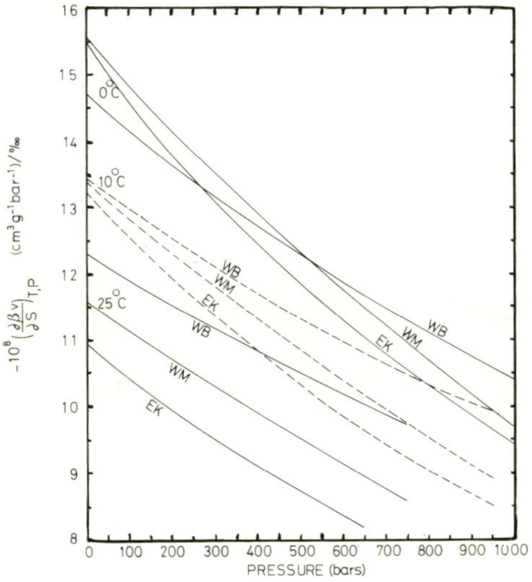

FIGURE 6.37

The salinity gradient of $(-\partial v_{sw}/\partial p)_T$ as a function of pressure and temperature.

EK Ekman (1908) eqs 6.81.3, 6.10.3, 6.10.4.

WB Wilson and Bradley (1968), eq 6.10.5.

WM Wang and Millero (1973), eq 6.10.6.

Ekman's 1-bar values are in close agreement with the binary derived results (Figure 6.36 curve B). The Tait-Gibson equation (eq 6.10.4) with Ekman's data, was used to calculate the values of $\bar{K}_s^{(p)}$ over the usual P-T ranges. The results are shown in Figure 6.38. The WM values are summarised in Figure 6.39 in the form of the deviations from the EK results.

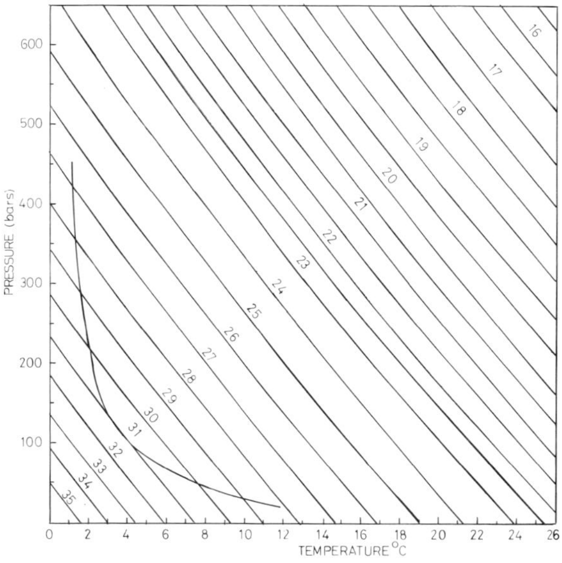

FIGURE 6.38

The mean partial molal compressibility of seasalt,

$m_s\bar{K}_s/(10^3+m_sM_t)$ ml per g seawater bar^{-1} x10^7

Data of Ekman, eq 6.10.4

S = 35‰

P-T characteristics of a south-west Pacific station superimposed, G3/62/65, Lat. 33 52S, Long. 153 27E (24 February, 1965), CSIRO (1965).

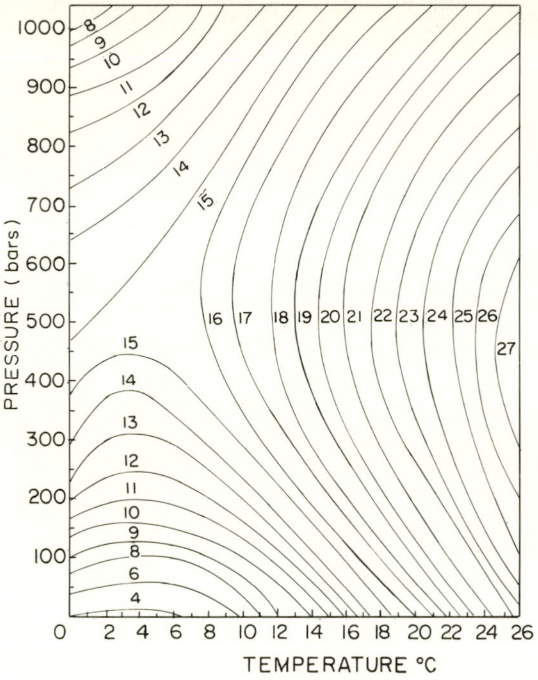

FIGURE 6.39

Deviations of $m_s\bar{K}_s$ values calculated from the WM and TG equations of state, i.e. $m_s(\bar{K}_s(\mathrm{WM}) - \bar{K}_s(\mathrm{EK}))/(10^3+m_sM_t)$ in ml per gram seawater $\mathrm{bar}^{-1} \times 10^8$ See eqs 6.10.4, and (6.10.3, 6.10.6), and Table 6.22. S = 35‰

The values of $\bar{K}_s^{(p)}$ follow the same P-T trends as the other partial molal quantities, i.e. the values for the average P-T profile show a region of effective stability around 1000 metres.

The discrepancies of the two data sets, EK and WM are largely due to the salinity gradient, as shown in Figure 6.37.

The concentration gradient of the mean apparent molal compressibility $(I\partial\phi_K/\partial I)$ or $S(1-c)\partial\phi_K/\partial S$ is of particular interest, as it is used in the calculation of the partial molal compressibility of electrolytes in the seawater medium. This quantity is given by $\bar{K}_s-\phi_K$ and values at 1-bar, as a function of temperature are shown in Figure 6.40.

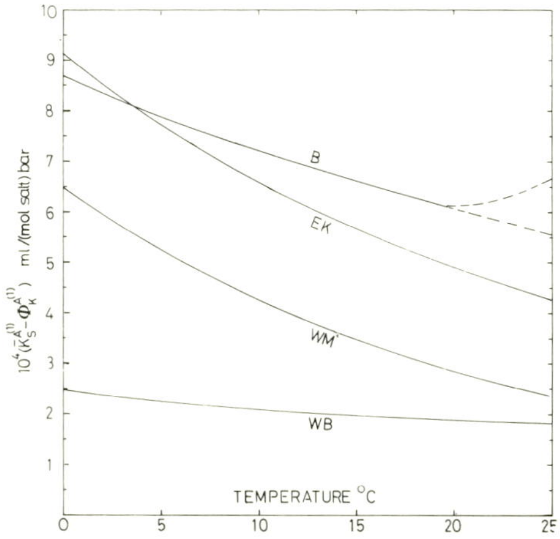

FIGURE 6.40

The concentration gradient of the mean apparent molal compressibility of seasalt $I\partial\phi_K/\partial I$ (or $S(1-c)\partial\phi_K/\partial S$) as a function of temperature, calculated from:

B binary solution data (Millero 1973)

EK Ekman (1908), eqs 6.11.2, 6.11.4.
WM Wang and Millero (1973), eq 6.10.2.
WB Wilson and Bradley (1968), eq 6.10.2.

Again the results for the EK data show closest agreement with the binary solution curve. Representative values at various pressures may be obtained from Table 6.22.

TABLE 6.22 The mean apparent molal compressibility and mean partial molal compressibility of seasalt as a function of pressure

S = 35‰, t = 5°C

$ml\,bar^{-1}g^{-1}$(seawater) x 10^7

P (bars)	$\bar{K}_s^{sw}$		ϕ_K^{sw}	
	EK	WM	EK	WM
1	32.6	32.9	36.6	35.6
150	29.6	30.5	33.1	33.0
250	27.8	29.0	31.0	31.3
350	26.1	27.5	29.1	29.6
450	24.6	26.1	27.4	28.1
550	23.3	24.8	25.8	26.6
750	20.8	22.2	23.0	23.7
850	19.8	21.0	21.8	22.4
950	18.8	19.9	20.7	21.1

N.B. $\bar{K}_s^{sw} = m_s\bar{K}_s/(10^3+m_sM_t)$, with $m_s = S/(1-c)M_t'$

for $\bar{K}_s^A$ (solution A) $M_t = 68.4148$, $M_t' = 68.0811$g

(see Table 4.16).

The differences in the EK and WM values of ($\bar{K}_s^{sw} - \phi_K^{sw}$), the mean value per gram of seawater, do not change appreciably with pressure at the low temperatures (Figure 6.41) i.e. in the seawater P-T range. The error function gradually changes its curvature as the temperature rises. It is as if the curves are pulled downwards at the high pressure end as the temperature increases.

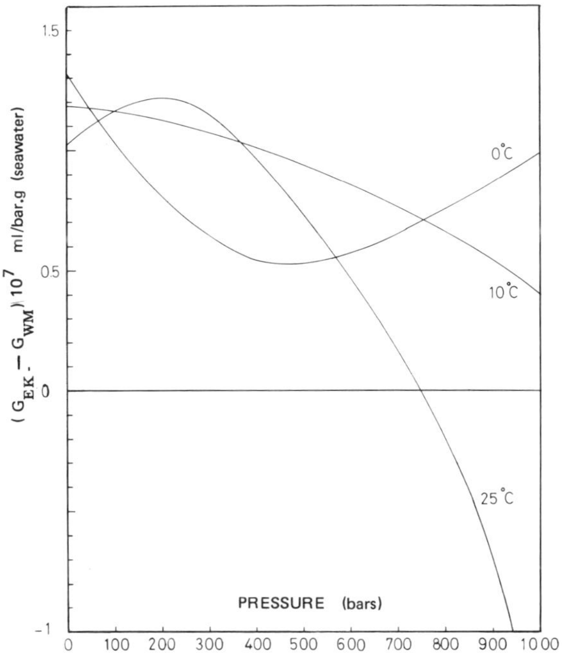

FIGURE 6.41

Differences in the values of $(\bar{K}_s^{sw} - \phi_K^{sw})$, or $\underline{G}$, calculated from the data of Ekman, and of Wang and Millero. eqs 6.11.2 and (6.10.2, 6.9.3, 6.10.3, 6.10.6) respectively. S = 35‰

6.11 The partial molal compressibilities of electrolytes in seawater, $(\bar{K}_i^{sw})$.

The general equation for $\bar{K}_i^{sw}$ (i.e. $(-\partial\bar{v}_i^{sw}/\partial p)_{T,Sal.}$), may be obtained by differentiating equation 4.4.8 with respect to pressure, viz.

$$\bar{v}_i^{sw} = \phi_{v_i} + (w_i/W)(\bar{V}_s - \Phi_v) + M \qquad (4.4.8)*$$

where M represents the mixing term

$$M = \Delta\Phi_v - (w_i/W)I(\partial\Delta\Phi_v/\partial I)_{p,T} + (I/W)(\partial\Delta\Phi_v/\partial m_i)_{m_j,P,T}$$

Since the pressure derivative of M is unknown and for most electrolytes is probably small, $\bar{K}_i^{sw(p)}$ can be represented by

$$-\bar{K}_i^{sw(p)} = -\phi_{k_i}^{(p)} + (w_i/W)(-\bar{K}_s + \phi_k)^{(p)} \qquad (6.11.1)$$

with $$(-\bar{K}_s + \Phi_K)^{(p)}/W = (\tfrac{A}{W}(1-c)/(B^* + p))(Q_1 + Q_2) \qquad (6.11.2)$$

and

$$Q_1 = (B - B^*)/S(B+p)$$

$$Q_2 = ((1-c)\partial B^*/\partial S)/(B^* + p)$$

$$\frac{A}{W} = 434.3 \times 0.315 M'_t/W = 6.860489 \times 10^3$$

(which is independent of recipe chosen for seawater)

$\bar{K}_i$ can change appreciably with pressure, as shown in Section 6.32 for the binary solutions, but can generally be represented by the simple function

$$\bar{K}_i^{(p)} = \bar{K}_i^{(1)} + B_k(p-1) \qquad (6.11.3)$$

with values of B_k of the order of 10^{-5} to 10^{-6} (ml mol^{-1}bar^{-2}) (Section 6.32, Table 6.8).

The value of B_k for the binary solution should be a good approximation for seawater. When data are unavailable, the values in Table 6.8 may be useful as a guide. In any case, the values of $\bar{K}^{sw(1)}$ are the important quantities and may be calculated from eq 6.11.1 and 6.11.2. Values of $(-\bar{K}_s + \phi_K)^{(1)}/W$ for a salinity of 35‰ are given as a function of temperature by

$$(-\bar{K}_s + \phi_K)^{(1)}/W = 10^{-4}(-6.725 + 0.21201t - 2.8030 \times 10^{-3} t^2) \text{ ml mol}^{-1}\text{bar}^{-1} \qquad (6.11.4)$$

(std.dev. $\pm$ 2x10^{-6})

using the data of Ekman, and a least squares fit.

with $w_i = \frac{1}{2}(z_+z_-)_i\nu_i$ (e.g. w_i = 1,3,4,6 for NaCl, Na_2SO_4, $MgSO_4$, $LaCl_3$ respectively)

The division by W of eq 6.11.4 naturally makes the result independent of any recipe chosen. (Tables 4.16, 4.17).

TABLE 6.23 The partial molal compressibilities of seawater components at S = 35‰, 1 bar.

$ml\,mol^{-1}bar^{-1} \times 10^4$

Salt	t^oC	0	5	10	15	25
NaCl[a]	$\bar{K}_i$	56.04	51.21	46.69	42.85	37.05
	$\bar{K}_i^{sw}$	56.97	52.40	48.15	44.45	38.72 (37.5)[b]
Na_2SO_4[c]	$\bar{K}_i$	163.40	149.38	136.68	126.53	109.84
	$\bar{K}_i^{sw}$	164.83	151.63	139.80	130.14	114.63 (110.8)
$NaHCO_3$	$\bar{K}_i$	48.71	43.78	38.77	34.63	28.14
	$\bar{K}_i^{sw}$	49.22	44.55	39.80	35.80	29.38
NaF	$\bar{K}_i$	86.03	79.30	72.99	67.15	56.84
	$\bar{K}_i^{sw}$	87.90	79.30	75.38	69.68	59.48
KCl	$\bar{K}_i$	48.31	43.25	38.86	35.02	29.57
	$\bar{K}_i^{sw}$	49.75	44.99	40.83	37.13	31.60 (30.3)
KBr	$\bar{K}_i$	41.28	35.82	31.24	27.10	20.96
	$\bar{K}_i^{sw}$	42.30	37.14	32.78	28.78	22.56
$MgCl_2$[d]	$\bar{K}_i$	113.52	104.50	96.43	89.76	79.99
	$\bar{K}_i^{sw}$	103.04	96.38	90.37	85.30	77.66 (69.6)
$CaCl_2$	$\bar{K}_i$	97.92	88.90	80.83	74.16	64.40
	$\bar{K}_i^{sw}$	87.44	80.78	74.77	69.70	62.07 (55)
$SrCl_2$	$\bar{K}_i$	98.52	89.50	81.43	74.76	65.01
	$\bar{K}_i^{sw}$	88.04	81.38	75.37	70.30	62.67

TABLE 6.23 (ctd)

a $B_k(25^oC) = 6.5x10^{-6}$ for eq $\bar{K}_i^{(p)} = \bar{K}_i^{(1)} + B_k(p-1)$

b bracketed values calculated using $(\bar{K}_s-\Phi_K)/W$ from "unweighted" binary solution data (Leyendekkers 1974)

c Binary solution data weighted to allow for interactions in seawater, Millero (1973)

$B_k(25^oC) = 6.5x10^{-6}$

d $B_k(25^oC) = 4.7x10^{-6}$

As noted previously, Section 6.2,the values of ϕ_{k_i}, (the apparent molal compressibility of the component i in the binary solution, at the same ionic strength as seawater) are calculated from

$$\phi_{k_i} = \phi^o_{k_i} + s'_k I_v^{1/2} \quad (6.2.2)*$$

It was shown that the molal ionic strength, I, could satisfactorily replace I_v, so that

$$\phi_{k_i} = \phi^o_{k_i} + s'_k \ (0.01994S/(1-c))^{1/2} \quad (6.11.5)$$

Values of $\phi^o_{k_i}$ and s'_k for a range of temperatures are given in Table 6A.1, Appendix 6.

Values of $\bar{K}_i^{sw(1)}$ and the corresponding $\bar{K}_i^{(1)}$ (binary solution) for seawater components are shown in Table 6.23 for a range of temperatures. Values for electrolytes at low concentration in seawater at 25°C are shown in Table 6.24.

The seawater values were calculated from eq 6.11.1 with $(\bar{K}_s-\phi_K)/W$ from eq 6.11.4 (Ekman's data). The binary solution values of this term are compared with Ekman's values in Table 6.25.

The overall agreement is good, as might be expected from Figure 6.36. These results suggest that for solutions that deviate from the "average" seawater composition, the summed binary values should give good estimates of the $(\bar{K}_s-\phi_K)/W$ term.

TABLE 6.24 The partial molal compressibility of electrolytes in seawater[a]

1 bar S = 35‰, t = 25°C

$10^4(-\bar{K}_s+\Phi_K)/W$ = $-3.8\ \mathrm{ml\,mol^{-1}bar^{-1}}$

Salt	$-\phi_{k_i}$ [b]	$-\bar{K}_i^{sw}$ [c]	Salt	$-\phi_{k_i}$	$-\bar{K}_i^{sw}$
	$10^4 \mathrm{ml\,mol^{-1}bar^{-1}}$			$10^4 \mathrm{ml\,mol^{-1}bar^{-1}}$	
$AgNO_3$	32.3	28.5	$LiClO_4$	22.4	18.6
$CaCO_3$	115.2	100.0 (97.0)	$LiNO_3$	28.9	25.1
$BaCl_2$	101.8	90.4	LiOH	63.4	59.6
CsBr	24.8	21.0	$MgSO_4$	123.2	108.0 (105.0)
$CsNO_3$	23.2	19.4	NaBr	40.2	36.4 (35.6)
HCl	5.8	2.0 (1.2)	$NaClO_3$	39.2	35.4
$KBrO_3$	40.5	36.7	$NaClO_4$	32.7	28.9
KCNS	22.8	19.0	Na_2CO_3	130.1	118.7 (116.5)
KF	62.3	58.5			
$KHCO_3$	28.6	24.8 (24.0)	NaOH	71.2	67.4
			NH_4Br	9.9	6.1
KI	23.5	19.7	NH_4Cl	16.4	12.6
KIO_3	64.2	60.4	NH_4ClO_4	4.6	0.8
KNO_3	30.0	26.2 (25.4)	NH_4NO_3	9.7	4.9
			$TlNO_3$	30.5	26.7
LiBr	29.6	25.8			

[a] The seawater concentration is taken as average (Table 4.16) so that the concentration of the added electrolyte should be < ca. $10^{-4}\mathrm{mol\,kg^{-1}}$

[b] At an ionic strength equivalent to $0.723\mathrm{mol\,kg^{-1}}$ (S-35‰)

[c] calculated from eq 6.11.1

bracketed values as for footnote b, Table 6.23.

TABLE 6.25 The weighted salinity gradient of the mean apparent molal compressibility of seasalt[a]

$10^4(\bar{K}_s - \Phi_K)/W$ $ml\,mol^{-1}bar^{-1}$

1 bar pressure, S = 35‰

t °C	Ekman's data eq 6.11.4[b]	Binary solution data eqs 6.9.1 and 6.10.1 Table 6A.1
0	6.73	6.39
5	5.74	5.81
10	4.89	5.34
15	4.18	4.89
20	3.61	4.52
25	3.18 (3.80)[c]	4.20 (4.55)[d]

a W = 1.3576 when $(\bar{K}_s^A - \phi_K^A)$ used. The value of $(\bar{K}_s - \phi_K)/W$ is independent of the solution chosen.

b For other salinites use equation 6.11.2 with p=1

c Using B instead of B*(S=0), see Figures 6.40, 6.32.

d From data listed by Parsons (1959). Otherwise values given by Millero (1973), with Na_2SO_4 weighted for the effect of mixing, are used.

See Appendix 6, Table 6A.1.

6.12 Optical properties and compressibility.

The optical properties of water are currently of interest because they are important for computing radiation transport through atmospheres containing water droplets and other aerosols, or through oceans containing hydrosols. They also have application in the development of optical remote sensing instruments for measuring the chemical and thermal quality or turbidity of environmental waters, and investigation of the optical properties and optical constants of aqueous solutions (e.g. seawater). Such properties can provide information on the structural properties of a solution. For example, Dobbins and Peck (1973) and Hawkes and Astheimer (1974) found anomalous results for the thermal coefficient of the refractive index of water near 35°C. The measurements were at very close intervals of temperature so that the refractive index was obtained as practically a continuous function of temperature (e.g. Dobbins and Peck obtained 3062 data points for the range $20 \leq t \leq 35^\circ C$). Such fine resolution greatly enhances the significance of the anomalies. Hawkes and Astheimer (1974), using an interferometer, found anomalous fringe behavior that indicated small convection currents, caused by local density (or temperature) inhomogeneities in the liquid. These results were not found for methanol or a petroleum oil, and did not arise from thermostat irregularities. As the authors point out, such inhomogeneity is consistent with local breakdown of association groups in water (like a change of phase) that could involve small latent heats. These results are very interesting in view of the compatible results outlined in Section 6.75.

It seems appropriate to include in this chapter some information on the optical properties related to the P-V-T properties of the solution. Hale and Querry (1973) reviewed the literature on the optical constants of water for the past 81 years, and determined the extinction coefficients and index of refraction at 25°C through a broad spectral region (200 nm to 200μm). However, the results of Eisenberg (1965) on the refractive index of water, developed on a thermodynamic basis will be found especially useful in the present context.

6.121 Refractive index.

WATER.

The relationship between the refractive index and the density of water (over the temperature and pressure ranges 0-60°C and 1-1000 bars, respectively) is accurately described by the equation (Eisenberg 1965)

$$f(n) = \frac{n^2-1}{n^2+2} = Ad_w^{B_n}EXP(-C_nT) \qquad (6.121.1)$$

where, n is the refractive index of water relative to vacuum, d_w is the density of water, T the absolute temperature. A, B_n, C_n are empirical constants dependent on the wavelength but independent of temperature and pressure. The differentiation of eq (6.121.1) relates B_n and C_n to the P-V-T properties of water, viz.

$$B_n = f'(n)\beta^{-1}\left(\frac{\partial n}{\partial p}\right)_T \qquad (6.121.1a)$$

$$C_n = -f'(n)\left(\frac{\partial n}{\partial T}\right)_p - \alpha B_n \qquad (6.121.1b)$$

or, $$C_n = -f'(n)\left(\frac{\partial n}{\partial T}\right)_V \qquad (6.121.1c)$$

with $f'(n) = 6n/(n^2+2)(n^2-1)$

As usual, β and α represent the isothermal compressibility and coefficient of expansion, respectively.

The values of A, B_n and C_n have been calculated from eq (6.121.1) by Eisenberg (1965). He used the seven-place refractive index data of Tilton and Taylor (1938) for 1-atm and covering various temperatures, and the 1-atm density data of Chappius (as evaluated by Tilton and Taylor, 1937).

The thirteen wavelengths covered ranged from 4046.6 to 7065.2Å. The values of A, B_n and C_n are summarised in Table 6.26 in the form of polynomial functions of the wavelength. The values for the Na-D and Hg-green lines and a few other wavelengths are given in Table 6.27, as examples.

TABLE 6.26

Parameters* for calculating the absolute refractive index, n, of water in the ranges
$1 \leqslant p \leqslant 1000$ bar
$0 \leqslant t \leqslant 100^{\circ}C$
$4046.6 \leqslant \lambda \leqslant 7065.2$ Å
Coefficients for the equation:

$$y = a_o + a_1\lambda + a_2\lambda^2 + a_3\lambda^3 + a_4\lambda^4 + a_5\lambda^5 + a_6\lambda^6$$

(with λ in Å, to 1 decimal place)

Coeff	y	A_t**	B_n	$10^5 C_n \deg^{-1}$
a_o		0.406728	0.348023	-14.7559
$10^3 a_1$		-0.164127	0.429044	23.4248
$10^7 a_2$		0.602333	-1.41505	-107.426
$10^{10} a_3$		-0.123469	0.237573	26.4192
$10^{14} a_4$		0.146135	-0.200986	-36.7200
$10^{19} a_5$		-0.937954	0.682538	271.635
$10^{23} a_6$		0.253552	-	-83.3143
std.dev.		1×10^{-7}	9×10^{-6}	1×10^{-4}

* for the equation: $f(n) = (n^2-1)/(n^2+2) = A d_w^{B_n} \mathrm{EXP}(-C_n T)$ $T^{\circ}K$

or, $f(n) = A_t d_w^{B_n} \mathrm{EXP}(-C_n t)$ $t^{\circ}C$

** $A = A_t \mathrm{EXP}(273.16 C_n)$

TABLE 6.27

Parameters* for calculating the refractive index, n, of water from

$(n^2-1)/(n^2+2) = A_t d_w^{B_n} \mathrm{EXP}(-C_n t)$

Absolute values

	Wavelength Å	A_t	B_n	$10^5 C_n$ deg^{-1}
Na-D	5892.6	0.2064709	0.88538	6.2037
Hg-green	5460.7	0.2073063	0.88410	6.3121
	4358.3	0.2105352	0.87898	6.5122
	7065.2	0.2047862	0.88796	5.8583

* Eisenberg (1965)

A number of interesting points follow from eq (6.121.1) and its derivatives.

At 4°C when $\alpha = 0$,

$$\left(\frac{\partial n}{\partial T}\right)_P = -C_n/f'(n) \qquad (6.121.1d)$$

whereas, when $\alpha = -C_n/B_n$

$$\left(\frac{\partial n}{\partial T}\right)_P = 0 \qquad (6.121.1e)$$

For liquids such as benzene and methanol C_n is zero (Eisenberg 1965, Reisler and Eisenberg 1965). It was concluded by these authors that C_n is a measure of the deviation of water from "normal" behavior, and may reflect a change with temperature of either the concentration of "ice-like" structures or the average polarizability of water molecules. The significance of C_n will be discussed in connection with the "internal pressure" of water in Section 6.123. Eisenberg (1965) has outlined the molecular significance of the constants B_n and C_n in terms of a model involving types of oscillators with given polarizabilities. This analysis showed that $C_n/(1-B_n)$ should not be very dependent on the wavelength. This was shown to be the case for eq (6.121.1). It was suggested that information on water structure may be obtained from the behavior of n, and hence B_n and C_n, under extreme conditions of temperature and pressure (i.e. t >100^oC and p> 2000 bars).

Change of refractive index with pressure. Eq (6.121.1) should be particularly useful for estimating the refractive index of water and seawater at pressures up to 1000 bars. Waxler et al (1964) measured the refractive index of water at wavelengths ranging from 6678 to 4678 over the pressure and temperature ranges 1-1110 bars and 1.6 to 53.4^oC. The experimental error was 0.0001. We have calculated values of n corresponding to the experimental values, using eq (6.121.1) and the polynomials listed in Table 6.26 (or the direct values of A, B_n and C_n). The results are summarised in Tables 6.28 and 6.29.

TABLE 6.28 Absolute refractive index of water *
Values calculated from eq (6.121.1)
Deviations, $10^5(n_{calc}-n_{exp})$, are underlined

Wavelength Å	6438.5	5085.8	4921.9	4799.9	4678.2
A_t	0.2056074	0.2081869	0.2086297	0.2089875	0.2093715
B_n	0.88666	0.88276	0.88208	0.88152	0.88092
p bars / 10^5C_n	6.0516	6.3943	6.4259	6.4477	6.4678
	t = 34.50^oC				
1	1.33021	1.33474	1.33552	1.33615	1.33684
	1	0	0	0	0
259.3	1.33382	1.33839	1.33918	1.33982	1.34051
	7	9	10	7	7
480.9	1.33676	1.34137	1.34216	1.34280	1.34350
	-3	4	3	6	5
799.2	1.34077	1.34542	1.34622	1.34687	1.34757
	8	4	9	11	10
1110.0	1.34447	1.34915	1.34996	1.35062	1.35132
	17	18	22	19	16

TABLE 6.28 (ctd)

Wavelength Å	6438.5	5085.8	4921.9	4799.9	4678.2
p bars			t = 24.80°C		
1	1.33136	1.33592	1.33670	1.33734	1.33802
	0	1	0	0	0
259.6	1.33506	1.33966	1.34045	1.34109	1.34178
	7	13	15	13	11
463.6	1.33783	1.34246	1.34326	1.34390	1.34460
	3	6	2	1	2
762.8	1.34170	1.34637	1.34718	1.34783	1.34853
	7	7	8	5	7
1108.6	1.34591	1.35062	1.35144	1.35210	1.35280
	10	9	11	9	9
			t = 7.64°C		
1	1.33264	1.33723	1.33803	1.33867	1.33935
	1	-1	0	0	0
256.1	1.33656	1.34119	1.34199	1.34264	1.34333
	11	11	13	12	8
497.8	1.34006	1.34474	1.34554	1.34619	1.34689
	4	7	10	5	1
730.7	1.34327	1.34798	1.34879	1.34945	1.35015
	3	-3	2	1	-58
1088.9	1.34792	1.35269	1.35351	1.35417	1.35488
	28	26	33	26	26
			t = 1.56°C		
1	1.33279	1.33740	1.33819	1.33883	1.33952
	0	1	-13	1	1
269.5	1.33706	1.34172	1.34252	1.34316	1.34386
	9	11	-2	11	12
507.4	1.34062	1.34532	1.34612	1.34677	1.34747
	15	15	-3	14	14
768.5	1.34432	1.34906	1.34987	1.35053	1.35123
	17	19	2	19	17
1049.7	1.34809	1.35287	1.35369	1.35436	1.35507
	24	32	10	28	28

λ = 6328 Å data from Stanley (1971)

p bars	t°C 1.03	5.03	10.02	15.01	20.01	24.98
1	0	0	-1	-1	-1	-2
345	-14	-8	-12	-6	-12	-13
689	-23	-9	-21	-15	-22	-20
1035	-25	-27	-29	-30	-30	-25
1379	-25	-37	-36	-35	-32	-25

* A_t, B_n and C_n from Table 6.26, density from eq 5.71.3, experimental data from Waxler et al (1964).

TABLE 6.29 Absolute refractive index of water
Deviations of calculated* and experimental** values

t^oC	p bars	λ Å	6678.1		5875.6		5015.7	
		A_t	0.2052736		0.2065009		0.2083698	
		B_n	0.88714		0.88532		0.88249	
		10^5C_n	5.9805		6.2084		6.4079	
			$10^5(n_{calc} - n_{exp})$					
1.56	1		0	(0)	1	(1)	0	(1)
	269.5		11	(12)	10	(11)	8	(8)
	507.4		11	(11)	13	(13)	13	(13)
	768.5		21	(22)	17	(17)	17	(17)
	1049.7		27	(27)	28	(28)	25	(25)
7.64	1		1	(1)	1	(1)	-1	(-1)
	256.1		10	(10)	12	(12)	13	(14)
	497.8		6	(6)	8	(8)	5	(5)
	730.7		-1	(-1)	-4	(-4)	3	(4)
	1088.9		33	(33)	29	(30)	30	(30)
24.80	1		0	(1)	0	(1)	0	(0)
	259.6		9	(9)	11	(12)	11	(11)
	463.6		10	(10)	0	(0)	1	(1)
	762.8		4	(4)	9	(10)	11	(11)
	1108.6		9	(9)	6	(6)	11	(11)
34.50	1		0	(0)	0	(1)	0	(1)
	259.3		4	(5)	9	(9)	7	(7)
	480.9		3	(4)	3	(3)	6	(7)
	799.2		10	(11)	9	(10)	8	(8)
	1110.0		19	(20)	19	(20)	21	(21)
54.34	1		0	(0)	0	(0)	0	(0)
	241.6		0	(1)	6	(6)	3	(3)
	489.7		1	(2)	5	(5)	4	(4)
	785.1		6	(6)	7	(7)	8	(8)
	1127.7		20	(20)	17	(17)	18	(18)

* From eq 6.121.1, coefficients A_t, B_n and C_n from Eisenberg (1965), as listed above. Bracketed values: using coefficients from Table 6.26. Density calculated from eqs 5.71.3, 6.3.2 and 6.75.2.

** Waxler, Weir and Schamp (1964), experimental error is 0.00010.

Except fcr pressures greater than 1000 bars, the calculated and experimental values generally agree, within the experimental errors. Values of n at a wavelength of 6328Å were also calculated and compared with the data of Stanley (1971). The accuracy of this data was given as 0.00006, however, the deviations ($n_{calc}-n_{exp}$) are generally higher than this and always negative. On the basis of the results for Waxler _et al_'s data it would appear that Stanley's values of n are somewhat high. The deviations are largely independent of temperature but increase linearly with pressure, at the rate of about 2.5×10^{-7} per bar. Stanley derived values of n from his experimental results via the equation

$$n(p,t) = n(1,t) + (\lambda/2t)dm + (n/3K)p$$

where dm represents the number of fringes passing a reference point due to the change in pressure, t is the thickness of the sample at a given temperature and pressure, λ the wavelength and K the bulk modulus of the sample. Errors in (n/3K) of around $2\text{-}3 \times 10^{-7}$ would give the observed deviations in n(p,t), Table 6.28. However, this is around the value for 1/3K. Errors in the second term may be the cause of the large deviations.

In general, it can be seen that eq (6.121.1), with the appropriate coefficients, gives an excellent summary of the values of the refractive index of water over the temperatures and pressures of oceanographic interest. It should be noted, however, that the trends of the deviations from Waxler _et al_'s data suggest a "water-structure" influence not fully accounted for by the equation. This is reflected in the tendency of the deviations to be relatively large at the two lowest temperatures and small at the higher temperatures.

SEAWATER.

The above results are of interest because Stanley (1971a) also measured the refractive index of seawater, using the same technique as for water. The errors could be larger in this case and the values of n(p,t) may be rather high for seawater. No other measurements at pressures above atmospheric appear to be available for comparison. Additional measurements on seawater,

TABLE 6.30 The absolute refractive index of seawater S = 35‰

t^oC	p bars	1	345	689	1035	1379
0	a	1.34015	1.34539	1.35025	1.35481	
	b	1.34482	1.35008	1.35507	1.35953	
	c	1.34127				
5	a	1.33977	1.34486	1.34962	1.35404	1.35814
	b	1.34455	1.34969	1.35450	1.35892	1.36314
	c	1.34100				
10	a	1.33935	1.34431	1.34896	1.35330	1.35738 (5)
	b	1.34422	1.34924	1.35394	1.35834	1.36240
	c	1.34069				
15	a	1.33899	1.34388	1.34844	1.35270	1.35668 (4)
	b	1.34379	1.34873	1.35332	1.35764	1.36166
	c	1.34029				
20	a	1.33850	1.34331	1.34780	1.35200	1.35592
	b	1.34327	1.34812	1.35268	1.35696	1.36095
	c	1.33978				
25	a	1.33795	1.34270	1.34713	1.35130	1.35520
	b	1.34272	1.34757	1.35208	1.35632	1.36019
	c	1.33924				
30	a	1.33737	1.34207	1.34648	1.35058	1.35444
	b	1.34215	1.34694	1.35138	1.35561	1.35946

a λ = 6328 Å - data of Stanley (1971a), mean values; the maximum
b λ = 5017 Å deviations ($x10^5$) from mean are given in brackets, values below $4x10^{-5}$ not shown. Experimental error given as 0.00006, but may be larger, see text.

c λ = 5893 Å values relative to air have been multiplied by 1.00029 to convert to absolute values. Relative values from Cox (1965 from data of Utterback et al 1934), experimental error estimated as 0.00005, cf. Table 6.31.

using a different technique would be useful as a check and to provide more extensive data.

The values of n for seawater at S = 35‰ and 1 bar, as given by Cox (1965, from the 1934 data of Utterback et al) and the data of Stanley (1971a) are listed in Table 6.30. Values at 1 bar for other salinities

(0-41‰) have been compiled by Cox (1965).

Mehu et al (1969) have summarised values over a range of wavelengths for salinities 8.83, 17.35, 26.36 and 35‰ and temperature 1-30°C. The values for S = 35‰ are given in Table 6.31.

TABLE 6.31 Absolute refractive index of seawater as a function of the temperature and wavelength.

S = 35‰, pressure = 1 bar

λ Å \ t°C	1	5	10	15	20	25	30
4047	1.35127	1.35106	1.35075	1.35034	1.34984	1.34928	1.34863
4358	1.34864	1.34845	1.34812	1.34770	1.34722	1.34665	1.34602
4678	1.34649	1.34630	1.34598	1.34554	1.34504	1.34449	1.34387
4800	1.34577	1.34558	1.34526	1.34484	1.34435	1.34378	1.34317
5085	1.34431	1.34412	1.34378	1.34336	1.34286	1.34233	1.34170
5461	1.34269	1.34248	1.34218	1.34174	1.34126	1.34071	1.34009
5770	1.34154	1.34137	1.34103	1.34062	1.34012	1.33957	1.33898
5791	1.34149	1.34130	1.34098	1.34054	1.34004	1.33950	1.33890
5893	1.34114	1.34097	1.34063	1.34020	1.33971	1.33915	1.33856
6438	1.33957	1.33938	1.33905	1.33863	1.33813	1.33758	1.33698

Calculated from Table 1, of Mehu and Johannin-Gilles (1969), which lists the data in the form of the specific refraction, $K = ((n_{sw}^2-1)/(n_{sw}^2+2))/d_{sw}$. The values of d_{sw} were calculated from eqs 6.8.1 and 5.71.1. Error in n_{sw} is about 0.00005.

The data of Utterback and Stanley were fitted to eq 6.121.1 which reproduces the data to $\pm 2\text{x}10^{-5}$ (i.e. less than the experimental error) using the coefficients listed in Table 6.32. One exception is the experimental value for wavelength 6328 Å at 0°C which has a deviation of $19\text{x}10^{-5}$. The experimental values of Utterback et al over the salinity range 5-41‰ are summarised in Table 6.33 in the form of the coefficients for eq 6.121.1. This set of data gives the most complete coverage of changes of n_{sw} with salinity.

Whilst eq 6.121.1 may be useful in studies of the solution structure, and for reproducing and interpolating 1 bar values, the coefficients are

TABLE 6.32

Coefficients for the equation

$$f(n)_{sw} = (n_{sw}^2 - 1)/(n_{sw}^2 + 2)$$

$$= A d_{sw}^{B_n} \mathrm{EXP}(-C_n T)$$

S = 35‰ p = 1 bar

Wavelength Å	A	B_n	$10^5 C_n \deg^{-1}$	t °C
5017	0.211964	0.78169	7.3434	0-30
5893*	0.210468	0.71654	7.8953	0-25
6328	0.207893	0.85014	5.6389	0-30

* gives the values of n_{sw} relative to air; to convert to absolute values, multiply by 1.00029.

N.B. T is the absolute temperature

d_{sw} is given by eq 6.8.1 (see Section 4.6 also) and eq 5.71.1

wavelengths 5017, 6328 - data of Stanley (1971a)
5893, data of Utterback *et al* (see Cox 1965).

not accurate enough to predict the high pressure values successfully. The deviations for pressures as low as 350 bars are 7×10^{-4}. This is unacceptable. In addition, as can be seen from Table 6.33, the values of A, B_n and C_n are not linear functions of salinity. Therefore, it may be simpler; and more appropriate, to treat the solute and solvent components separately. This could be done by using the equations developed by Eisenberg (1965) for water, and the Tammann and Gibson hypotheses. That is, the contribution of the water to the refractive index of seawater would be given by eq 6.121.1, using the coefficients A_t, B_n and C_n of water for each wavelength (Table 6.26) and the density of water under a pressure P_e (see Sections 4.7, 6.5.3, eq 4.62.5). The contribution of the solute could be expected to be a simple function of ψ_t, the specific volume of the melted solute in solution (Section 6.31) and hence a simple function of salinity and temperature. This treatment would probably be more useful for estimating the effect of pressure on n_{sw} over a wide range of wave-lengths. The water-in-seawater contribution would be expected to be accurate to 1×10^{-4} on the basis of the

TABLE 6.33

Coefficients for the equation

$$(n_{sw}^2-1)/(n_{sw}^2+2) = Ad_{sw}^{B_n}EXP(-C_nT)$$

$$0 \leqslant t \leqslant 25^oC$$

S‰	A	B_n	$10^5C_n deg^{-1}$
41	0.209373	0.78319	6.8732
40	0.209741	0.76315	7.2467
39	0.209327	0.79265	6.8204
38	0.210355	0.72525	7.8423
37	0.210513	0.71591	7.9971
36	0.210147	0.73572	7.5563
35	0.210468	0.71654	7.8952
34	0.210136	0.73649	7.5025
33	0.210538	0.70968	7.9215
32	0.210936	0.68372	8.3525
31	0.211265	0.66018	8.7029
30	0.210930	0.67935	8.2921
25	0.209170	0.82251	6.2229
20	0.208199	0.90474	4.7894
15	0.208887	0.88930	5.5472
10	0.209809	0.80784	6.6058
5	0.210818	0.67046	7.9709

T in ^oK, n_{sw} relative to air; multiply by 1.00029 to convert to absolute values.

A, B_n and C_n are independent of temperature

d_{sw} is given by eqs 6.8.1 (see also Section 4.6) and 5.71.1

Above coefficients are based on the data of Utterback et al (1934) as given by Cox (1965).

pure water results. The value for the solute contribution should not be very sensitive to pressure, since ψ_t is independent of pressure.

Analyses along these lines were recently carried out (Leyendekkers 1976). The refractive index of seawater (under a pressure p bars) was shown to be given by:

$$f(n_{sw}^{(p)}) = f(n_{wsw}^{(p)}) + \Delta f(n) \qquad (6.121.2)$$

where $f(n_{wsw})$, the refractive index function for the water-in-seawater, is given by:

$$f(n_{wsw}^{(p)}) = A_t(1/\psi_w^{(p)})^{B_n}EXP(-C_nt) \qquad (6.121.2a)$$

from eq (6.121.1) with $(1/\psi_w)$ representing the density of the water, and

$$\psi_w^{(p)} = v_w^{(1)}(1-0.3151\log\frac{(B^*+p)}{(B+1)}) \quad \text{from } (4.62.5)^*$$

with $B = 2671.8 + 19.454t - 0.27028t^2 + 0.0009798t^3$ bars (Section 6.75)

and $B^* = 2670.8 + 6.8966S + (19.39 - 0.0703S)t - 0.223t^2$ from eq 4.61.7*.

A_t, B_n and C_n are given in Table 6.26 and $v_w^{(1)}$ in Chapter 5, Section 5.71.

The contribution of the solute is given by:

$$\Delta f(n) = b_o + b_1\lambda + (b_2 + b_3\lambda)S/\psi_t \qquad (6.121.2b)$$

with $\psi_t = 0.54214 - 0.0021896t$, from Sections 6.31 and 4.61.

The b coefficients are given in Table 6.33a for four salinities (35, 26.4, 17.3 and 8.8 ppt), and covering the ranges of temperature, pressure and wavelength as shown, the data are from Mehu *et al* (1969).

TABLE 6.33a Coefficients for eq 6.121.2b

S‰	Coeff	$10^3 b_o$	$10^7 b_1$	$10^5 b_2$	$10^9 b_3$	$\pm 10^5 \sigma$
34.998		-3.0195	1.47483	7.69745	-3.06555	3
26.356		-1.2237	-0.534862	5.46151	0.122601	3
17.348		-2.9938	2.99067	11.8276	-9.41295	2
8.829		-0.45880	-0.107134	5.95281	-0.433293	3

valid for the ranges: $4047\ \text{Å} \leq \lambda \leq 6438\ \text{Å}$
$0 \leq t \leq 30^oC$
$1 \leq p \leq 1000$ bars

The absolute values of the refractive index of seawater for the Na-D line (5892.6Å), Utterback's data, for the salinity range 5-39‰ and temperature range 0-25°C, are given by eq 6.121.2, with

$$\Delta f(n) = a_o + a_1 S/\psi_t \qquad (6.121.2c)$$

where, $a_o = 3.11905\text{x}10^{-5} + 6.91429\text{x}10^{-7}t$

and $a_1 = 2.56670\text{x}10^{-5} + 1.29817\text{x}10^{-7}t$ (with $\sigma \pm 1\text{x}10^{-5}$)

and t represents the temperature in ^{o}C.

These equations should enable prediction of the refractive index of seawater to within 0.00010 for the normal seawater range of temperature and pressure (Leyendekkers 1976). The value of n_{sw} is simply obtained from the function, via

$$n_{sw}^{(p)} = \left[(1+2f(n_{sw}^{(p)}))/(1-f(n_{sw}^{(p)}))\right]^{\frac{1}{2}}$$

6.122 <u>Light Scattering</u> - Rayleigh Ratio

The derivative $\left(\frac{\partial n}{\partial p}\right)_T \beta^{-1}$ is of primary importance in the calculation of the Rayleigh ratio of pure liquids in light scattering experiments. The pressure coefficient of n is, from eq (6.121.1a),

$$n\left(\frac{\partial n}{\partial p}\right)_T = \beta B_n(n^2+2)(n^2-1)/6 \tag{6.122.1}$$

The Rayleigh ratio $R_u(90)$ in the direction perpendicular to the unpolarized incident beam is given by, (Cohen and Eisenberg 1965, Eisenberg and Kauzmann 1969):

$$R_u(90) = \frac{2\Pi^2 n^2}{\lambda^4} \frac{kT}{\beta}\left(\frac{\partial n}{\partial p}\right)_T^2 f \tag{6.122.2}$$

with λ = wavelength of incident light

n = refractive index for λ

β = isothermal compressibility

$\left(\frac{\partial n}{\partial p}\right)_T$ = isothermal piezo-optic coefficient

The quantity, f, is given by:

$$f = (6 + 6\rho_u)/(6 - 7\rho_u) \tag{6.122.2a}$$

and is known as the Cabannes factor, which allows for the optical anisotropy of the molecules. ρ_u is called the depolarisation ratio. Eq (6.122.2) is known as the Einstein-Smoluchowski-Cabannes equation. All the experimental quantities in this expression have been measured so that $R_u(90)$ as measured directly, can be compared with the calculated value. Cohen and

Eisenberg (1965) found an agreement within 4% for the two sets of values.

Litan (1968) showed, on the basis of the theory of light scattering from water, that eq (6.122.2) should underestimate the amount of scattering. The discrepancy, however, is within the range of the present experimental error. Litan derived an expression for the excess scattering, over that calculated by eq (6.122.2) and found it to be less than 1.5% of the value calculated by the conventional approach. For the model used, this excess light scattering was due to fluctuations in "iceberg" concentration.

We have calculated values of $R_u(90)$ at 25°C from

$$R_u(90) = 2\Pi^2 kTB_n^2\beta((n^2+2)(n^2-1))^2 f/36\lambda^4 \qquad (6.122.3)$$

from eqs (6.122.1) and (6.122.2), with

$k = 1.38049\text{x}10^{-22}\ \text{cm}^3\text{bardeg}^{-1}$

$\beta = 45.52\text{x}10^{-6}\ \text{bar}^{-1}$

n from eq (6.121.1), using the A_t, B_n and C_n values from Table 6.27.

B_n from Table 6.27.

The values of f were experimental (Table 6.34). The calculated values are compared with the experimental values in Table 6.34. The predicted values are generally lower than the experimental ones, in accord with the results of Litan (1968). The experimental values of Cohen and Eisenberg appear to be too low. The excess scattering (1.5% of calculated value) is $0.015\text{x}10^{-6}$ cm^{-1} and $0.037\text{x}10^{-6}\text{cm}^{-1}$ for the wavelengths 5460.7 and 4358.3, respectively.

TABLE 6.34 Rayleigh ratio* of water at 25°C, 1 bar

λ Å	n	ρ_u Exp.	$R_u(90)\text{x}10^6\text{cm}^{-1}$ calc.	Exp.
5460.7	1.3343358	0.076	0.928	0.865**
		0.116	1.015	1.08
				1.05
		0.108	0.997	1.0
4358.3	1.3400737	0.087	2.428	2.32**
		0.100	2.499	2.45
		0.108	2.545	2.60
		0.083	2.407	2.89

* from Eisenberg and Kauzmann (1969). ** values of Cohen and Eisenberg (1965).

6.123 <u>The thermal pressure coefficient and the thermodynamic "internal pressure" of water.</u>

The thermal pressure coefficient $(\partial p/\partial T)_V$ and the thermodynamic internal pressure $(\partial U/\partial V)_T$ are two useful functions for characterising the nature of the liquid state. Their relationship is summarised by:

$$(\partial U/\partial V)_T = T(\partial p/\partial T)_V - p \qquad (1.8.10)*$$

and, from Section 2.2

$$(\partial p/\partial T)_V = -(\partial V/\partial T)_p/(\partial V/\partial p)_T$$

$$= \alpha/\beta \qquad (2.2.6)*$$

Thus, the thermodynamic internal pressure for water and seawater (at constant composition) can be calculated from

$$(\partial U/\partial V)_T = (T\alpha/\beta - p) \qquad (6.123.1)$$

Since, for water,

$$(\alpha v)^{(p)} = \left[0.315v_w^{(1)}/2.303\right] \frac{dB}{dt}\frac{(p-1)}{(B+1)(B+p)}$$

$$+ (\alpha v)^{(1)}\left[1 - 0.315\log\frac{(B+p)}{(B+1)}\right] \qquad (5.71.6)*$$

and

$$(\beta v)^{(p)} = 0.315v_w^{(1)}/2.303(B+p) \qquad (6.3.1)*$$

then

$$(\alpha/\beta)^{(p)} = \frac{(p-1)dB}{(B+1)dt}$$

$$+ (2.303/0.315)\alpha^{(1)}(B+p)\left[1-0.315\log\frac{(B+p)}{(B+1)}\right] \qquad (6.123.2)$$

Values of $(\partial U/\partial V)_T$ at 0° and 25^oC were calculated from eqs (6.123.1) and (6.123.2), using

$$\alpha^{(1)} = \left[(1-d_w)/d_w\right]\left[2/(t-a_1) - a_5/(t+a_3)(t+a_2)\right] \qquad (5.71.4)*$$

$$\text{with } d_w = 1 - \left[(t-a_1)^2(t+a_2)/a_4(t+a_3)\right] \qquad (5.71.3)*$$

where $a_1 = 3.9863$, $a_2 = 288.9414$, $a_3 = 68.12963$,

$a_4 = 508929.2$, $a_5 = 220.81177$, and

$$B = 2671.8 + 19.454t - 0.27028t^2 + 0.0009798t^3 \qquad (6.75.2)*$$

from Section 6.75.

The values for $(\partial U/\partial V)_T$ for a range of pressures are summarised in Table 6.35. The "structural" internal pressure, Z, is also shown. The method of estimating this will now be outlined. It is of interest to note here that at $0^\circ C$ Z is the dominant component of the thermodynamic internal pressure at low external pressures (Table 6.35).

TABLE 6.35 The thermal pressure coefficient and the thermodynamic internal pressure of water.

pressure unit: bar

p	t = 0°C: $\left(\frac{\partial p}{\partial T}\right)_V$* bar deg^{-1}	t = 0°C: $\left(\frac{\partial U}{\partial V}\right)_T$ eq 6.123.1	t = 0°C: Z** λ=5893Å	t = 25°C: $\left(\frac{\partial p}{\partial T}\right)_V$ bar deg^{-1}	t = 25°C: $\left(\frac{\partial U}{\partial V}\right)_T$ eq 6.123.1	t = 25°C: Z λ=5893Å
1	-1.327	-363.60	-373.93	5.648	1683.00	-458.96
50	-0.992	-320.91	-379.84	5.854	1695.47	-465.41
100	-0.649	-277.34	-385.85	6.065	1708.13	-471.98
200	0.036	-190.13	-397.82	6.484	1733.28	-485.06
300	0.722	-102.87	-409.73	6.903	1758.18	-498.07
400	1.408	- 15.54	-421.57	7.321	1782.85	-511.03
500	2.094	71.86	-433.35	7.739	1807.30	-523.92
600	2.780	159.30	-445.07	8.155	1831.53	-536.75
700	3.466	246.81	-456.73	8.571	1855.55	-549.52
800	4.153	334.37	-468.33	8.987	1879.35	-562.24
900	4.840	421.98	-479.88	9.401	1902.96	-574.90
1000	5.527	509.65	-491.37	9.815	1926.37	-587.51

* calculated from α/β , see text

** " structural" internal pressure from eq 6.123.8, see Table 6.36 for the effect of a change in the wavelength.

From eqs (6.121.1a and b):

$$\alpha/\beta = -\left[\left(\frac{\partial \ln f(n)}{\partial T}\right)_p + c_n\right] / \left(\frac{\partial \ln f(n)}{\partial p}\right)_T \qquad (6.123.3)$$

and, since $\left(\frac{\partial \ln f(n)}{\partial T}\right)_p = f'(n)\left(\frac{\partial n}{\partial T}\right)_p$

and

$$\left(\frac{\partial \ln f(n)}{\partial p}\right)_T = f'(n)\left(\frac{\partial n}{\partial p}\right)_T$$

with $f'(n) = (6n/(n^2+2)(n^2-1))$

$$\alpha/\beta = -\left[\frac{(\partial n/\partial T)_p}{(\partial n/\partial p)_T} + C_n/(f'(n)(\partial n/\partial p)_T)\right] \qquad (6.123.4)$$

The thermodynamic internal pressure may be thought of as made up of two terms (apart from p), viz:

$$(\partial U/\partial V)_T = -\frac{T(\partial n/\partial T)_p}{(\partial n/\partial p)_T} - \frac{TC_n}{f'(n)(\partial n/\partial p)_T} - p \qquad (6.123.5)$$

As noted above, when $\alpha = -C_n/B_n$, $(\partial n/\partial T)_p = 0$, and the first term of eq (6.123.5) vanishes. Also, for "normal" liquids, such as benzene and methanol, C_n is zero. The special contribution of structural factors in water may therefore be considered as reflected in the second term of eq (6.123.5). If Z, say, represents this term,

$$Z = -TC_n(n^2+2)(n^2-1)/6n(\partial n/\partial p)_T \qquad (6.123.6)$$

or, in terms of β, from eq (6.122.1)

$$Z = -TC_n/\beta B_n \qquad (6.123.7)$$

Since C_n and B_n are independent of temperature and pressure the "structural" internal pressure is inversely proportional to β for a given temperature and wavelength. For the wavelength range $4047\text{Å} \leq \lambda \leq 7065\text{Å}$, the value of $(C_n/B_n)\text{x}10^5$ ranges from 6.6 to 7.5 deg^{-1}.

For the Na-D line, $C_n/B_n = 7.0068\text{x}10^{-5}\text{deg}^{-1}$, at 25°C, 1 bar, the value of β is $45.52\text{x}10^{-6}\ \text{bar}^{-1}$, so that

$$Z = -298.15\text{x}70.068/45.52 \quad \text{bar}$$

$$= -458.9 \text{ bar}$$

From the Tait equation (Section 6.3),

$$\beta^{-1} = \left[(B+p)(1-0.315\log\frac{(B+p)}{(B+1)})\right]/0.4343\text{x}0.315 \quad \text{bar}$$

So that,

$$Z^{(p)} = -TC_n(B+p)(1-0.315\log\frac{(B+p)}{(B+1)})/B_n 0.4343\text{x}0.315 \qquad (6.123.8)$$

and at 1 bar

$$Z^{(1)} = -TC_n(B+1)/B_n 0.4343\text{x}0.315 \qquad (6.123.8\text{a})$$

The Tait parameter B is a function of temperature but independent of pressure. It is presumed to represent the difference between the expansive pressure, due to thermal energy, and the cohesive pressure, due to temperature-independent attraction between the molecules (Harned and Owen 1958). Eq (6.123.8) illustrates its relationship with the thermodynamic internal pressure of water.

Some values of Z for 0°C and 25°C and from 1-1000 bars are listed in Table 6.36, see Table 6.35 also.

TABLE 6.36 The "structural" thermodynamic internal pressure of water at various wavelengths

Z bars (eq 6.123.8)

λ = 5892.6Å, B_n = 0.88538, 10^5C_n = 6.2037deg^{-1}		λ= 7065.2Å, B_n = 0.88796, 10^5C_n = 5.8583deg^{-1}		λ= 4046.6Å, B_n = 0.87642, 10^5C_n = 6.5413deg^{-1}		p bars
25°C	0°C	25°C	0°C	25°C	0°C	
-458.96	-373.93	-432.15	-352.08	-488.89	-398.31	1
-465.41	-379.84	-438.22	-357.65	-495.76	-404.60	50
-471.98	-385.85	-444.41	-363.31	-502.75	-411.01	100
-485.06	-397.82	-456.72	-374.58	-516.68	-423.76	200
-498.07	-409.73	-468.98	-385.79	-530.55	-436.44	300
-511.03	-421.57	-481.17	-396.94	-544.34	-449.05	400
-523.92	-433.35	-493.31	-408.03	-558.08	-461.60	500
-536.75	-445.07	-505.39	-419.07	-571.74	-474.08	600
-549.52	-456.73	-517.42	-430.05	-585.35	-486.51	700
-562.24	-468.33	-529.39	-440.97	-598.90	-498.87	800
-574.90	-479.88	-541.31	-451.84	-612.38	-511.17	900
-587.51	-491.37	-553.19	-462.67	-625.81	-523.41	1000

REFERENCES Chapter 6.

Amagat E.N. 1893.

Elasticité et dilatibilité des fluides.
Ann. Chim. Phys. 6: 68-136, 29: 505-574.

Barlow A.J., and E. Yazgan 1967.

Pressure dependence of the velocity of sound in water as a function of temperature.
Br.J.Appl.Phys. 18: 645-651.

Barnartt S. 1952.

The velocity of sound in electrolyte solutions.
J.Chem.Phys. 20: 278-279.

Bradshaw A. and K.E.Schleicher 1970.

Direct measurement of thermal expansion of seawater under pressure.
Deep-Sea Res. 17: 691-706.

Caldwell D.R. and B.E. Tucker 1970.

Determination of thermal expansion of seawater by observing onset of convection.
Deep-Sea Res. 17: 707-719.

Cohen G. and H. Eisenberg 1965.

Light scattering of water, deuterium oxide, and other pure liquids.
J.Chem.Phys. 43(11): 3881-3887.

Cox R.A. 1965.

The physical properties of seawater, Chapter 3 of "Chemical Oceanography 1" eds J.P.Riley and G.Skirrow. Academic Press (p108, Table VI).

Cox R.A. and N.D. Smith 1959.

The specific heat of seawater.
Proc.R.Soc(A). 252: 51-62.

Crease J. 1962.

The specific volume of seawater under pressure as determined by recent measurements of sound velocity.
Deep-Sea Res. 9: 209-213.

CSIRO 1968.

Aust.Oceanogr. Cruise Rep. 26
Oceanographic Observations in the Pacific Ocean in 1963, Cruise G3/63.

CSIRO 1969.

Aust.Oceanogr. Cruise Rep. 44
Oceanographic Observations in the Pacific Ocean in 1965, Cruise G3/65.

Chapter 6.

Dayantis J. 1972.

No. 16 Considérations sur l'équation de Tait et son obtention à partir de l'equation de van der Waals. J.de Chim.Phys. 69(1): 94-99.
N.B. this reference is incorrectly cited in Chemical Abstracts.

Del Grosso V.A. 1952.

The velocity of sound in seawater at zero depth. Rept.U.S.N.Res.Lab. 4002, 39pp.

Del Grosso V.A. and C.W. Mader 1972.

Speed of sound in pure water. J.Acoust.Soc.Amer. 52(5): 1442-1446.

Del Grosso V.A. and C.W. Mader 1973.

Another search for anomalies in the temperature dependence of the speed of sound in pure water. J.Acoust.Soc.Amer. 53(2): 561-563.

Diaz Pena M. and M.L. McGlashan 1959.

An apparatus for the measurement of the isothermal compressibility of liquid. Trans.Faraday Soc. 55: 2018-2024.

Dobbins H.M. and E.R. Peck 1973.

Change of refractive index of water as a function of temperature. J.Opt.Soc.Am. 63(3): 318-320.

Dorsey N.E. 1940.

Properties of ordinary water-substance. American Chemical Society Monograph Series 81, Reinhold, New York, 673pp.

Duedall I.W. and S. Paulowich 1973.

A bellows-type differential compressimeter for determining the difference between the compressibilities of two seawater solutions to 900 bars. Rev.Sci.Instrum. 44: 120-127.

Eckart C. 1958.

Properties of water. J.Am.Sci. 256: 225-240.

Eisenberg H. 1965.

Equation for the refractive index of water. J.Chem.Phys. 43(11): 3887-3892.

Eisenberg D. and W. Kauzmann 1969.

"The structure and properties of water". Oxford University Press, London 296pp.

Chapter 6.

Ekman V.W. 1908.

Die Zusammendrueckbarkeit des Meerwassers.
Publ.Circonst.Perm.Explor.Mer. 43: 1-47.

Emmet R.T. and F.J. Millero 1974.

High pressure density measurements of seawater - preliminary results.
J.Geophys.Res. 79(24): 3463-3472.

Erlander Stig. R. 1968a

The structure of water and its relationship to hydrocarbon-water interactions.
J.Macromol.Sci. - Chem. A2(3): 595-621.

Erlander Stig. R. 1968b

Determination of the structure and effective dielectric constant of hydrated ions.
J.Macromol.Sci. - Chem. A2(4): 833-858.

Fine R.A. and F.J. Millero 1973.

Compressibility of water as a function of temperature and pressure.
J.Chem.Phys. 59(10): 5529-5536.

Fine R.A., Dong-Ping Wang, and F.J. Millero 1974.

The equation of state of seawater.
J.Mar.Res. 32(3): 433-456.

Fisher F.H. 1962.

The effect of pressure on the equilibrium of magnesium sulfate.
J.Phys.Chem. 66: 1607-1611.

Fisher F.H. 1975.

Dissociation of Na_2SO_4 from ultrasonic absorption reduction in $MgSO_4$-NaCl solutions.
J.Sol.Chem. 4(3): 237-240.

Fisher F.H. and A.P. Fox 1975.

$NaSO_4^-$ ion pairs in aqueous solutions at pressures up to 2000 atm.
J.Sol.Chem. 4(3): 225-236.

Fofonoff N.P. 1962.

Physical properties of seawater, in "The Sea", vol. 1, edited by M.N.Hill, pp3-30, Interscience, New York.

Foster K.R. and E.D. Finch 1974.

An acoustical technique for measuring the temperature of maximum density of dilute aqueous solutions.
J.Phys.Chem. 78(22): 2305-2306.

Chapter 6.

Franks, F., Editor 1972.

"Water, a comprehensive treatise", Vol. 1, Plenum Press, New York, N.Y.

Gibson R.E. 1934.

The influence of concentration on the compressions of aqueous solutions of certain sulfates and a note on the representation of the compressions of aqueous solutions as a function of pressure. J.Am.Chem.Soc. 56: 4-14.

Gibson R.E. 1935.

The influence of the concentration and nature of the solute on the compressions of certain aqueous solutions. J.Am.Chem.Soc. 57: 284-293.

Ginell Robert 1961.

Derivation of the Tait equation and its relation to the structure of liquids. J.Chem.Phys. 34(4): 1249-1252.

Hagler, A.T., H.A. Scheraga, and G.Nemethy 1973.

Current status of the water-structure problem; application to proteins. Ann.N.Y.Acad.Sci. 204: 51-78.

Hale G.M. and M.R. Querry 1973.

Optical constants of water in the 200-nm to 200-μm wavelength region. Applied Optics. 12(3): 555-563.

Harned H.S. and B.B Owen 1958.

"The physical chemistry of electrolytic solutions" 3rd ed, A.C.S. Monograph, Reinhold, New York, 803pp.

Hawkes J.B. and R.W. Astheimer 1974.

Optical properties of water in the neighborhood of 35°C. J.Opt.Soc.Am. 64(1): 105-106.

Hawkins D.T. 1975.

A bibliography on the physical and chemical properties of water: 1969-1974. J.Sol.Chem. 4(8): 621-744.

Hayward A.T.J. 1967.

Compressibility equations for liquids: A comparative study. Brit.J.Appl.Phys. 18: 965-977.

Hirschfelder J.O., C.F. Curtiss, and R.B. Bird 1964.

"Molecular theory of gases and liquids". John Wiley & Sons, Inc., New York. 1249pp.

Chapter 6.

Horne R.A., Editor 1972.

"Water and aqueous solutions,"
Wiley-Interscience, New York. 837pp.

Kearns E.R. 1966.

Compressibilities of some dilute aqueous solutions.
Ph.D. Thesis, Yale Univ., University Microfilms Inc, Ann Arbor, Michigan, Ord No. 66-4902.

Kell G.S. 1970.

Isothermal compressibility of liquid water at 1 atm.
J.Chem.Eng.Data. 15(1): 119-122.

Kell G.S. and E. Whalley 1965.

The PVT properties of water. I. Liquid water in the temperature range 0 to 150°C and at pressures up to 1 kb.
Phil.Trans.Roy.Soc.A. 258: 565-614.

Kell G.S. and E. Whalley 1975.

Reanalysis of the density of liquid water in the range 0-150°C and 0-1 kbar.
J.Chem.Phys. 62(9): 3496-3503.

Knudsen M.H.C. 1901.

Hydrographische Tabellen, G.E.C. Grad., Copenhagen.

Lee J. 1966.

The apparent and partial molal volumes of electrolytes in water and in aqueous sodium-chloride solution.
Thesis, Yale Univ., Ann Arbor, Mich, Univ. Microfilms, 66-4906, Abstr. B,27,131.

Lentz, B.R., A.T. Hagler, and H.A. Scheraga 1974.

Structure of liquid water. II. Improved statistical thermodynamic treatment and implications of a cluster model.
J.Phys.Chem. 78(15): 1531-1550.

Lepple F.K. and F.J. Millero 1971.

The isothermal compressibility of seawater near one atmosphere.
Deep-Sea Res. 18: 1233-1254.

Leyendekkers J.V. 1974.

Prediction of the partial molal volumes of electrolytes in seawater.
Mar.Chem. 2(2): 89-110.

Leyendekkers J.V. 1975.

The effect of pressure on the chemical potentials of seawater components, on ionic species in seawater, and on calcite saturation levels.
Mar.Chem. 3: 23-41.

Chapter 6.

Leyendekkers J.V. 1976.

Prediction of the refractive index of seawater as a function of temperature, pressure, salinity and wavelength.
Mar.Chem. 4: Sept-Dec.

Litan A. 1968.

Fluctuation theory of light scattering from pure water.
J.Chem.Phys. 48(3): 1059-1063.

Li Yuan-Hui 1967.

Equation of state of water and sea water.
J.Geophys.Res. 72(10): 2665-2678.

Lown D.A., H.R. Thirsk, and Lord Wynne-Jones 1968.

Effect of pressure on ionization equilibria in water at 25°C.
Trans.Faraday Soc. 64: 2073-2080.

Macdonald J.R. 1969.

Review of some experimental and analytical equations of state.
Rev.Mod.Phys. 41(2): 316-349.

Mehu A. et A. Johannin-Gilles 1969.

Variation de la réfraction spécifique de l'eau de mer étalon de Copenhague et de ses dilutions en fonction de la longueur d'onde, de la température et de la chlorinité.
Deep-Sea Res. 16: 605-611.

Millero F.J. 1973.

Theoretical estimates of the isothermal compressibility of seawater.
Deep-Sea Res. 20: 101-105.

Millero F.J. 1974.

Sea water as a multicomponent electrolyte solution, in The Sea - Ideas and Observations, vol. 5, E. Goldberg, editor, Chap. 1, pp1-55, Interscience, New York.

Millero F.J., R.W. Curry, and W.J. Drost-Hansen 1969.

Isothermal compressibility of water at various temperatures.
J.Chem.Eng.Data 14: 422-425.

Millero F.J. and T. Kubinski 1975.

Speed of sound in seawater as a function of temperature and salinity at 1 atm.
J.Acoust.Soc.Am. 57(2): 312-319.

Chapter 6.

Millero F.J. and W.L. Masterton 1974.

Volume change for the formation of magnesium sulfate ion pairs at various temperatures.
J.Phys.Chem. 78(13): 1287-1294.

Millero F.J., G.K. Ward, F.K. Lepple, and E.V. Hoff 1974.

Isothermal compressibility of aqueous sodium chloride, magnesium chloride, sodium sulfate, and magnesium sulfate solutions from 0 to 45° at 1 atm.
J.Phys.Chem. 78(16):1636-1643.

Olofsson Gerd and L.G. Hepler 1975.

Thermodynamics of ionization of water over wide ranges of temperature and pressure.
J.Sol.Chem. 4(2): 127-143.

Owen B.B. and H.L. Simons 1957.

Standard partial molal compressibilities by ultrasonics. I. Sodium chloride and potassium chloride at 25°C.
J.Phys.Chem. 61: 479-482.

Perron G., J.E. Desnoyers, and F.J. Millero 1974.

Apparent molal volumes and heat capacities of alkaline earth chlorides in water at 25°C.
Can.J.Chem. 52(22): 3738-3741.

Pushin N.A. and E.V. Grebenshchikov 1923.

The adiabatic cooling of water and the temperature of its maximum density as a function of pressure.
J.Chem.Soc. (London). 123: 2717-2725.

Read A.J. 1975.

The first ionization constant of carbonic acid from 25 to 250°C and to 2000 bar.
J.Sol.Chem. 4(1): 53-70.

Reisler E. and H. Eisenberg 1965.

Refractive indices and piezo-optic coefficients of deuterium oxide, methanol, and other pure liquids.
J.Chem.Phys. 43(11): 3875-3880.

Stanley E.M. 1971.

Refractive index of pure water for wavelength of 6328Å at high pressure and moderate temperature.
J.Chem.Eng.Data. 16(4): 454-457.

Stanley E.M. 1971a

The refractive index of seawater as a function of temperature, pressure and two wavelengths.
Deep-Sea Res. 18: 833-840.

Chapter 6.

Sweers H.E. 1971.

A comparison of methods used to calculate sigma-t, specific volume anomaly and dynamic height. Mar.Technol.Soc.J. 5: 7-25.

Tait P.G. 1888.

Report on some of the physical properties of fresh water and seawater. Rep.Sci.Results Voy.H.M.S. Challenger,Phys.Chem.2: 1-76.

Tilton L.W. and J.K. Taylor 1937.

Accurate representation of refractivity and density of distilled water as a function of temperature. J.Res.Nat.Bur.Stand. 18: 205-214.

Tilton L.W. and J.K. Taylor 1938.

Refractive index and dispersion of distilled water for visible radiation, at temperatures 0 to 60°C. J.Res.Natl.Bur.Std. 20: 419-477.

Utterback C.L., T.G. Thompson and B.A. Thomas 1934.

Refractivity-chlorinity temperature relationship of ocean waters. J.Cons.Perm.Int.Explor.Mer. 9: 35-38.

Vedam R. and G. Holton 1968.

Specific volumes of water at high pressures obtained from ultrasonic-propagation measurements. J.Acoust.Soc.Am. 43: 108-116.

Wang Dong-Ping and F.J. Millero 1973.

Precise representation of the P-V-T properties of water and seawater determined from sound speeds. J.Geophys.Res. 78(3): 7122-7128.

Waxler R.M., C.E. Weir, and H.W. Schamp, Jr. 1964.

Effect of pressure and temperature upon optical dispersion of benzene, carbon tetrachloride and water. J.Res.Nat.Bur.Stand. 68A(5): 489-498.

Weres Oleh and S.A. Rice 1972.

A new model of liquid water. J.Am.Chem.Soc. 94(26): 8983-9002.

Wilson W. 1959.

Speed of sound in distilled water as a function of temperature and pressure. J.Acoust.Soc.Am. 31: 1067-1072.

Wilson W. 1960.

Speed of sound in seawater as a function of T, P, and salinity. J.Acoust.Soc.Am. 38: 641-644.

Chapter 6.

Wilson W. and D. Bradley 1968.

Specific volume of seawater as a function of temperature, pressure, and salinity.
Deep-Sea Res. 15: 335-363.

Winnick Jack and Sung J. Cho 1971.

PVT behavior of water at negative pressures.
J.Chem.Phys. 55(5): 2092-2097.

APPENDIX 6.

TABLE 6A.1 Parameters for the equation

$$\phi_{k_i} = \phi^o_{k_i} + s'_k I_v^{1/2} \qquad \text{eq (6.2.2)*}$$

For seawater components (Solution A, Table 4.16)

at various temperatures, and 1 bar. $cm^3 mol^{-1} bar^{-1}$

t^oC	i	NaCl	Na_2SO_4	$NaHCO_3$	NaF	KCl	KBr	$MgCl_2$	$CaCl_2$	$SrCl_2$
0	$-10^4\phi^o_k$	79.0	228.2*	70.4	111.8	72.8	64.5	142.6	127.0	127.6
		81.4	230.8					147.8		
	$10^4 s'_k$	18.0	50.8*	17.0	20.2	19.2	18.2	22.8	22.8	22.8
		21.0	54.2					26.6		
5	$10^4\phi^o_k$	72.0	207.8*	63.3	102.9	65.7	57.0	131.8	116.2	116.8
	$10^4 s'_k$	16.3	45.8*	15.3	18.5	17.6	16.6	21.4	21.4	21.4
10	$-10^4\phi^o_k$	65.7	190.0*	56.5	94.8	59.4	50.5	122.2	106.6	107.2
	$10^4 s'_k$	14.9	41.8*	13.9	17.1	16.1	15.1	20.2	20.2	20.2
15	$-10^4\phi^o_k$	60.2	175.0*	50.7	87.3	53.9	44.7	114.0	98.4	99.0
		60.1	168.4					113.9		
	$10^4 s'_k$	13.6	38.0*	12.6	15.8	14.8	13.8	19.0	19.0	19.0
		14.9	33.3					19.0		
20	$-10^4\phi^o_k$	55.5	162.6*	45.7	80.4	49.2	39.7	107.2	91.6	92.2
	$10^4 s'_k$	12.5	35.4*	11.5	14.7	13.6	12.6	18.0	18.0	18.0
25	$-10^4\phi^o_k$	51.6	152.8*	41.4	74.3	45.2	35.3	101.6	86.0	86.6
		51.3	145.4					100.4		
	$10^4 s'_k$	11.5	33.8*	10.5	13.7	12.4	11.4	17.0	17.0	17.0
		12.2	26.9					16.0		
30	$-10^4\phi^o_k$	48.4	145.8*	37.9	68.2	42.1	31.9	97.6	82.0	82.6
		48.5	139.2					96.2		
	$10^4 s'_k$	10.8	33.0*	9.8	13.0	11.3	10.3	16.4	16.4	16.4
		11.3	25.8					14.7		
35	$-10^4\phi^o_k$	46.0	141.4*	35.2	63.9	39.7	29.2	94.8	79.2	79.8
	$10^4 s'_k$	10.2	33.0*	9.2	12.4	10.3	9.3	15.6	15.6	15.6
40	$-10^4\phi^o_k$	44.4	139.8*	33.2	59.7	38.2	27.3	93.4	77.8	78.4
	$10^4 s'_k$	11.8	34.4*	8.8	12.0	9.1	8.1	15.2	15.2	15.2

from the compilation of the ionic values given by Millero (1973).

*The values for SO_4^{2-} were derived from Na_2SO_4 and $MgSO_4$ data, using a weighting technique (Millero 1973). Underlined values - see Table 6.2c, Section 6.22.

TABLE 6A.1a Ionic compressibility parameters ϕ_k^o and s_k' values based on $\phi_k^o(H+) = s_k'(H+) = 0$ at all temperatures

t^oC	0	5	10	15	20	25	30	35	40
$-10^4\ \phi_k^o(Cl-)$	26.0	20.8	16.4	12.9	10.2	8.3	7.2	6.9	7.5
$10^4 s_k'(Cl-)$	3.0	3.0	3.0	3.0	3.0	3.0	3.0	3.0	3.0

The other ionic values are obtained by the principle of additivity from Table 6A.1, e.g.

$$0^oC \quad \phi_k^o\ (Mg^{2+}) = \phi_k^o(MgCl_2) - 2\phi_k^o(Cl-)$$
$$= -(142.6 - 52.0)x10^{-4}$$
$$= -90.6x10^{-4}\ cm^3mol^{-1}bar^{-1}$$

See also Figs. 6.1 and 6.2, as a guide for estimating values for those ions for which no data are available.

N.B. See Table 6.1 for the effect of using I(mol/kg) instead of I_v (mol/litre).

TABLE 6A.2 Parameters for the equation

$$\phi_{k_i} = \phi^o_{k_i} + s_k c^{1/2} \qquad \text{eq(6.2.1)*}$$

$cm^3 mol^{-1} bar^{-1}$ at 25°C, 1 bar

i	$-10^4\phi^o_k$	$10^4 s_k$
$AgNO_3$	38.2	6.9
$BaCl_2$	115	26.8
$BeSO_4$	93	
$CaCl_2$	(87)	(26.0)
	86^x	29.4^x
$CdSO_4$	127	
$CeCl_3$	176	
CsBr	32.7	9.3
CsCl	41.3	11.5
CsI	22.1	7.5
$CsNO_3$	31.4	9.7
Cs_2SO_4	119	
$CuSO_4$	132	
HCl	8.3	3.0
KAcetate	47	
KBr	39.6	8.9
	35.3^x	11.4^x
$KBrO_3$	49.0	10.6
KCNS	26.9	4.8
KCl	45.2	12.4
	45.2^x	12.4^x
$KClO_3$	37.2	9.0
K_2CrO_4	139	
KF	75.4	15.4

i	$-10^4\phi^o_k$	$10^4 s_k$
$KHCO_3$	35.	
	35.0^x	11.4^x
KI	29.0	6.5
KIO_3	78.0	16.2
KNO_3	38.2	9.9
KOH	81	
K_2SO_4	139	
	140^x	61.7^x
LiBr	35.6	7.1
LiCl	42.0	9.1
$LiClO_4$	25.2	3.3
LiI	25.1	5.1
$LiNO_3$	34.0	6.0
LiOH	77.9	17.0
Li_2SO_4	146	
	133.6^x	50.2^x
$MgCl_2$	100.4^x	27.7^x
$MgSO_4$	153	35.2
	135.7^x	35.8^x
NH_4Br	12.8	3.4
NH_4Cl	22.0	6.6
NH_4ClO_4	3.1	-1.8
NH_4I	2.7	1.0
NH_4NO_3	11.7	2.4
$(NH_4)_2SO_4$	91	
	93.6^x	41.6^x

i	$-10^4\phi^o_k$	$10^4 s_k$
NaBr	46.8	8.0
Na_2CO_3	171	48.8
NaCl	51.6	11.4
	51.3^x	12.2^x
$NaClO_3$	45.7	7.6
$NaClO_4$	36.8	4.8
NaF	82.2	16.2
	74.3^x	13.7^x
$NaHCO_3$	41.4^x	10.5^x
NaI	35.7	4.9
$NaNO_3$	45.6	9.3
NaOH	89.0	20.9
Na_2SO_4	154	
	145.4^x	46.6^x
RbBr	36.7	9.0
RbCl	45.3	11.2
RbI	25.9	6.5
$RbNO_3$	35.1	9.6
$TlNO_3$	35.1	5.4
$ZnSO_4$	140	
Me_4NCl	17.2	6.5
Me_4NI	-2.1	9.0
Et_4NCl	20.1	4.8
Et_4NI	0.2	10.0

Data from the compilation by Parsons (1959)

x from Tables 6A.1 and 6A.1a - these values are considered more accurate than the older data

( ) at 30°C and 100 bars

TABLE 6A.3 The theoretical limiting law slope s_k^L for eq 6.21.2

$10^4 s_k^L/(w_i^{1/2})$ $cm^3 l.mol^{-2}bar^{-1}$

Charge type t°C	1:1	1:2 or 2:1	2:2	3:1 or 1:3
0	1.4	4.2	5.6	8.4
15	2.9	8.7	11.6	15.6
30	4.2	12.6	16.8	25.2
45	5.3	15.9	21.2	31.8

from Millero, Ward _et. al_ (1974)

TABLES 6A.4 and 6A.5

The values in Table 6A.4 are based on the summation of the values for the salts in the binary solutions, at the same ionic strength as seawater of a salinity of 35‰, i.e. 0.723 mol/kg.

$$\Phi_k^A = \Sigma m_i \phi_{k_i} / \Sigma m_i \qquad (6.9.1)^*$$

$$\bar{K}_s^A = \Sigma m_i \bar{K}_i / \Sigma m_i \qquad (6.10.1)^*$$

where A indicates a solution of salts according to recipe A, Table 4.16 (i.e. NaCl, Na_2SO_4, $NaHCO_3$, NaF, KCl, KBr, $MgCl_2$, $CaCl_2$, $SrCl_2$, H_3BO_3), the contribution of the last component has been neglected.

The calculation of the mean apparent molal compressibility, Φ_k, and the mean partial molal compressibility, $\bar{K}_s$, of the solute in seawater can be most conveniently made from the Tait-Gibson set of equations based on the data of Ekman (1908) (as given by Li, 1967). (Table 6A.5).

TABLES 6A.4 and 6A.5 (continued)

Mean apparent molal compressibility of seasalt.

$$\Phi_k^{(p)} = 434.3 \text{x} 0.315 M'_t v_w^{(1)} (1-c)(B-B^*)/S(B^*+p)(B+p) \qquad (6.9.4)^*$$

and $\Phi_k^{(1)} = \Phi_k^{(p)}(p=1)$

For solution A, $M'_t = 68.0811$ g

$v_w^{(1)}$ is given by eq 5.71.1. Table 5.18.

$B^* = 2670.8 + 6.8966S + (19.39-0.0703S)t - 0.223t^2$

$B = B^*(S=0)$

$\partial B^*/\partial S = 6.8966 - 0.0703t$, S represents the salinity (‰), $c = 10^{-3}S$.

Also, $\Phi_k^{(p)} = \Phi_k^{(1)}(B^*+1)(B+1)/(B^*+p)(B+p)$

Mean partial molal compressibility of seasalt.

$$\bar{K}_s^{(p)} = \frac{-(434.3 \text{x} 0.315 M'_t v_w^{(1)})(1-c)^2}{(B^*+p)^2} \frac{\partial B^*}{\partial S} \qquad (6.10.4)^*$$

$\bar{K}_s^{(1)} = \bar{K}_s^{(p)}(p=1)$

Also, $\bar{K}_s^{(p)} = \bar{K}_s^{(1)} \frac{(B^*+1)^2}{(B^*+p)}$

To obtain the total solute effect for the seawater (a solution containing 1 kg of water) multiply Φ_k or $\bar{K}_s$ by m_s, i.e. $S/(1-c)M'_t$

TABLE 6A.4 Estimates of the mean apparent and mean partial molal compressibilities of seasalt* from binary solution data. 1 bar pressure, S = 35‰

	ml/(mol salts).bar		ml bar^{-1}	
$t^{o}C$	$-10^4\Phi_k^A$	$-10^4\bar{K}_s^A$	$-10^4 m_s\Phi_k$	$-10^4 m_s\bar{K}_s$
0	77.18	68.50	41.11	36.49
5	70.54	62.65	37.58	33.37
10	64.49	57.25	34.36	30.50
15	59.35	52.71	31.62	28.08
20	54.96	48.82	29.28	26.01
25	51.41	45.71 44.3^x	27.39	24.35
30	48.50	43.09	25.84	22.96

x $\Sigma\bar{K}_i^{sw}$ from Table 6.23

	ml/(g seawater).bar		ml/mol(ions).bar	
$t^{o}C$	$-10^7 m_s\Phi_k/w_{sw}$	$-10^7 m_s\bar{K}_s/w_{sw}$	$-10^4\Phi_k/\Sigma\nu_i f_i$	$-10^4\bar{K}_s/\Sigma\nu_i f_i$
0	39.67	35.21	35.42	31.44
5	36.26	32.20	32.37	28.75
10	33.15	29.42	29.60	26.27
15	30.51	27.09	27.24	24.19
20	28.25	25.09	25.23	22.41
25	26.42	23.49	23.59	20.98
30	24.93	22.15	22.26	19.78

* The calculations are for a seawater solution containing 1 kg of water and Σm_i moles of salts (the individual salts, i, are according to solution A, Table 4.16)

$w_{sw} = (10^3+\Sigma m_i M_i)$ g, M_i is the molecular weight of the salt, i,

$m_s = \Sigma m_i$, $f_i = m_i/m_s$.

The data for the binary solutions were taken from Millero (1973)

TABLE 6A.5 The mean apparent molal compressibility and mean partial molal compressibility * of "seasalt" (the solute in seawater).

at 1 bar pressure

$ml\,mol^{-1}bar^{-1}$

S‰	30		35		40	
t°C	$-10^4\Phi_k^A$	$-10^4\bar{K}_s^A$	$-10^4\Phi_k^A$	$-10^4\bar{K}_s^A$	$-10^4\Phi_k^A$	$-10^4\bar{K}_s^A$
0	81.02 (78.77)	72.94 (73.35)	79.65 (77.82)	70.49 (71.28)	78.31 (76.84)	68.14 (69.25)
5	72.31 (70.07)	65.49 (65.77)	71.15 (69.31)	63.40 (64.08)	70.02 (68.53)	61.40 (62.40)
10	64.98 (63.48)	59.17 (60.06)	64.00 (62.87)	57.38 (58.63)	63.03 (62.23)	55.66 (57.22)
15	58.74 (58.30)	53.75 (55.57)	57.89 (57.81)	52.20 (54.35)	57.06 (57.28)	50.71 (53.14)
20	53.36 (54.12)	49.04 (51.92)	52.63 (53.72)	47.70 (50.87)	51.90 (53.28)	46.40 (49.81)
25	48.67 (50.68)	44.91 (48.89)	48.03 (50.35)	43.74 (47.97)	47.40 (49.98)	42.60 (47.04)

* see also Tables 6.22, 6A.4

Values based on the data of Ekman (1908), eqs 6.9.3, 6.10.4

Bracketed values based on the "acoustic" equation of Wang and Millero (1973), eqs 6.9.3, 6.10.3, 6.10.6 and Table 4.18.

REFERENCES Appendix 6

Ekman V.W. 1908.

Die Zusammendrueckbarkeit des Meerwassers.
Publ.Circnst.perm.Explor.Mer. 43: 1-47.

Li Yuan-Hui 1967.

Equation of state of water and sea water.
J.Geophys.Res. 72(10): 2665-2678.

Millero F.J. 1973.

Theoretical estimates of the isothermal compressibility of sea water.
Deep-Sea Res. 20: 101-105.

Millero F.J., G.K. Ward, F.K. Lepple and E.V.Hoff 1974.

Isothermal compressibility of aqueous sodium chloride, magnesium chloride, sodium sulfate and magnesium sulfate solutions from 0 to 45° at 1 Atm.
J.Phys.Chem. 78(16): 1636-1643.

Parsons R. 1959.

"Handbook of electrochemical constants", Butterworths, London, 113pp.

Wang Dong-Ping and F.J. Millero 1973.

Precise representation of the P-V-T properties of water and seawater determined from sound speeds.
J.Geophys.Res. 78(3): 7122-7128.

CHEMICAL INDEX

The salts and ions are listed alphabetically according to the first letter of the formula. See also the Tables in Appendix 3.

D

E

F

G

H

SUBJECT INDEX